D0712654

Design Through Verilog HDL

Design Through Verilog HDL

T. R. Padmanabhan
B. Bala Tripura Sundari

IEEE PRESS

A JOHN WILEY & SONS, INC., PUBLICATION

For general information on our other products and services please contact our Customer Care Department within the U.S. at 877-762-2974, outside the U.S. at 317-572-3993 or fax 317-572-4002.

Wiley also publishes its books in a variety of electronic formats. Some content that appears in print, however, may not be available in electronic format.

Library of Congress Cataloging-in-Publication Data:

Padmanabhan, T. R.
    Design through Verilog HDL / T. R. Padmanabhan, B. Bala Tripura Sundari.
        p. cm.
    Includes bibliographical references and index.
    ISBN 0-471-44148-1 (cloth)
    1. Verilog (Computer hardware description language)   I. Tripura Sundari, B. Bala.   II. Title.

TK7885.7.P37 2003
621.39'2–dc22   2003057671

To my parents

B. Bala Tripura Sundari

To Ravi and Chandra

T.R. Padmanabhan

CONTENTS

PREFACE .... *xi*
ACKNOWLEDGEMENTS .... *xiii*

1 INTRODUCTION TO VLSI DESIGN   **1**
1.1 INTRODUCTION *1*
1.2 CONVENTIONAL APPROACH TO DIGITAL DESIGN *1*
1.3 VLSI DESIGN *3*
1.4 ASIC DESIGN FLOW *4*
1.5 ROLE OF HDL *9*

2 INTRODUCTION TO VERILOG   **11**
2.1 VERILOG AS AN HDL *11*
2.2 LEVELS OF DESIGN DESCRIPTION *11*
2.3 CONCURRENCY *13*
2.4 SIMULATION AND SYNTHESIS *14*
2.5 FUNCTIONAL VERIFICATION *14*
2.6 SYSTEM TASKS *16*
2.7 PROGRAMMING LANGUAGE INTERFACE (PLI) *16*
2.8 MODULE *16*
2.9 SIMULATION AND SYNTHESIS TOOLS *22*
2.10 TEST BENCHES *27*

3 LANGUAGE CONSTRUCTS AND CONVENTIONS IN VERILOG   **31**
3.1 INTRODUCTION *31*
3.2 KEYWORDS *31*
3.3 IDENTIFIERS *32*
3.4 WHITE SPACE CHARACTERS *33*
3.5 COMMENTS *33*
3.6 NUMBERS *34*
3.7 STRINGS *36*
3.8 LOGIC VALUES *38*
3.9 STRENGTHS *39*
3.10 DATA TYPES *40*
3.11 SCALARS AND VECTORS *41*
3.12 PARAMETERS *42*

3.13 MEMORY 43
3.14 OPERATORS 43
3.15 SYSTEM TASKS 44
3.16 EXERCISES 46

4 GATE LEVEL MODELING – 1   **47**
4.1 INTRODUCTION 47
4.2 AND GATE PRIMITIVE 47
4.3 MODULE STRUCTURE 50
4.4 OTHER GATE PRIMITIVES 51
4.5 ILLUSTRATIVE EXAMPLES 51
4.6 TRI-STATE GATES 64
4.7 ARRAY OF INSTANCES OF PRIMITIVES 66
4.8 ADDITIONAL EXAMPLES 69
4.9 EXERCISES 79

5 GATE LEVEL MODELING – 2   **81**
5.1 INTRODUCTION 81
5.2 DESIGN OF FLIP-FLOPS WITH GATE PRIMITIVES 81
5.3 DELAYS 91
5.4 STRENGTHS AND CONTENTION RESOLUTION 102
5.5 NET TYPES 109
5.6 DESIGN OF BASIC CIRCUITS 115
5.7 EXERCISES 124

6 MODELING AT DATA FLOW LEVEL   **127**
6.1 INTRODUCTION 127
6.2 CONTINUOUS ASSIGNMENT STRUCTURES 127
6.3 DELAYS AND CONTINUOUS ASSIGNMENTS 133
6.4 ASSIGNMENT TO VECTORS 135
6.5 OPERATORS 136
6.6 ADDITIONAL EXAMPLES 150
6.7 EXERCISES 157

7 BEHAVIORAL MODELING — 1   **159**
7.1 INTRODUCTION 159
7.2 OPERATIONS AND ASSIGNMENTS 160
7.3 FUNCTIONAL BIFURCATION 161
7.4 INITIAL CONSTRUCT 164
7.5 ALWAYS CONSTRUCT 168
7.6 EXAMPLES 170
7.7 ASSIGNMENTS WITH DELAYS 184
7.8 wait CONSTRUCT 192
7.9 MULTIPLE ALWAYS BLOCKS 195

7.10 DESIGNS AT BEHAVIORAL LEVEL *197*
7.11 BLOCKING AND NONBLOCKING ASSIGNMENTS *201*
7.12 THE case STATEMENT *205*
7.13 SIMULATION FLOW *214*
7.14 EXERCISES *217*

8 BEHAVIORAL MODELING II   **219**
8.1 INTRODUCTION *219*
8.2 if AND if–else CONSTRUCTS *219*
8.3 assign–deassign CONSTRUCT *225*
8.4 repeat CONSTRUCT *236*
8.5 for LOOP *238*
8.6 THE disable CONSTRUCT *244*
8.7 while LOOP *249*
8.8 forever LOOP *254*
8.9 PARALLEL BLOCKS *258*
8.10 force–release CONSTRUCT *261*
8.11 EVENT *266*
8.12 EXERCISES *268*

9 FUNCTIONS, TASKS, AND USER-DEFINED PRIMITIVES   **273**
9.1 INTRODUCTIUON *273*
9.2 FUNCTION *273*
9.3 TASKS *286*
9.4 USER-DEFINED PRIMITIVES (UDP) *292*
9.5 EXERCISES *302*

10 SWITCH LEVEL MODELING   **305**
10.1 INTRODUCTION *305*
10.2 BASIC TRANSISTOR SWITCHES *305*
10.3 CMOS SWITCH *318*
10.4 BIDIRECTIONAL GATES *328*
10.5 TIME DELAYS WITH SWITCH PRIMITIVES *333*
10.6 INSTANTIATIONS WITH STRENGTHS AND DELAYS *334*
10.7 STRENGTH CONTENTION WITH TRIREG NETS *334*
10.8 EXERCISES *337*

11 SYSTEM TASKS, FUNCTIONS, AND COMPILER DIRECTIVES   **339**
11.1 INTRODUCTION *339*
11.2 PARAMETERS *339*
11.3 PATH DELAYS *348*
11.4 MODULE PARAMETERS *371*
11.5 SYSTEM TASKS AND FUNCTIONS *373*
11.6 FILE-BASED TASKS AND FUNCTIONS *383*

11.7 COMPILER DIRECTIVES *385*
11.8 HIERARCHICAL ACCESS *393*
11.9 GENERAL OBSERVATIONS *404*
11.10 EXERCISES *405*

12 QUEUES, PLAS, AND FSMS   **407**
12.1 INTRODUCTION *407*
12.2 QUEUES *407*
12.3 PROGRAMMABLE LOGIC DEVICES (PLDs) *414*
12.4 DESIGN OF FINITE STATE MACHINES *418*
12.5 EXERCISES *433*

APPENDIX A (Keywords and Their Significance) *443*
APPENDIX B (Truth Tables of Gates and Switches) *447*
REFERENCES *449*
INDEX *451*

PREFACE

Verilog has rapidly become a widely accepted language for VLSI design. The language is well-structured and defined to cater to the steady increase in the size of ICs to be designed without sacrificing the advantages associated with design at the "grass roots" level. A designer aspiring to master the language in its versatility should become familiar with the various constructs in it, practice their use in real applications, and use them in combinations to be successful.

Describing a design using Verilog is only half the story: Writing Test benches, testing a design for all its desired functions, and identifying the faults and removing them remain equally challenging tasks. This book is an attempt to address these issues effectively. The constructs in Verilog are discussed through apt illustrative examples. Equal importance is given to design description and test benches. The examples have been tested with popular and commonly used simulation packages and the results reproduced. In many of the cases the tested designs have been synthesized, and the synthesized circuit has also been reproduced. "Seeing is believing": Seeing a design available as a software routine, transformed to a circuit, will add a lot to the confidence level of novices who use the book. flip-flops, counters, registers, coders, decoders, mux, demux *etc.*, have been considered at different levels of design; this should help in clarifying the perspectives regarding levels, need, and significance.

Place and significance of Verilog in VLSI design have been brought out in Chapters 1 and 2. Basics of the language, its conventions, *etc.*, are dealt with in Chapters 2 and 3. Chapters 4 and 5 form an introduction to design through Verilog. It is done at the gate level, which may be the most comfortable for the beginner. Any design, however involved it may be, can be completely realized in terms of the gate primitives of Verilog. We hope that the illustrative examples considered and the exercises at the end of the chapters, impart such a confidence to a designer. Chapter 6 is devoted to design at the data flow level. Continuous assignments using operators linking operands, which allow designs to be described more compactly but still close enough to the circuit level, form the theme of this chapter. Behavioral level design is discussed in Chapters 7 and 8. Mastery at this level – akin to the C language – is essential for a successful designer working at the system level. Functions and tasks, which facilitate structuring of designs and their orderly description, form the theme of Chapter 9. The switch primitives in Verilog constitute the link with actual VLSI implementation although their mastery is not essential to many of the designers with their higher level activities. Chapter 10 is devoted exclusively to switch level design; since it stands out from the main text flow so far, its discussion is consciously deferred to this stage. Chapter 11 forms an introduction to the system tasks and functions in Verilog and their use in typical environments. Chapter 12 deals with design using PLDs and FSMs. Though subdued, the treatment is enough to give the necessary lead to more comprehensive designs.

All the chapters have enough exercises at the end. Some help mastery of the material in the chapter, through practice; others are structured to stimulate the users to explore avenues of their own. The step-by-step build-up of a processor in Chapter 12 is of this type.

All simulation results presented in the text as part of illustrative examples, have been obtained using the "Modelsim" software of Mentor Graphics. All synthesis results wherever presented, have been obtained using the "Leonardo Spectrum" software of Mentor Graphics. These have been reproduced by courtesy of Mentor Graphics.

Users' views and suggestions are welcome; for this purpose, the website *www.aitec.amrita.edu/publications* may be accessed.

T. R. PADMANABHAN
B. BALA TRIPURA SUNDARI

July 2003

ACKNOWLEDGEMENTS

Many of our acquaintances and associates have contributed to the fruition of this venture. K.N.C. Eswaran is responsible for all the delicate and subtle touches with Word. Our colleagues — Subha, Sathyapriya, and Rajagopal — have made many useful suggestions. Anand Srinivasan helped with simulation in his own way. Ajai Narendran of the Systems Wing of our Institute has been helpful in many ways. Our families — Krishna Sudarshan, Saketh, Srikanth, Ravi, Chandra, and Uma — have put up with our transient oddities. Brahmachari Abhayamrita Chaitanya — Chief Operating Officer of Amrita Vishwa Vidyapeetham — made the Institute facilities, especially the VLSI laboratory, available for us. Dr. N. Narayana Pillai, Dean (Students), and Prof. R. Sundararajan of our Institute have been of great encouragement to us. Ms Christina Kuhnen, Associate Acquisitions Editor at IEEE Press, has been quite helpful throughout; she has effectively bridged the distance between New York and Coimbatore. The painstaking efforts of the Referees to wade through the manuscript, understand the matter and their constructive suggestions have conspicuously contributed to the book in its present form. We give our sincere thanks to all of them.

Our obeisance goes to *Mata Amritanandamayi Devi* for her commitment to societal transformation through quality education; this is a humble attempt to add another brick to the edifice being built by her.

1

INTRODUCTION TO VLSI DESIGN

1.1   INTRODUCTION

The word digital has made a dramatic impact on our society. More significant is a continuous trend towards digital solutions in all areas – from electronic instrumentation, control, data manipulation, signals processing, telecommunications *etc.*, to consumer electronics. Development of such solutions has been possible due to good digital system design and modeling techniques.

1.2   CONVENTIONAL APPROACH TO DIGITAL DESIGN

Digital ICs of SSI and MSI types have become universally standardized and have beenaccepted for use. Whenever a designer has to realize a digital function, he uses a standard set of ICs along with a minimal set of additional discrete circuitry. Consider a simple example of realizing a function as

$$Q_{n+1} = Q_n + (A\ B)$$

Here Q_n, A, and B are Boolean variables, with Q_n being the value of Q at the nth time step. Here $A\ B$ signifies the logical AND of A and B; the '+' symbol signifies the logical OR of the logic variables on either side. A circuit to realize the function is shown in Figure 1.1. The circuit can be realized in terms of two ICs – an A-O-I gate and a flip-flop. It can be directly wired up, tested, and used.

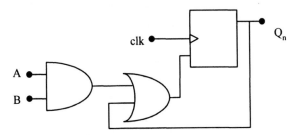

Figure 1.1 A simple digital circuit.

1

With comparatively larger circuits, the task mostly reduces to one of identifying the set of ICs necessary for the job and interconnecting; rarely does one have to resort to a microlevel design [Wakerly]. The accepted approach to digital design here is a mix of the top-down and bottom-up approaches as follows [Hill & Peterson]:

- Decide the requirements at the system level and translate them to circuit requirements.
- Identify the major functional blocks required like timer, DMA unit, register-file *etc.*, say as in the design of a processor.
- Whenever a function can be realized using a standard IC, use the same –for example programmable counter, mux, demux, *etc.*
- Whenever the above is not possible, form the circuit to carry out the block functions using standard SSI – for example gates, flip-flops, *etc.*
- Use additional components like transistor, diode, resistor, capacitor, *etc.*, wherever essential.

Once the above steps are gone through, a paper design is ready. Starting with the paper design, one has to do a circuit layout. The physical location of all the components is tentatively decided; they are interconnected and the 'circuit-on-paper' is made ready. Once a paper design is done, a layout is carried out and a net-list prepared. Based on this, the PCB is fabricated, and populated and all the populated cards tested and debugged. The procedure is shown as a process flowchart in Figure 1.2.

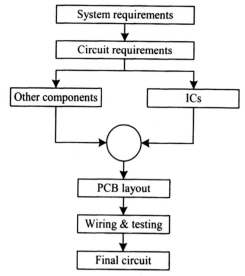

Figure 1.2 Sequence of steps in conventional electronic circuit design.

At the debugging stage one may encounter three types of problems:

- *Functional mismatch*: The realized and expected functions are different. One may have to go through the relevant functional block carefully and locate any error logically. Finally the necessary correction has to be carried out in hardware.
- *Timing mismatch*: The problem can manifest in different forms. One possibility is due to the signal going through different propagation delays in two paths and arriving at a point with a timing mismatch. This can cause faulty operation. Another possibility is a race condition in a circuit involving asynchronous feedback. This kind of problem may call for elaborate debugging. The preferred practice is to do debugging at smaller module stages and ensuring that feedback through larger loops is avoided: It becomes essential to check for the existence of long asynchronous loops.
- *Overload*: Some signals may be overloaded to such an extent that the signal transition may be unduly delayed or even suppressed. The problem manifests as reflections and erratic behavior in some cases (The signal has to be suitably buffered here.). In fact, overload on a signal can lead to timing mismatches.

The above have to be carried out after completion of the prototype PCB manufacturing; it involves cost, time, and also a redesigning process to develop a bugfree design.

1.3   VLSI DESIGN

The complexity of VLSIs being designed and used today makes the manual approach to design impractical. Design automation is the order of the day. With the rapid technological developments in the last two decades, the status of VLSI technology is characterized by the following [Wai-kai, Gopalan]:

- A steady increase in the size and hence the functionality of the ICs.
- A steady reduction in feature size and hence increase in the speed of operation as well as gate or transistor density.
- A steady improvement in the predictability of circuit behavior.
- A steady increase in the variety and size of software tools for VLSI design.

The above developments have resulted in a proliferation of approaches to VLSI design. We briefly describe the procedure of automated design flow [Rabaey, Smith MJ]. The aim is more to bring out the role of a Hardware Description Language (HDL) in the design process. An abstraction based model is the basis of the automated design.

1.3.1   Abstraction Model

The model divides the whole design cycle into various domains (see Figure 1.3). With such an abstraction through a division process the design is carried out in different layers. The designer at one layer can function without bothering about the layers above or below. The thick horizontal lines separating the layers in the figure signify the compartmentalization. As an example, let us consider design at the gate level. The circuit to be designed would be described in terms of truth tables and state tables. With these as available inputs, he has to express them as Boolean logic equations and realize them in terms of gates and flip-flops. In turn, these form the inputs to the layer immediately below. Compartmentalization of the approach to design in the manner described here is the essence of abstraction; it is the basis for development and use of CAD tools in VLSI design at various levels.

The design methods at different levels use the respective aids such as Boolean equations, truth tables, state transition table, *etc.* But the aids play only a small role in the process. To complete a design, one may have to switch from one tool to another, raising the issues of tool compatibility and learning new environments.

1.4   ASIC DESIGN FLOW

As with any other technical activity, development of an ASIC starts with an idea and takes tangible shape through the stages of development as shown in Figure 1.4 and shown in detail in Figure 1.5. The first step in the process is to expand the idea in terms of behavior of the target circuit. Through stages of programming, the same is fully developed into a design description – in terms of well defined standard constructs and conventions.

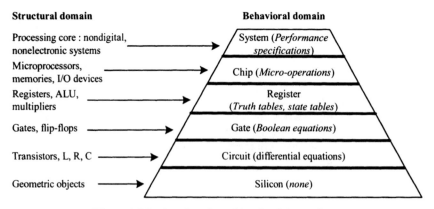

Figure 1.3 Design domain and levels of abstraction.

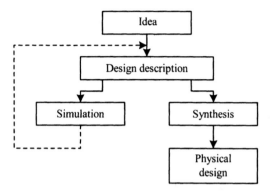

Figure 1.4 Major activities in ASIC design.

The design is tested through a simulation process; it is to check, verify, and ensure that what is wanted is what is described. Simulation is carried out through dedicated tools. With every simulation run, the simulation results are studied to identify errors in the design description. The errors are corrected and another simulation run carried out. Simulation and changes to design description together form a cyclic iterative process, repeated until an error-free design is evolved.

Design description is an activity independent of the target technology or manufacturer. It results in a description of the digital circuit. To translate it into a tangible circuit, one goes through the physical design process. The same constitutes a set of activities closely linked to the manufacturer and the target technology

1.4.1   Design Description

The design is carried out in stages. The process of transforming the idea into a detailed circuit description in terms of the elementary circuit components constitutes design description. The final circuit of such an IC can have up to a billion such components; it is arrived at in a step-by-step manner.

The first step in evolving the design description is to describe the circuit in terms of its behavior. The description looks like a program in a high level language like C. Once the behavioral level design description is ready, it is tested extensively with the help of a simulation tool; it checks and confirms that all the expected functions are carried out satisfactorily. If necessary, this behavioral level routine is edited, modified, and rerun – all done manually. Finally, one has a design for the expected system – described at the behavioral level. The behavioral design forms the input to the synthesis tools, for circuit synthesis. The behavioral constructs not supported by the synthesis tools are replaced by data flow and gate level constructs. To surmise, the designer has to develop synthesizable codes for his design.

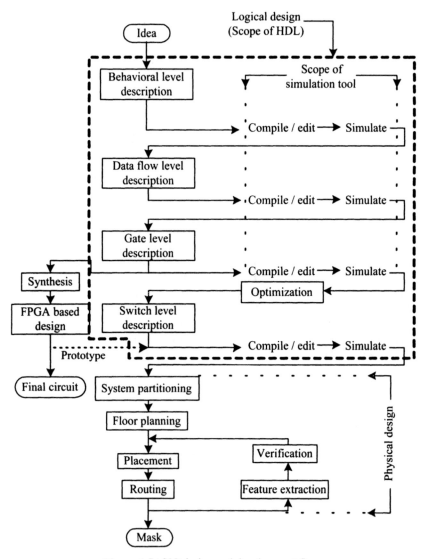

Figure 1.5ASIC design and development flow.

The design at the behavioral level is to be elaborated in terms of known and acknowledged functional blocks. It forms the next detailed level of design description. Once again the design is to be tested through simulation and iteratively corrected for errors. The elaboration can be continued one or two steps further. It leads to a detailed design description in terms of logic gates and transistor switches.

1.4.2   Optimization

The circuit at the gate level – in terms of the gates and flip-flops – can be redundant in nature. The same can be minimized with the help of minimization tools. The step is not shown separately in the figure. The minimized logical design is converted to a circuit in terms of the switch level cells from standard libraries provided by the foundries. The cell based design generated by the tool is the last step in the logical design process; it forms the input to the first level of physical design [Micheli].

1.4.3   Simulation

The design descriptions are tested for their functionality at every level – behavioral, data flow, and gate. One has to check here whether all the functions are carried out as expected and rectify them. All such activities are carried out by the simulation tool. The tool also has an editor to carry out any corrections to the source code. Simulation involves testing the design for all its functions, functional sequences, timing constraints, and specifications.   Normally testing and simulation at all the levels – behavioral to switch level – are carried out by a single tool; the same is identified as "scope of simulation tool" in Figure 1.5.

1.4.4   Synthesis

With the availability of design at the gate (switch) level, the logical design is complete.   The corresponding circuit hardware realization is carried out by a synthesis tool. Two common approaches are as follows:

- The circuit is realized through an FPGA [Oldfield]. The gate level design description is the starting point for the synthesis here. The FPGA vendors provide an interface to the synthesis tool. Through the interface the gate level design is realized as a final circuit. With many synthesis tools, one can directly use the design description at the data flow level itself to realize the final circuit through an FPGA. The FPGA route is attractive for limited volume production or a fast development cycle.
- The circuit is realized as an ASIC. A typical ASIC vendor will have his own library of basic components like elementary gates and flip-flops. Eventually the circuit is to be realized by selecting such components and interconnecting them conforming to the required design. This constitutes the physical design. Being an elaborate and costly process, a physical design may call for an intermediate functional verification through the FPGA route. The circuit realized through the FPGA is tested as a prototype. It provides another opportunity for testing the design closer to the final circuit.

1.4.5   Physical Design

A fully tested and error-free design at the switch level can be the starting point for a physical design [Baker & Boyce, Wolf]. It is to be realized as the final circuit using (typically) a million components in the foundry's library. The step-by-step activities in the process are described briefly as follows:

- *System partitioning*: The design is partitioned into convenient compartments or functional blocks. Often it would have been done at an earlier stage itself and the software design prepared in terms of such blocks. Interconnection of the blocks is part of the partition process.
- *Floor planning*: The positions of the partitioned blocks are planned and the blocks are arranged accordingly. The procedure is analogous to the planning and arrangement of domestic furniture in a residence. Blocks with I/O pins are kept close to the periphery; those which interact frequently or through a large number of interconnections are kept close together, and so on. Partitioning and floor planning may have to be carried out and refined iteratively to yield best results.
- *Placement*: The selected components from the ASIC library are placed in position on the "Silicon floor." It is done with each of the blocks above.
- *Routing*: The components placed as described above are to be interconnected to the rest of the block: It is done with each of the blocks by suitably routing the interconnects. Once the routing is complete, the physical design cam is taken as complete. The final mask for the design can be made at this stage and the ASIC manufactured in the foundry.

1.4.6   Post Layout Simulation

Once the placement and routing are completed, the performance specifications like silicon area, power consumed, path delays, *etc.*, can be computed. Equivalent circuit can be extracted at the component level and performance analysis carried out. This constitutes the final stage called "verification." One may have to go through the placement and routing activity once again to improve performance.

1.4.7   Critical Subsystems

The design may have critical subsystems. Their performance may be crucial to the overall performance; in other words, to improve the system performance substantially, one may have to design such subsystems afresh. The design here may imply redefinition of the basic feature size of the component, component design, placement of components, or routing done separately and specifically for the subsystem. A set of masks used in the foundry may have to be done afresh for the purpose.

1.5   ROLE OF HDL

An HDL provides the framework for the complete logical design of the ASIC. All the activities coming under the purview of an HDL are shown enclosed in bold dotted lines in Figure 1.4. Verilog and VHDL are the two most commonly used HDLs today. Both have constructs with which the design can be fully described at all the levels. There are additional constructs available to facilitate setting up of the test bench, spelling out test vectors for them and "observing" the outputs from the designed unit.

IEEE has brought out Standards for the HDLs, and the software tools conform to them. Verilog as an HDL was introduced by Cadence Design Systems; they placed it into the public domain in 1990. It was established as a formal IEEE Standard in 1995. The revised version has been brought out in 2001. However, most of the simulation tools available today conform only to the 1995 version of the standard.

Verilog HDL used by a substantial number of the VLSI designers today is the topic of discussion of the book.

2

INTRODUCTION TO VERILOG

2.1   VERILOG AS AN HDL

Verilog has a variety of constructs as part of it. All are aimed at providing a functionally tested and a verified design description for the target FPGA or ASIC. The language has a dual function – one fulfilling the need for a design description and the other fulfilling the need for verifying the design for functionality and timing constraints like propagation delay, critical path delay, slack, setup, and hold times [Smith DJ, Wai-Kai].

Verilog as an HDL has been introduced here and its overall structure explained. A widely used development tool for simulation and synthesis has been introduced; the brief procedural explanation provided suffices to try out the Examples and Exercises in the text.

2.2   LEVELS OF DESIGN DESCRIPTION

The components of the target design can be described at different levels with the help of the constructs in Verilog.

2.2.1   Circuit Level

At the circuit level, a switch is the basic element with which digital circuits are built. Switches can be combined to form inverters and other gates at the next higher level of abstraction. Verilog has the basic MOS switches built into its constructs, which can be used to build basic circuits like inverters, basic logic gates, simple 1-bit dynamic and static memories. They can be used to build up larger designs to simulate at the circuit level, to design performance critical circuits. Figure 2.1 shows the circuit of an inverter suitable for description with the switch level constructs of Verilog.

2.2.2   Gate Level

At the next higher level of abstraction, design is carried out in terms of basic gates.
All the basic gates are available as ready modules called "Primitives." Each such
primitive is defined in terms of its inputs and outputs. Primitives can be
incorporated into design descriptions directly. Just as full physical hardware can
be built using gates, the primitives can be used repeatedly and judiciously to build
larger systems. Figure 2.2 shows an AND gate suitable for description using the
gate primitive of Verilog. The gate level modeling or structural modeling as it is
sometimes called is akin to building a digital circuit on a bread board, or on a
PCB. One should know the structure of the design to build the model here. One
can also build hierarchical circuits at this level. However, beyond 20 to 30 of such
gate primitives in a circuit, the design description becomes unwieldy; testing and
debugging become laborious.

2.2.3   Data Flow

Data flow is the next higher level of abstraction. All possible operations on signals
and variables are represented here in terms of assignments. All logic and algebraic
operations are accommodated. The assignments define the continuous functioning
of the concerned block. At the data flow level, signals are assigned through the
data manipulating equations. All such assignments are concurrent in nature. The
design descriptions are more compact than those at the gate level. Figure 2.3
shows an A-O-I relationship suitable for description with the Verilog constructs at
the data flow level.

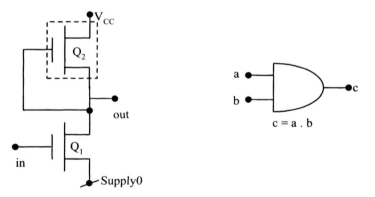

Figure 2.1 A simple Inverter circuit at the   **Figure 2.2** A simple AND gate represented
switch level.   at the gate level.

2.2.4 Behavioral Level

Behavioral level constitutes the highest level of design description; it is essentially at the system level itself [Bhaskar]. With the assignment possibilities, looping constructs and conditional branching possible, the design description essentially looks like a "C" program. The statements involved are "dense" in function. Compactness and the comprehensive nature of the design description make the development process fast and efficient. Figure 2.4 shows an A-O-I gate expressed in pseudo code suitable for description with the behavioral level constructs of Verilog.

2.2.5 The Overall Design Structure in Verilog

The possibilities of design description statements and assignments at different levels necessitate their accommodation in a mixed mode. In fact the design statements coexisting in a seamless manner within a design module is a significant characteristic of Verilog. Thus Verilog facilitates the mixing of the above-mentioned levels of design. A design built at data flow level can be instantiated to form a structural mode design. Data flow assignments can be incorporated in designs which are basically at behavioral level.

2.3 CONCURRENCY

In an electronic circuit all the units are to be active and functioning concurrently. The voltages and currents in the different elements in the circuit can change simultaneously. In turn the logic levels too can change. Simulation of such a circuit in an HDL calls for concurrency of operation. A number of activities – may be spread over different modules – are to be run concurrently here. Verilog simulators are built to simulate concurrency. (This is in contrast to programs in the normal languages like C where execution is sequential.) Concurrency is achieved by proceeding with simulation in equal time steps. The time step is kept small enough to be negligible compared with the propagation delay values. All the activities scheduled at one time step are completed and then the simulator

$e = \overline{a.b + c.d}$	If $(a, b, c$ or d changes) Compute e as $e = \overline{a.b + c.d}$

Figure 2.3 An A-O-I gate represented as a data flow type of relationship.

Figure 2.4 An A-O-I gate in pseudo code at behavioral level.

advances to the next time step and so on. The time step values refer to simulation time and not real time. One can redefine timescales to suit technology as and when necessary and carry out test runs.

In some cases the circuit itself may demand sequential operation as with data transfer and memory-based operations. Only in such cases sequential operation is ensured by the appropriate usage of sequential constructs from Verilog HDL.

2.4   SIMULATION AND SYNTHESIS

The design that is specified and entered as described earlier is simulated for functionality and fully debugged. Translation of the debugged design into the corresponding hardware circuit (using an FPGA or an ASIC) is called "synthesis." The tools available for synthesis relate more easily with the gate level and data flow level modules [Smith MJ]. The circuits realized from them are essentially direct translations of functions into circuit elements. In contrast many of the behavioral level constructs are not directly synthesizable; even if synthesized they are likely to yield relatively redundant or wrong hardware. The way out is to take the behavioral level modules and redo each of them at lower levels. The process is carried out successively with each of the behavioral level modules until practically the full design is available as a pack of modules at gate and data flow levels (more commonly called the "RTL level").

2.5   FUNCTIONAL VERIFICATION

Testing is an essential ingredient of the VLSI design process as with any hardware circuit. It has two dimensions to it – functional tests and timing tests. Both can be carried out with Verilog. Often testing or functional verification is carried out by setting up a "test bench" for the design. The test bench will have the design instantiated in it; it will generate necessary test signals and apply them to the instantiated design. The outputs from the design are brought back to the test bench for further analysis. The input signal combinations, waveforms and sequences required for testing are all to be decided in advance and the test bench configured based on the same.

The test benches are mostly done at the behavioral level. The constructs there are flexible enough to allow all types of test signals to be generated.

In the process of testing a module, one may have to access variables buried inside other modules instantiated within the master module. Such variables can be accessed through suitable hierarchical addressing.

2.5.1   Test Inputs for Test Benches

Any digital system has to carry out a number of activities in a defined manner. Once a proper design is done, it has to be tested for all its functional aspects. The system has to carry out all the expected activities and not falter. Further, it should not malfunction under any set of input conditions. Functional testing is carried out to check for such requirements. Test inputs can be purely combinational, periodic, numeric sequences, random inputs, conditional inputs, or combinations of these. With such requirements, definition and design of test benches is often as challenging as the design itself.

As the circuit design proceeds, one develops smaller blocks and groups them together to form bigger circuit units. The process is repeated until the whole system is fully built up. Every stage calls for tests to see whether the subsystem at that layer behaves in the manner expected. Such testing calls for two types of observations:

- Study of signals within a small unit when test inputs are given to the whole unit.
- Isolation of a small element and doing local test to facilitate debugging.

Verilog has constructs to accommodate both types of observation through a hierarchical description of variables within.

2.5.2   Constructs for Modeling Timing Delays

Any basic gate has propagation delays and transmission delays associated with it. As the elements in the circuit increase in number, the type and variety of such delays increase rapidly; often one reaches a stage where the expected function is not realized thanks to an unduly large time delay. Thus there is a need to test every digital design for its performance with respect to time. Verilog has constructs for modeling the following delays:

- Gate delay
- Net delay
- Path delay
- Pin-to-pin delay

In addition, a design can be tested for setup time, hold time, clock-width time specifications, *etc.* Such constructs or delay models are akin to the finite delay time, rise time, fall time, path or propagation delays, *etc.*, associated with real digital circuits or systems. The use of such constructs in the design helps simulate realistic conditions in a digital circuit. Further, one can change the values of delays in different ways. If a buffer capacity is increased, its associated delays can be reduced. If a design is to migrate to a better technology, the delay values can be rescaled. With such testing, one can estimate the minimum frequency of operation, the maximum speed of response, or typical response times.

2.6   SYSTEM TASKS

A number of system tasks are available in Verilog. Though used in a design description, they are not part of it. Some tasks facilitate control and flow of the testing process. The values of signals in a module can be displayed in the course of simulation. The tasks available for the purpose display them in desired formats. Reading data from specified files into a module and writing back into files are also possible through other tasks. Timescale can be changed prior to simulation with the help of specific tasks for the purpose.

A set of system functions add to the flexibility of test benches: They are of three categories:

- Functions that keep track of the progress of simulation time
- Functions to convert data or values of variables from one format to another
- Functions to generate random numbers with specific distributions.

There are other numerous system tasks and functions associated with file operations, PLAs, *etc.*

2.7   PROGRAMMING LANGUAGE INTERFACE (PLI)

PLI provides an active interface to a compiled Verilog module. The interface adds a new dimension to working with Verilog routines from a C platform. The key functions of the interface are as follows:

- One can read data from a file and pass it to a Verilog module as input. Such data can be test vectors or other input data to the module. Similarly, variables in Verilog modules can be accessed and their values written to output devices.
- Delay values, logic values, *etc.*, within a module can be accessed and altered.
- Blocks written in C language can be linked to Verilog modules.

2.8   MODULE

Any Verilog program begins with a keyword – called a "**module.**" A **module** is the name given to any system considering it as a black box with input and output terminals as shown in Figure 2.5. The terminals of the module are referred to as 'ports'. The ports attached to a module can be of three types:

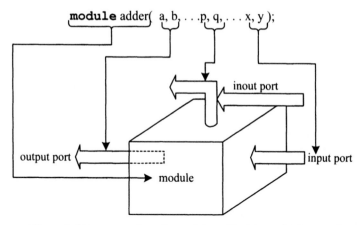

Figure 2.5 Representation of a module as black box with its ports.

- **input** ports through which one gets entry into the module; they signify the input signal terminals of the module.
- **output** ports through which one exits the module; these signify the output signal terminals of the module.
- **inout** ports: These represent ports through which one gets entry into the module or exits the module; These are terminals through which signals are input to the module sometimes; at some other times signals are output from the module through these.

Whether a module has any of the above ports and how many of each type are present depend solely on the functional nature of the module. Thus one module may not have any port at all, another may have only input ports, while a third may have only output ports, and so on.

All the constructs in Verilog are centered on the module. They define ways of building up, accessing, and using modules. The structure of modules and the mode of invoking them in a design are discussed here.

A module comprises a number of "lexical tokens" arranged according to some predefined order. The possible tokens are of seven categories:

- White spaces
- Comments
- Operators
- Numbers
- Strings
- Identifiers
- Keywords

The rules constraining the tokens and their sequencing will be dealt with as we progress. For the present let us consider modules. In Verilog any program which forms a design description is a "module." Any program written to test a design description is also a "module." The latter are often called as "stimulus modules" or "test benches." A module used to do simulation has the form shown in Figure 2.6. Verilog takes the active statements appearing between the "**module**" statement and the "**endmodule**" statement and interprets all of them together as forming the body of the module. Whenever a module is invoked for testing or for incorporation into a bigger design module, the name of the module ("**test**" here) is used to identify it for the purpose.

A digression into design using SSI ICs is in order here. Consider the IC 7430, an eight input NAND gate. In any design using it, the IC can be looked up on as a black box with eight input leads and one output lead (Figure 2.7a). Three aspects characterize the IC – its function, its input leads, and its output lead. Other ICs may have more output leads. A NAND gate module is defined in an analogous manner in terms of its function, input leads and the output lead. The module used to describe the circuit here also follows the earlier format; that is, the "**module**" statement signifies the beginning of the module, the "**endmodule**" statement signifies the end of the module. However, the initial statement "**module**" has to be more elaborate with the input and the output ports forming part of it (see Figure 2.7b).

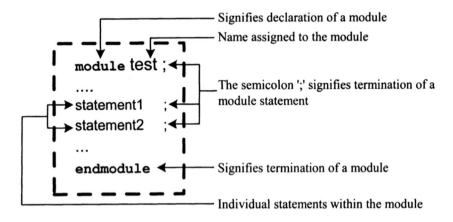

Figure 2.6 Structure of a typical simulation module.

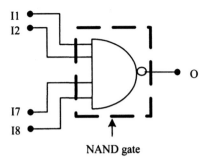

NAND gate

Figure 2.7(a) Eight input NAND gate (IC 7430). Gate proper with terminals.

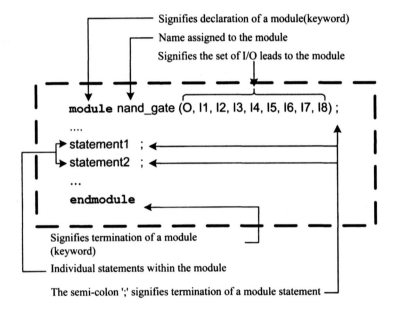

Signifies declaration of a module(keyword)

Name assigned to the module

Signifies the set of I/O leads to the module

module nand_gate (O, I1, I2, I3, I4, I5, I6, I7, I8) ;

....

statement1 ;

statement2 ;

...

endmodule

Signifies termination of a module
(keyword)

Individual statements within the module

The semi-colon ';' signifies termination of a module statement

Figure 2.7(b) Eight input NAND gate (IC 7430). Structure of the gate module.

The same type of IC – 7430 – may be repeatedly used in a circuit. Each time it is used, a different name is assigned to it in the design sheet. Part of such a typical design sheet will look as in Figure 2.8. The associated table (Table 2.1) allows us to identify each type of IC to be used and put in its proper place. An automated design description can use a module defined above, repeatedly in a number of places as in the circuit of Figure 2.8. Each such use is an "instantiation." A typical instantiation of the module defined above has the form shown in Figure 2.9. The following observations are in order here:

Table 2.1 Partial list of IC numbers and their types for a typical design

IC No	IC1	IC2	IC3	...	IC9	...
IC type	7430	7430		...	7405	...

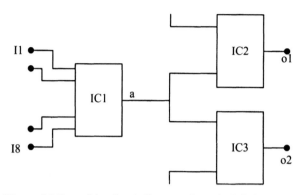

Figure 2.8 Part of the circuit diagram of a typical digital circuit.

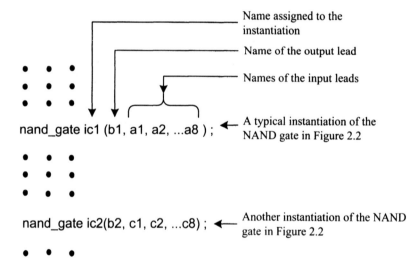

Figure 2.9 Instantiations of module nand_gate in another module.

- The designer has defined a specific function within a module; the module is assigned the name "nand_gate."
- The nand_gate can be invoked (instantiated) by him in a design as many times as desired.
- Each instantiation has to be assigned a separate identifier name by him (called "IC1", "IC2", *etc.*).

- As part of the instantiation declaration, the input and output terminals are to be defined. The convention followed is to stick to the same order as in the module declaration. It is further illustrated in Figure 2.9.

Some modules may have a large number of ports. Sticking to the order of the ports in an instantiation is likely to cause (human) errors. An alternative (and sometimes more convenient) form of instantiation is also possible – shown in Figure 2.10. The terminal identifications are explicit (though elaborate) here. Further one need not stick to the order of the ports as they appear in the module definition. With such a form of port assignments, the possibility of errors is considerably reduced.

The following aspects of the modules and their instantiation are noteworthy:

- Each module can be defined only once.
- Module definitions are to be done independently. One module cannot be defined inside another – they cannot be nested.
- Any module can be instantiated inside another any number of times. Each instantiation has to be done with a separate name assigned to it.

The various constructs and features available in Verilog are discussed in the following chapters. However, certain conventions and constructs essential for the progress of the book at this stage are discussed in Chapter 3.

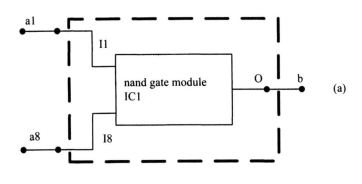

nand_gate ic1(O(b), I8(a8), ...   I1(a1));   (b)

Figure 2.10 (a) A typical circuit block and (b) its instantiation.

2.9   SIMULATION AND SYNTHESIS

A variety of Software tools related to VLSI design is available. We discuss here
two of them directly relevant to us – Modelsim and Leonardo Spectrum of Mentor
Graphics. Modelsim has been used to simulate the designs. Simulation results
presented for the variety of examples discussed in the book have been obtained
using it. Leonardo Spectrum has been used to obtain the synthesized circuits
presented. We would like to draw the attention of the readers to the following in
this context:

- Only the essential aspects of the tools are presented – those essential for
  the progress of the book.
- For more details of the tools and the variety of facilities they offer, one
  can refer to the respective user manuals and the Help menus.
- Tools from other sources are similar in essentials. Any of them can be
  used.

2.9.1   Use of Modelsim SE 5.5

The procedure to invoke the tool and use it is briefly described here. The tool can
be used to prepare a source file, edit and compile it, and simulate the compiled
version.

Editing and Compilation

- Open the Modelsim Window. We get the following menus listed at the top:

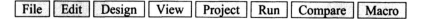

File | Edit | Design | View | Project | Run | Compare | Macro

- Click on "View." We get the following menus:

All
Hide Workspace
Sources
Structure
Variables
Signals
List
Process
Wave

Data flow
Data sets
New
Other

- Click on "Source." The "Source" window opens with the following set of menus listed at the top:

File	Edit	Object	Options	Window

- Click on "File" option. We get the following options:

New
Open
Use source
Source directory
Properties
Save
Save as
Compile
Close

- Click on "New." We get the following options:

VHDL
Verilog
Others

- Click on "Verilog." A "Source_edit-new.v" opens.
  The Verilog design can be keyed in. It forms the source file. The source file considered in various examples in the book can be created in this manner (*e.g.*, Example 4.2 and Figure 4.4).
- Click on "File" option. We get a pull down menu.
- Click on "Save as."
  Select a Directory of your choice. Give a suitable filename with extension ".v" (Say "demo.v"). Click on "Save" and save the file. The source (design) file has been created and saved. Now it is ready for compilation.

- Click on "Compile." "Compile HDL Source Files" window opens. File name "demo" is displayed. Library "Work" is displayed. The selected file (demo.v) will be compiled and loaded into Work. The lines of display in the main window confirm this.
- If the source file has any syntax or logical errors, compilation will not take place. The errors will be indicated in the main window. The source file can be opened (by clicking on the main menu) and edited. Once again compilation can be attempted. The procedure has to be repeated iteratively until all the errors in the source file have been removed and compilation is successfully completed.

Simulation

- In the main window click on "Design" pulldown menu.
- In the options displayed, click on "Load Design." The following options are displayed at the top:–

- Select "Design" and click on it. A small window appears on the screen. "Library: Work" is displayed, implying that the working library is open. The module name "demo" is displayed under it. In the normal course the names of all the compiled files will be listed alphabetically one below the other. The specific file to be simulated is to be selected by clicking on the same.
- The "Load" button below gets highlighted. Click on it. The design gets loaded and is ready for simulation run.
- Click the "Run" menu in the Modelsim main window. Select 100 ns runtime.
- The design runs for 100 ns (by default) and the output list appears in the main window. The listing can be selected, copied, and pasted to another file. The simulation results for the various examples in the book have been obtained in this manner. If necessary, the time duration of simulation can be altered in the main window.

Observing Waveforms

Simulation results can alternately be viewed as waveforms with the following procedure:

- In the main Modelsim window click on "Signals." The signals window opens with the following options displayed at the top:

- Click on the "View" pulldown menu. We get the options as shown below:

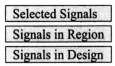

```
Wave
List
Log
Filter
```

- Amongst the options available, click on "Wave." We get the following options:

```
Selected Signals
Signals in Region
Signals in Design
```

- Select "Signals in Design." The "Waveform Window" opens and shows the signals in the design. The Window has a "Run" option.
- Click on "Run" to run the design and get the waveforms displayed.
  The waveforms shown as simulated outputs for different examples in the book have been obtained in this manner.

One can practice simulation of a few examples given in the book. Subsequently options available at the different stages can be tried, and the tool with its full versatility can be mastered.

2.9.2   Synthesis

Conversion of the code into hardware logic and fitting it into an FPGA or ASIC to realize the circuit is termed "Synthesis." We have used the Mentor Graphics Synthesis tool called "Leonardo Spectrum" for the purpose. The synthesis procedure is briefly described here:

- Double click on "Leonardo Spectrum 2000.1b."
- The Main Window named "Examplar Logic – Leonardo Spectrum Level 3"opens with a pulldown menu as follows:

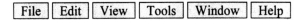

```
File    Edit    View    Tools    Window    Help
```

- Click on "File". A pulldown menu opens with options such as the following:

- Select "New." A window named "untitled" opens. We can type in a new program and save it as a file with a name assigned to it (Say "name.v") in a directory of our choice. The procedure is similar to that followed above to create and save a new file with extension ".v" (signifying that it is a Verilog file). The file is now ready for synthesis. However, it is always preferable to simulate a file and be fully satisfied with at the simulation stage itself before synthesizing it.
- Click on the "Tools" menu on the main window. A set of options appear on the screen.
- Select "Quick Set up." A window of the type shown in Figure 2.11 appears. All the settings necessary to complete the synthesis can be carried out with it.
- Click on "Open files." Select the Verilog source file to be synthesized. It will be visible under "Input" in the figure.
- Under "Technology" select "FPGA." Select a device of (say) Xilinx – for example, XC4000XL. The selected Xilinx device name is displayed under 'Device'.
- Select a "Clock Frequency" – say 10 MHz.
- Click on the "Run Flow" button. The synthesis program runs and completes the synthesis. Summarized results will be displayed on the screen.
- If the coding is correct and synthesizable, the display "Ready" appears highlighted at the bottom left-hand corner. If not, error details will be displayed. The program may be rectified and synthesis attempted again. Icons for "RTL Schematic", "Gate Level Schematic" and "Critical Path Schematic" at the top become active.
- We can click on each of them in succession. The circuit schematic can be viewed at the RTL level or the gate level. The critical path can be viewed – it represents the path that takes the maximum time of operation on a pin-to-pin basis. It sets the upper limit to the speed of operation of the circuit.

The synthesized circuits shown for the different examples in the book have been obtained in this manner. The device selected to synthesize the design, is called the "Target Device." One can select any other suitable target device of Xilinx or other FPGA vendors like Actel, Altera, Cypress, Lattice, Lucent, Quicklogic, *etc.*

The program generates a summary of the synthesis activity and displays it as a "Sum File." It gives a report on the utilization of the "Target Device" by the

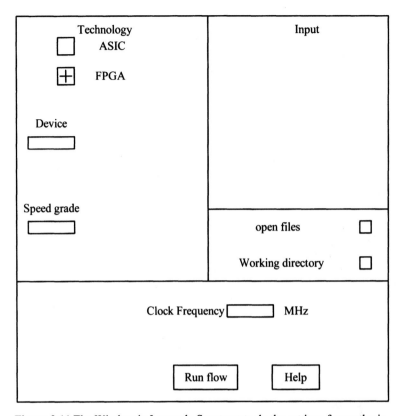

Figure 2.11 The Window in Leonardo Spectrum to do the settings for synthesis.

design that was synthesized. It also generates and displays some timing information like "Critical Path Timing."

2.10   TEST BENCHES

Any digital circuit that has been designed and wired goes through a testing process before being declared as ready for use. Testing involves studying circuit behavior under simulated conditions for the following:

- Check and ensure that all functions are carried out as desired. It is the test for the static behavior of the circuit. A set of logic input values are applied at selected points and the logic values at another set of points observed.
- Check and ensure that all the functional sequences are carried out as desired. It is one of the tests for the dynamic behavior of the circuit. It may call for the

generation of specific input sequences with respect to time, applying them to the circuit and observing selected outputs.
- Check for the timing behavior: One tests for the propagation and other types of delays here. A variety of tests may have to be carried out. It may involve observation of variations in the signals at selected points, measuring the time delay between specified events, measuring pulse widths, and so on.

Verilog has the provision for all the above. One sets up a "test bench" in software and caries out a simulated test. The facilities required to set up test benches are discussed in detail in Chapters 7 and 8. However, the need to test the designs in Chapters 4 to 6 warrants a brief introduction to them here; only the essentials are discussed. Further, the "test benches" up to Chapter 7 are kept simple and easily understandable.

Simulated testing is a time-based activity. It is usually carried out in simulated time. With any simulation tool the simulation progresses through equal simulation time steps. The time step can be 1 fs, 1 ps, 1 ns and so on. In the text the default value is taken as 1 ns. In some cases it is mentioned explicitly; in other situations it is implicit, *that is,* whenever 'time step' is mentioned, it implies 1ns of simulation time. If required, the simulation time step can be altered (see Chapter 11).

Consider the group of statements below reproduced from the test bench of Figure 4.1:

```
Initial
Begin
        a1 = 0;
        a2 = 0;
#3      a1 = 1;
#1      a1 = 0;
#2      a2 = 1;
#4      a1 = 1;
#3      a2 = 0;
#1      a2 = 1;
end
and g1(b, a1, a2);
initial $monitor ( $time, "a1 = %b, a2 = %b, b = %b" a1, a2, b);
#100 $finish;
```

The **keyword** initial is followed by a sequence of statements between the keywords **begin** and **end**. Usually the **initial** banner signifies a setting done on a once or a "once for all" basis. The "# 3" implies a time delay or wait time of 3 time steps in simulation. Thus the sequence implies the following:

- At 0 simulation time the logic variables a1 and a2 are assigned the logic level 0.

- With a delay of 3 ns a1 is reassigned the logic value of 1.
- With a further delay of 1 ns – that is, at the 4th ns - a1 is reverted to the logic level 0.
- Similarly at the 6th, 10th, 13th and 14th ns values of simulation time, further changes are made to a1 and a2.
- Note that every time value specified here is an increment in simulation time.

The values of a1 and a2 are not changed beyond the 14th ns. The statement

initial # 100 $**finish**;

implies that the simulation is to be continued up to the 100th ns of simulation time and then stopped.

The above constitutes the generation of the test sequence for testing. Such test signals are applied to the designed circuit through instantiation; the statement

and g1(b, a1, a2);

implies as much. The statement

initial $monitor ($**time**, "a1 = %b, a2 = %b, b = %b"' a1, a2, b);

monitors a1, a2, and a3 for changes; whenever any of them changes, all of them are sampled and the sampled values displayed.

Summarizing testing constitutes three activities:

- Generation of the test signals – under the "**initial**" banner
- Application of the test signal to the circuit under test – through instantiation
- Observing selected signal values – through the $**monitor** statement

Many of the test benches for the subsequent examples are also structured in a similar fashion. Changes are kept to the minimum to ensure focus on the example concerned. As and when such changes are made, the same is explained.

3

LANGUAGE CONSTRUCTS AND CONVENTIONS IN VERILOG

3.1   INTRODUCTION

The constructs and conventions make up a software language. A clear understanding and familiarity of these is essential for the mastery of the language. Verilog has its own constructs and conventions [IEEE, Sutherland]. In many respects they resemble those of C language [Gottfried]. We discuss the constructs and conventions essential to the progress of the book. More of these follow in the ensuing chapters.

Any source file in Verilog (as with any file in any other programming language) is made up of a number of ASCII characters. The characters are grouped into sets — referred to as "lexical tokens." A lexical token in Verilog can be a single character or a group of characters. Verilog has 7 types of lexical tokens — operators, keywords, identifiers, white spaces, comments, numbers, and strings. Operators are introduced in Chapter 6. All the other tokens are discussed here. Some other aspects of Verilog essential to the progress of the book are also discussed subsequently.

3.1.1   Case Sensitivity

Verilog is a case-sensitive language like C. Thus sense, Sense, SENSE, sENse,... *etc.*, are all treated as different entities / quantities in Verilog.

3.2   KEYWORDS

The keywords define the language constructs. A keyword signifies an activity to be carried out, initiated, or terminated. As such, a programmer cannot use a keyword for any purpose other than that it is intended for. All keywords in

Verilog are in small letters and require to be used as such (since Verilog is a case-sensitive language). All keywords appear in the text in New Courier Bold-type letters.

Examples

module ← signifies the beginning of a module definition.
endmodule ← signifies the end of a module definition.
begin ← signifies the beginning of a block of statements.
end ← signifies the end of a block of statements.
if ← signifies a conditional activity to be checked
while ← signifies a conditional activity to be carried out.

A list of keywords in Verilog with the significance of each is given in Appendix A.

3.3   IDENTIFIERS

Any program requires blocks of statements, signals, *etc.*, to be identified with an attached nametag. Such nametags are identifiers. It is good practice for us to use identifiers, closely related to the significance of variable, signal, block, *etc.*, concerned. This eases understanding and debugging of any program.

e.g., clock, enable, gate_1, . . .

There are some restrictions in assigning identifier names. All characters of the alphabet or an underscore can be used as the first character. Subsequent characters can be of alphanumeric type, or the underscore (_), or the dollar ($) sign – for example

name, _name. Name, name1, name_$, . . . ← all these are allowed as identifiers
name aa ← not allowed as an identifier because of the blank ("name" and "aa" are interpreted as two different identifiers)
$name ← not allowed as an identifier because of the presence of "$" as the first character.
1_name ← not allowed as an identifier, since the numeral "1" is the first character
@name ← not allowed as an identifier because of the presence of the character "@".
A+b ← not allowed as an identifier because of the presence of the character "+".

An alternative format makes it is possible to use any of the printable ASCII characters in an identifier. Such identifiers are called "escaped identifiers"; they have to start with the backslash (\) character. The character set between the first backslash character and the first white space encountered is treated as an identifier. The backslash itself is not treated as a character of the identifier concerned.

Examples

\b=c
\control-signal
\&logic
\abc // Here the combination "abc" forms the identifier.

It is preferable to use the former type of identifiers and avoid the escaped identifiers; they may be reserved for use in files which are available as inputs to the design from other CAD tools.

3.4   WHITE SPACE CHARACTERS

Blanks (\b), tabs (\t), newlines (\n), and formfeed form the white space characters in Verilog. In any design description the white space characters are included to improve readability. Functionally, they separate legal tokens. They are introduced between keywords, keyword and an identifier, between two identifiers, between identifiers and operator symbols, and so on. White space characters have significance only when they appear inside strings.

3.5   COMMENTS

It is a healthy practice to comment a design description liberally – as with any other program. Comments are incorporated in two ways. A single line comment begins with "//" and ends with a new line – for example

module d_ff (Q, dp, clk); //This is the design description of a D flip-flop.

//Here Q is the output.

// dp is the input and clk is the clock.

One can incorporate multiline comments also without resorting to "//" at every line. For such multiline comments "/*" signifies the beginning of a comment and "*/" its end. All lines appearing between these two symbol combinations are together treated as a single block comment – for example

module d_ff (Q, dp, clk);

/* This module forms the design description of a d_flip_flop wherein
   Q is the output of the flip-flop ,
   dp is the data input and
   clk the clock input*/

Multiline comments cannot be nested. For example, the following comment is not valid.

/*The following forms the design description of a D flip-flop /*which can be modified to form other types of flip-flops*/ with clock and data inputs.*/

A valid alternative can be as follows: -

/*The following forms the design description of a D flip-flop (which can be modified to form other types of flip-flops) with clock and data inputs.*/

3.6   NUMBERS

Frequently numbers need to be specified in a design description. Logic status of signal lines, buses, delay values, and numbers to be loaded in registers are examples. The numbers can be of integer type or real type.

3.6.1   Integer Numbers

Integers can be represented in two ways. In the first case it is a decimal number – signed or unsigned; an unsigned number is automatically taken as a positive number. Some examples of valid number representations of this category are given below:

2
25
253
−253

The following are invalid since nondecimal representations are not permissible.

2a
B8
−2a
−B8

Normally the number is taken as 32 bits wide. Thus all the following numbers are assigned 32 bits of width:

2
25

253
–2
–25
–253

If a design description has a number specified in the form given here, the circuit synthesizer program will assign 32 bits of width to it and to all the related circuits. Hence all such number specifications – despite their simplicity – may be avoided in design descriptions. Number representation in this form may preferably be restricted to test benches.

The alternate form of number representation is more specific – though elaborate. The number can be specified in binary, octal, decimal, or hexadecimal form. The representation has three tokens with an optional sign preceding it. Figure 3.1 shows typical number representations with the significance of each field explained separately.

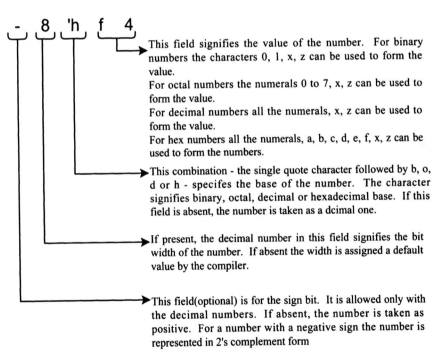

Figure 3.1 Representation of a number in Verilog: One can use capital letters instead of small letters in the last two fields.

Observations:

- The characters used to specify the base number, the sign or the magnitude can be in either case (Thus A, B, C, D, E, or F can be used in place of a, b, c, d, e, or f, respectively, to specify the concerned hex digit. **X** or **Z** can be used in place of **x** or **z** value, respectively).
- The single quote character in the base field has to be immediately followed by the character representing the base. Intervening white spaces are not allowed. However, such white spaces can precede the magnitude field.
- Negative numbers are represented in 2's complement form.
- The question mark character – "?" – can be used in place of **z**. The underscore character can be used anywhere after the first character. It adds to the readability. It is normally ignored.
- If the number size is smaller than the size specified, the size is made up by padding 0's to the left. However, if the leftmost bit is a **x** or **z**, the same is padded to the left.
- Left truncation and right extension can often be confusing. It is preferable to specify the numbers fully.

Table 3.1 shows the format of specifications of the integer type numbers along with illustrative examples.

3.6.2   Real Numbers

Real numbers can be specified in decimal or scientific notation. The decimal notation has the form

-a.b

where a, b, the negative sign, and the decimal point have the usual significance. The fields a and b must be present in the number. A number can be specified in scientific notation as

4.3e2

where 4.3 is the mantissa and 2 the exponent. The decimal equivalent of this number is 430. Other examples of numbers represented in scientific notation are –4.3e2, –4.3e–2, and 4.3e–2. The representations are common.

3.7   STRINGS

A string is a sequence of characters enclosed within double quotes. A string must be contained on a single line; that is, it cannot be carried over to two lines with a

Table 3.1 Different ways of number representations in Verilog

Representation	Remarks
33 'd33	Both of these represent decimal numbers of unspecified size – normally interpreted by Verilog as 32 bitwide, *i.e.*, 0000 0000 0000 0000 0000 0000 0010 0001
9'd439 9'D439 9'D4_39	All these represent 3 digit decimal numbers. D & d both specify decimal numbers. "_" (underscore) is ignored
9'b1_1011__1x01 9'b11011x01 9'B11011x01	All these represent binary numbers of value 11011x01. B & b specify binary numbers. "_" is ignored. x signifies the concerned bit to be of unknown value.
9'o123 9'O123 9'o1x3 9'o12z	All these represent 9-bit octal numbers. The binary equivalents are 001 010 011, 001 010 011, 001 xxx 011, 001 010 zzz respectively. z signifies the concerned bits to be in the high impedance state.
'o213	An octal number of unspecified size having octal value 213.
8'ha5 8'HA5 8'hA5 8'ha_5	All these are 8 bit-wide-hex numbers of hex value a5h. The equivalent binary value is 1010 0101.
11'hb0	A 11 bit number with a hex assignment. Its value is 000 1011 0000. The number of bits specified is more than that indicated in the value field. Enough zeros are padded to the left as shown.
9'hza	A hex number of 9 bits. Its value is taken as zzzzz 1010.
5'hza	A 5-bit hex number. Its value is taken as z 1010.
5'h?a	A 5-bit hex number. Its value is taken as z 1010. '?' is another representation for 'z'.
-5'h1a -3'b101	Negative numbers. Negative numbers are represented in 2's complement form.
-4'd7	A 4 bit negative number. Its value in 2's complement form is 7. Thus the number is actually $-(16 - 7) = -9$.

carriage return. Special characters are specified by preceding them with the "\" character. Verilog treats a string as a sequence of ASCII characters – for example,

"This is a string"

"This string is one \t with a gap in between"

"This is called a \"string\"".

When a string of ASCII characters as above is an operand in an expression, it is treated as a binary number. This binary number is formed by replacing each ASCII character by 8 bits – a 0 bit followed by the 7-bit ASCII equivalent – and treating the resulting binary sequence as a single binary number. For example, the statement (with P defined as a 32-bit vector beforehand)

P = "numb"

assigns the binary value

0110 1110 0111 0101 0110 1101 0110 0010

to P (0110 1110, 0111 0101, 0110 1101 and 0110 0010 are the 8-bit equivalents of the letters n, u, m, and b, respectively).

3.8   LOGIC VALUES

Signal lines, logic values appearing on signal lines, *etc.*, can normally take two logic levels:

1 ← signifies the 1 or high or true level
0 ← signifies the 0 or low or false level.

Two additional levels are also possible – designated as **x** and **z**. Here **x** represents an unknown or an uninitialized value. This corresponds to the don't-care case in logic circuits. **z** represents / signifies a high impedance state. This is possible when a signal line is tri-stated or left floating. The following are noteworthy here:

- When a variable in an expression is in the **z** state, the effect is the same as it having **z** value. But when an input to a gate is in the **z** state (see Chapter 4), it is equivalent to having the **x** value.
- The MOS switches discussed in Chapter 10 form an exception to the above. If the input to a MOS switch is in the **z** state, its output too remains at the **z** state.
- With a few exceptions all data types in Verilog can take on all the 4 logic values or levels. The **event** (see Section 8.11) is an exception to this. It cannot store any value. The **trireg** cannot take on the **z** value (see Chapter 5).

A logic state can have a "strength" associated with it. It is a quantitative representation of the internal impedance value of the corresponding hardware circuit; a change in the internal impedance is reflected as a corresponding change in the strength level. Whenever the logic values from two sources are combined, there can be a conflict and the resulting contention has to be resolved. The strength values are discussed below. Details of contention and its resolution are discussed in Chapter 5.

3.9   STRENGTHS

The logic levels are also associated with strengths. In many digital circuits, multiple assignments are often combined to reduce silicon area or to reduce pin-outs. To facilitate this, one can assign strengths to logic levels. Verilog has eight strength levels – four of these are of the driving type, three are of capacitive type and one of the hi-Z type. Details are given in Table 3.2 (see also Section 5.4).

When a signal line is driven simultaneously from two sources of different strength levels, the stronger of the two prevails. A few illustrative examples are considered here.

- If a signal line **a** is driven by two sources – **b** at 1 level with strength "**strong1**" and **c** at level 0 with strength "**pull0**"– a will take the value 1.

3.2 Details of strengths in Verilog

Strength name	Strength level (signifies inverse of source impedance)	Specification keyword	Abbreviation	Element modeled
Supply drive	7	**Supply1**	**Su1**	Power supply connection
		Supply0	**Su0**	
Strong drive	6	**Strong1**	**St1**	Default gate and assign output strength
		Strong0	**St0**	
Pull drive	5	**Pull1**	**Pu1**	Gate and assign output strength
		Pull0	**Pu0**	
Large capacitor	4	**Large1**	**La1**	Size of trireg net capacitor
		Large0	**La0**	
Weak drive	3	**Weak1**	**We1**	Gate and assign output strength
		Weak0	**We0**	
Medium capacitor	2	**Medium1**	**Me1**	Size of trireg net capacitor
		Medium0	**Me0**	
Small capacitor	1	**Small1**	**Sm1**	Size of trireg net capacitor
		Small0	**Sm0**	
High impedance	0	**Highz1**	**Hi1**	Tri-stated line
		Highz0	**Hi0**	

- If a signal line a is driven by two sources – b at 1 level with strength "**pull1**" and c at level 0 with strength "**strong0**," a will take the value 0.
- If a signal line a is driven by two sources – b at 1 level with strength "**strong**1" and c at level 0 with strength "**strong0**," a will take the value x (indeterminate).
- If a signal line a is driven by two sources – b at 1 level with strength "**weak1**" and c at level 0 with strength "**large0**," a will take the value 0. (Note that **large** signifies a capacitive drive on a tri-stated line whereas **weak** signifies a gate / assigned output drive with a high source impedance; despite this, due to the higher strength level, the **large** signal prevails.)

The significance of strengths is further explained in Chapter 5.

3.10   DATA TYPES

The data handled in Verilog fall into two categories:

(i) Net data type

(ii) Variable data type

The two types differ in the way they are used as well as with regard to their respective hardware structures. Data type of each variable or signal has to be declared prior to its use. The same is valid within the concerned block or module.

3.10.1   Nets

A net signifies a connection from one circuit unit to another. Such a net carries the value of the signal it is connected to and transmits to the circuit blocks connected to it. If the driving end of a net is left floating, the net goes to the high impedance state. A net can be specified in different ways.

wire: It represents a simple wire doing an interconnection. Only one output is connected to a wire and is driven by that.

tri: It represents a simple signal line as a wire. Unlike the wire, a tri can be driven by more than one signal outputs.

Functionally, **wire** and **tri** are identical. Distinct nomenclatures are provided for the convenience of assigning roles. Other types of nets are discussed in Chapter 5.

3.10.2   Variable Data Type

A variable is an abstraction for a storage device. It can be declared through the
keyword **reg** and stores the value of a logic level: 0, 1, **x**, or **z**. A net or wire
connected to a **reg** takes on the value stored in the **reg** and can be used as input
to other circuit elements. But the output of a circuit cannot be connected to a **reg**.
The value stored in a **reg** is changed through a fresh assignment in the program.
time, **integer**, **real**, and **realtime** are the other variable types of data;
these are dealt with later.

3.11   SCALARS AND VECTORS

Entities representing single bits — whether the bit is stored, changed, or
transferred — are called "scalars." Often multiple lines carry signals in a cluster –
like data bus, address bus, and so on. Similarly, a group of **reg**s stores a value,
which may be assigned, changed, and handled together. The collection here is
treated as a "vector." Figure 3.2 illustrates the difference between a scalar and a
vector. wr and rd are two scalar nets connecting two circuit blocks circuit1 and
circuit2. b is a 4-bit-wide vector net connecting the same two blocks. b[0], b[1],
b[2], and b[3] are the individual bits of vector b. They are "part vectors."

A vector **reg** or net is declared at the outset in a Verilog program and hence
treated as such. The range of a vector is specified by a set of 2 digits (or
expressions evaluating to a digit) with a colon in between the two. The
combination is enclosed within square brackets.

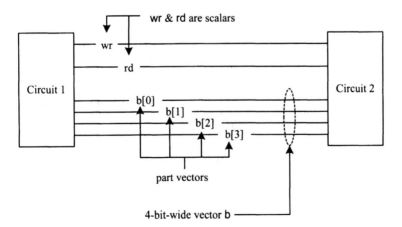

Figure 3.2 Illustration of scalars and vectors.

Examples:

wire[3:0] a;   /* a is a four bit vector of net type; the bits are designated as a[3], a[2], a[1] and a[0]. */

reg[2:0] b;   /* b is a three bit vector of **reg** type; the bits are designated as b[2], b[1] and b[0]. */

reg[4:2] c;   /* c is a three bit vector of **reg** type; the bits are designated as c[4], c[3] and c[2]. */

wire[-2:2] d ;   /* d is a 5 bit vector with individual bits designated as d[-2], d[-1], d[0], d[1] and d[2]. */

Whenever a range is not specified for a net or a **reg**, the same is treated as a scalar – a single bit quantity. In the range specification of a vector the most significant bit and the least significant bit can be assigned specific integer values. These can also be expressions evaluating to integer constants – positive or negative.

Normally vectors – nets or **regs** – are treated as unsigned quantities. They have to be specifically declared as "**signed**" if so desired.

Examples

wire signed[4:0] num;   // num is a vector in the range -16 to +15.

reg signed [3:0] num_1;   // num_1 is a vector in the range -8 to +7.

3.12   PARAMETERS

In some designs, certain parameter values are not committed at the outset. Proportionality constants, frequency-scaling levels, number of taps in digital filters, *etc.*, are typical examples. There are also situations where the size of the design is left open and decided at a later stage. Bus width, LIFO depth, and memory size are such quantities which may be committed later. All such constants can be declared as parameters at the outset in a Verilog module, and values can be assigned to them; for example,

parameter word_size = 16;

parameter word_size = 16, mem_size = 256;

Such parameter assignments are made at compiler time. The parameter values cannot be changed (normally) at runtime. However, a parameter that has been assigned a value in a module definition can have its value changed at runtime – that is, when the module is used at runtime in some other design (*i.e.*, instantiated) or when it is tested.   Such modifications are carried out through a "**defparameter**" statement.   The parameter assignment done as part of parameter declaration can have the appropriate constant on the right-hand side of the assignment statement, as was the case above. The assignment can also have algebraic expressions on the right hand side. Such expressions can involve constants and other parameters declared already; for example

Parameter word_size = 16, factor = word_size/2;

3.13   MEMORY

Different types and sizes of memory, register file, stack, *etc.*, can be formed by extending the vector concept. Thus the declaration

Reg [15:0] memory[511:0];

declares an array called "memory"; it has 512 locations. Each location is 16 bits wide. The value of any chosen location can be assigned to a selected register or *vice versa*; this constitutes memory reading or writing [see Example 8.10]. The index used to refer a memory location can be a number or an algebraic expression which reduces to an integral value – positive, zero, or negative. As an example, consider the assignment statement

B = mem[(p-q)/2];

The simulator first evaluates (p - q)/2 (which should be an integer): Let it reduce to 3. Then the data stored at mem[3] is assigned to B. Stack pointer, program counter, index register, *etc.*, can be implemented through the above concept. Different types of memory addressing like indirect, indexed, *etc.*, can also be accommodated. Page addressing can be accomplished by a slight adaptation of the concept.

3.14   OPERATORS

Verilog has a number of operators akin to the C language. These are of three types:

1. Unary: the unary operator is associated with a single operand. The operator precedes the operand – for example, ~a.
2. Binary: the binary operator is associated with two operands. The operator appears between the two operands – for example, a&b.
3. Ternary: the ternary operator is associated with three operands. The two operators together constitute a ternary operation. The two operators separate the three operands – for example
    a?b:c   // Here the operators "?" and ":" together define an operation.

Operators are discussed in detail in Chapter 6.

3.15   SYSTEM TASKS

During the simulation of any design, a number of activities are to be carried out to monitor and control simulation. A number of such tasks are provided / available in Verilog. Some other tasks serve other functions. However, a few of these are used commonly; these are described here. The "$" symbol identifies a system task. A task has the format

`$<keyword>`

3.15.1 $display

When the system encounters this task, the specified items are displayed in the formats specified and the system advances to a new line. The structure, format, and rules for these are the same as for the "printf" / "scanf" function in C. Refer to a standard text in "C" language for the text formatting codes in common usage [Gottfried].

Examples

`$display` ("The value of a is : a = , %d", a);
Execution of this line results in printing the value of a as a decimal number (specified by "%d"). The string present within the inverted commas specifies this. Thus if a has the value 3.5, we get the display

The value of a is : a = 3.5.

After printing the above line, the system advances to the next line.

`$display;`   /* This is a display task without any arguments. It advances output to a new line. */

3.15.2 $monitor

The `$monitor` task monitors the variables specified whenever any one of those specified changes. During the running of the program the monitor task is invoked and the concerned quantities displayed whenever any one of these changes. Following this, the system advances to the next line. A monitor statement need appear only once in a simulation program. All the quantities specified in it are continuously monitored. In contrast, the `$display` command displays the quantities concerned only once – that is, when the specific line is encountered during execution. The format for the `$monitor` task is identical to that of the `$display` task.

Examples

$monitor ("The value of a is : a = , %d", a);

With the task, whenever the value of a changes during execution of a program, its new value is printed according to the format specified. Thus if the value of a changes to 2.4 at any time during execution of the program, we get the following display on the monitor.

The value of a is: a = 2.4.

3.15.3   Tasks for Control of Simulation

Two system tasks are available for control of simulation:

$finish task, when encountered, exits simulation. Control is reverted to the Operating System. Normally the simulation time and location are also printed out by default as part of the exit operation.

$stop task, suspends simulation; if necessary the simulation can be resumed by user intervention. Thus with the stop task, the simulator is in an interactive mode. In contrast with $finish, simulation has to be started afresh.

3.16   EXERCISES

1.   Run the Verilog program in Figure 3.3. Observe the output.

```
module fancy2;
integer i,j;
initial repeat(5)
begin
#1      j=0;
        while(j<=10)
        begin
                j=j+1;
                for(i=0;i<=j;i=i+1) $write(" b");
                $display("*");
        end
#1      while(j>=0)
```

continued

continued

```
        begin
                for(i=0;i<=j;i=i+1) $write(" c");
                $display("*");
                j=j-1;
        end
end
initial #12 $stop;
endmodule
```

Figure 3.3 A simple Verilog module.

2.   In Exercise 3.1 above, delete b and c in the write statement lines. Rerun the program.

3.   Try other combinations of I and j values and repeat the run.

4.   Run the Verilog program in Figure 3.4.

5.   In the program of Figure 3.4 replace the "**always**" statement by "**initial**" statement and run the program.

6.   In the program of Figure 3.4 replace the "a=a+7" statement by "a=a-7" statement and run the program.


```
module fancy3;
reg[11:0]a;
always
begin
#0      $display("See this:        ah=%d, ad=%h, ao=%o, ab=%b",a,a,a,a);
#1      $display("How about this? ah=%0d, ad=%0h, ao=%0o, ab=%0b",a,a,a,a);
        a=a+7;
end
initial
begin
        a=0;
        #10 $stop;
end
endmodule
```

Figure 3.4 Another simple Verilog module.

4

GATE LEVEL MODELING – 1

4.1   INTRODUCTION

Digital designers are normally familiar with all the common logic gates, their symbols, and their working. Flip-flops are built from the logic gates. All other functionally complex and more involved circuits can also be built using the basic gates. All the basic gates are available as "Primitives" in Verilog. Primitives are generalized modules that already exist in Verilog [IEEE]. They can be instantiated directly in other modules. Further design description using gate primitives is quite close to the actual circuits (design description using the switch primitives dealt with in Chapter 10 are still closer). We describe features of gate level primitives, ways of working with them, and ways of building more involved circuits with them [Palnitkar, Lee]. In this process some of the commonly used features of Verilog are also brought out.

4.2   AND GATE PRIMITIVE

The AND gate primitive in Verilog is instantiated with the following statement:
and g1 (O, I1, I2, . . ., In);
Here '**and**' is the keyword signifying an AND gate. g1 is the name assigned to the specific instantiation. O is the gate output; I1, I2, *etc*., are the gate inputs. The following are noteworthy:

- The AND module has only one output. The first port in the argument list is the output port.
- An AND gate instantiation can take any number of inputs — the upper limit is compiler-specific.
- A name need not be necessarily assigned to the AND gate instantiation; this is true of all the gate primitives available in Verilog.

4.2.1   Example 4.1

Figure 4.1 shows the stimulus program for testing the AND gate g1.   The output obtained by stimulating the program is shown in Figure 4.2.   Some explanation regarding the simulation program is in order here.

- The module test_and has no port. It instantiates the AND module once.
- The test input sequence is specified within the **initial** block – the sequence of statements between the **begin** and **end** statements together form this block.
- The keyword "**initial**" signifies the settings done initially — that is, only once for the whole routine.
- The first set of statements within the **initial** block
        a1 = 0;
        a2 = 0;
        make
        a1 = a2 = 0
        at zero simulation time.
- After 3 time steps, a1 is set to one but a2 remains at 0. The expression "#3" means "after 3 time steps". Subsequent changes in a1 and a2 also can be explained in the same manner.

```
module test_and;
reg a1, a2;
wire b;
Initial
Begin
        a1 = 0;
        a2 = 0;
  #3    a1 = 1;
  #1    a1 = 0;
  #2    a2 = 1;
  #4    a1 = 1;
  #3    a2 = 0;
  #1    a2 = 1;
end
and g1(b, a1, a2);
initial $monitor ( $time, "a1 = %b, a2 = %b, b = %b" a1, a2, b);
initial #100 $finish;
endmodule
```

Figure 4.1 A module to instantiate the AND gate primitive and test it.

```
 0 a1  =  0 a2  =  0 b  =  0
 3 a1  =  1 a2  =  0 b  =  0
 4 a1  =  0 a2  =  0 b  =  0
 6 a1  =  0 a2  =  1 b  =  0
10 a1  =  1 a2  =  1 b  =  1
13 a1  =  1 a2  =  0 b  =  0
14 a1  =  1 a2  =  1 b  =  1
```

Figure 4.2 The output obtained by running the module of Figure 4.1.

- The program displays the variable values – that is, the values of o, a1, and a2 whenever any one of these changes. This is evident from the printout on the monitor, which has been reproduced in Figure 4.2.
- A pair of variables a1 and a2 are declared in the program, and the values stored in them are given as inputs to the AND gate instantiation.
- Any variable not declared in the module is by default taken as a net of wire type; it is also taken as a scalar. The same is true of all modules in Verilog.
- The term $time in the $monitor statement signifies the running time of the program. Here it causes the value of time at the instant of capturing the data for display, to be displayed.
- The statement

    #100 $finish;

    signifies that the program will stop simulation and exit the operating system at the end of 100 time steps.

4.2.2   Truth Table of AND Gate Primitive

The truth table for a two-input AND gate is shown in Table 4.1. It can be directly extended to AND gate instantiations with multiple inputs. The following observations are in order here:

Table 4.1 Truth table of AND gate primitive

		Input 1			
		0	1	x	z
Input 2	0	0	0	0	0
	1	0	1	x	x
	x	0	x	x	x
	z	0	x	x	x

- If any one of the inputs to the AND gate instantiation is in the 0 state, its output is also in the 0 state. It is irrespective of whether the other inputs are at the 0, 1, **x** or **z** state.
- The output is at 1 state if and only if every one of the inputs is at 1 state.
- For all other cases the output is at the **x** state.
- Note that the output is never at the **z** state – the high impedance state. This is true of all other gate primitives as well.

4.3   MODULE STRUCTURE

Figure 4.1 shows a typical module. In a general case a module can be more elaborate. A lot of flexibility is available in the definition of the body of the module. However, a few rules need to be followed:

- The first statement of a module starts with the keyword **module**; it may be followed by the name of the module and the port list if any (see Section 2.8).
- All the variables in the ports-list are to be identified as **inputs**, **outputs**, or **inouts**. The corresponding declarations have the form shown below:

    - **Input a1, a2;**
    - **Output b1, b2;**
    - **Inout c1, c2;**

- The port-type declarations here follow the module declaration mentioned above.
- The ports and the other variables used within the body of the module are to be identified as nets or registers with specific types in each case. The respective declaration statements follow the port-type declaration statements. Examples:

    wire a1, a2, c;
    reg b1, b2;

    The type declaration must necessarily precede the first use of any variable or signal in the module.

- The executable body of the module follows the declaration indicated above.
- The last statement in any module definition is the keyword "**endmodule**".
- Comments can appear anywhere in the module definition.

4.4   OTHER GATE PRIMITIVES

All other basic gates are also available as primitives in Verilog. Details of the facilities and instantiations in each case are given in Table 4.2. The following points are noteworthy here:

- In all cases of instantiations, one need not necessarily assign a name to the instantiation. It need be done only when felt necessary – say for clarity of circuit description.
- In all the cases the output port(s) is (are) declared first and the input port(s) is (are) declared subsequently.
- The buffer and the inverter have only one input each. They can have any number of outputs; the upper limit is compiler-specific. All other gates have one output each but can have any number of inputs; the upper limit is again compiler-specific.

4.4.1   Truth Table

Extending the concepts of Section 4.2.2, one can form the truth tables of all other gate primitives. The basic features of each are given in Table 4.3. The truth tables themselves are given in Appendix B.

4.5   ILLUSTRATIVE EXAMPLES

The examples considered here illustrate the use of gate primitives in designs. Further, they bring out how one can build fairly large designs by judiciously combining smaller modules in a repeated fashion [Bignel, Sedra].

Table 4.2 Basic gate primitives in Verilog with details

Gate	Mode of instantiation	Output port(s)	Input port(s)
AND	and ga (o, i1, i2, . . . i8);	o	i1, i2, . .
OR	or gr (o, i1, i2, . . . i8);	o	i1, i2, . .
NAND	nand gna (o, i1, i2, . . . i8);	o	i1, i2, . .
NOR	nor gnr (o, i1, i2, . . . i8);	o	i1, i2, . .
XOR	xor gxr (o, i1, i2, . . . i8);	o	i1, i2, . .
XNOR	xnor gxn (o, i1, i2, . . . i8);	o	i1, i2, . .
BUF	buf gb (o1, o2, ... i);	o1, o2, o3, . .	i
NOT	not gn (o1, o2, o3, . . . i);	o1, o2, o3, . .	i

Table 4.3 Rules for deciding the output values of gate primitives for different input combinations

Type of gate	0 output state	1 output state	**x** output state
AND	Any one of the inputs is zero	All the inputs are at one	All other cases
NAND	All the inputs are at one	Any one of the inputs is zero	
OR	All the inputs are at zero	Any one of the inputs is one	
NOR	Any one of the inputs is one	All the inputs are at zero	
XOR	If every one of the inputs is definite at zero or one, the output is zero or one as decided by the XOR or XNOR function		If any one of the inputs is at **x** or **z** state, the output is at **x** state
XNOR			
BUF	If the only input is at 0 state	If the only input is at 1 state	All other cases of inputs
NOT	If the only input is at 1 state	If the only input is at 0 state	

4.5.1   Example 4.2

The commonly used A-O-I gate is shown in Figure 4.3 for a simple case. The module and the test bench for the same are given in Figure 4.4. The circuit has been realized here by instantiating the AND and NOR gate primitives. The names of signals and gates used in the instantiations in the module of Figure 4.4 remain the same as those in the circuit of Figure 4.3. The module aoi_gate in the figure has input and output ports since it describes a circuit with signal inputs and an output. The module aoi_st is a stimulus module. It generates inputs to the aoi_gate module and gets its output. It has no input or output ports.

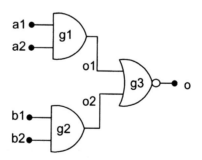

Figure 4.3 A typical A-O-I gate circuit.

```
/*module for the aoi-gate of figure 4.3 instantiating
the gate primitives - fig4.4*/
module aoi_gate(o,a1,a2,b1,b2);
input a1,a2,b1,b2;// a1,a2,b1,b2 form the input
//ports of the module
output o;//o is the single output port of the module
wire o1,o2;//o1 and o2 are intermediate signals
//within the module
and g1(o1,a1,a2); //The AND gate primitive has two
and g2(o2,b1,b2);// instantiations with assigned
//names g1 & g2.
nor g3(o,o1,o2);//The nor gate has one instantiation
//with assigned name g3.
endmodule

//Test-bench for the aoi_gate above
module aoi_st;
reg a1,a2,b1,b2;
//specific values will be assigned to a1,a2,b1,
// and b2 and these connected
//to input ports of the gate insatntiations;
//hence these variables are declared as reg
wire o;
initial
begin
        a1 = 0;
        a2 = 0;
        b1 = 0;
        b2 = 0;
    #3 a1 = 1;
    #3 a2 = 1;
    #3 b1 = 1;
    #3 b2 = 0;
    #3 a1 = 1;
    #3 a2 = 0;
    #3 b1 = 0;
end
initial #100 $stop;//the simulation ends after
//running for 100 tu's.
initial $monitor($time   ,   "   o = %b , a1 = %b ,
  a2 = %b , b1 = %b ,b2 = %b ",o,a1,a2,b1,b2);
aoi_gate gg(o,a1,a2,b1,b2);
endmodule
```

Figure 4.4 Module for the AOI gate of Figure 4.3 and a test bench for the same.

The A-O-I gate module has three instantiations – two of these being AND gates and the third a NOR gate; this conforms to the circuit of A_O_I gate in Figure 4.3. Within the aoi_gate module, all signals are of type net. The aoi_ gate module in Figure 4.4 is instantiated once in the module aoi_st for testing. Any such instantiation of a user-defined module in another module has to be assigned a name. (As mentioned earlier, this is not mandatory with the instantiation of gate primitives available in Verilog.) The instantiation is given the name gg here. Note that all the inputs to the instantiation of aoi_gate in the test bench are fed through **reg**s.

The aoi_gate and aoi_st are compiled and run. Different combinations of values are assigned to a1, a2, b1, and b2 in the test bench at regular intervals of 3 time steps. At all such time steps at least one of the signals included in the monitor statement changes. Hence all the signal values are displayed on the monitor at three time step intervals. The results of running the test bench are reproduced in Figure 4.5, which confirms this.

The module aoi_gate has been synthesized and the synthesized circuit shown in Figure 4.6; the figure does not warrant any detailed explanation.

Both the modules can do with some elegant simplification. First consider the stimulus module aoi_st in Figure 4.4. All the four inputs can be clubbed together and treated as a "vector" input. Often this may be possible to be identified with a four-bit-wide bus in a system. It makes the vector representation all the more meaningful. With this, the variables together can be declared as a single vector. The value taken by the vector can be defined with relevant time delays. To accommodate such a change, the AOI module of Figure 4.4 is recast in Figure 4.7. The compactness achieved here is carried over to the instantiation of the module for its test bench aoi_st2, which is also shown in the figure.

The AOI gate itself (aoigate2 in Figure 4.7) has been made compact on two counts: All the four inputs have been clubbed together and treated as a four-bit vector. Further, the two and gate instantiations are clubbed together into one statement. Note the format of the statement – a comma separates the two instantiations, and as usual a semicolon signifies the end of the statement. In any set of instantiations, all similar instantiations in a module can be combined in this manner. The module aoigate2 has an input/output port since it describes a circuit with signal inputs and outputs. aoi_st2 is a stimulus module. It generates inputs

```
#     0    o = 1 ,  a1 = 0 ,  a2 = 0 ,  b1 = 0 ,b2 = 0
#     3    o = 1 ,  a1 = 1 ,  a2 = 0 ,  b1 = 0 ,b2 = 0
#     6    o = 0 ,  a1 = 1 ,  a2 = 1 ,  b1 = 0 ,b2 = 0
#     9    o = 0 ,  a1 = 1 ,  a2 = 1 ,  b1 = 1 ,b2 = 0
#    18    o = 1 ,  a1 = 1 ,  a2 = 0 ,  b1 = 1 ,b2 = 0
#    21    o = 1 ,  a1 = 1 ,  a2 = 0 ,  b1 = 0 ,b2 = 0
```

Figure 4.5 Results of running the aoi_st test bench of Figure 4.3.

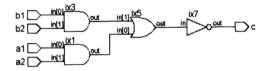

Figure 4.6 Synthesized version of the module aoi_gate of Figure 4.4.

to the module from within the stimulus module and gets its output. It has no input or output port. In a more general case one may have a number of modules defined at different levels, which are repeatedly instantiated in bigger modules. The stimulus module may be at the apex. It may carry out the stimulus activity by generating the inputs to the other ports in the hierarchy and receiving their outputs.

```
module aoi_gate2(o,a);
input [3:0]a;//A is a vector of 4 bits width
output o;// output o is a scalar
wire o1,o2;//these are intermediate signals
and (o1,a[0],a[1]),(o2,a[2],a[3]);
nor (o,o1,o2);/*The nor gate has one instantiation
with assigned name g3.*/
endmodule

module aoi_st2;
reg[3:0] aa;
aoi_gate2 gg(o,aa);
initial
        begin
        aa = 4'b000;//a being a vector, all its
 #3     aa = 4'b0001;//bit components are
 #3     aa = 4'b0010;//assigned values at one go.
 #3     aa = 4'b0100;//Similarly their changes are
 #3     aa = 4'b1000;//combined in the assignments
 #3     aa = 4'b1100;
 #3     aa = 4'b0110;
 #3     aa = 4'b0011;
        end
initial
$monitor( $time , " aa = %b , o = %b " , aa,o);
initial #24 $stop;
endmodule
```

Figure 4.7 Another realization of the A-I-O gate with the input declared as a vector; the test bench for the module is also shown in the figure.

The stimulus module need not necessarily have a port; aoi_st in Figure 4.4 and aoi_st2 in Figure 4.7 are typical examples. The results of running the test bench aoi_st2 of Figure 4.7 are shown in Figure 4.8.

To facilitate involved design descriptions, some additional flexibility is available in Verilog.

- Signals at the ports can be identified by a hierarchical name. Such addressing may become useful when displaying them in the stimulus module.
- Signal instantiations illustrated above specify inputs and outputs in the same sequence as was done in the definition. The procedure is simple and acceptable in situations with only a few numbers of inputs and outputs. But in modules with a comparatively large number of inputs and outputs, sticking to the sequence and keeping track of it becomes strenuous. In such situations the instantiation can be done by identifying the inputs and outputs on a one-to-one basis [see Section 2.8]. Thus the instantiation of the aoi_gate2 in the test bench of Figure 4.7 can be described alternately as

aoigate2 gg (.o(o), .a[1](aa[1]), .a[2](aa[2]), .a[3](aa[3]), .a[4](aa[4]));

Here one need not stick to the same order of assignment of the ports as in the module definition. Thus the instantiation entered as

aoigate2 gg (.a[1](aa[1]), .o(o),.a[2](aa[2]), .a[4](aa[4]), a[3](aa[3]));

is equally valid.

4.5.2   Example 4.3: 4-to-16 Decoder

Decoder design using gates can be described in various ways. Here we define a 2-to-4 decoder module and instantiate it repeatedly and judiciously to realize a 4-to-16 decoder. The procedure is not necessarily the best or most elegant.

```
#              0      aa = 0000 , o = 1
#              3      aa = 0001 , o = 1
#              6      aa = 0010 , o = 1
#              9      aa = 0100 , o = 1
#             12      aa = 1000 , o = 1
#             15      aa = 1100 , o = 0
#             18      aa = 0110 , o = 1
#             21      aa = 0011 , o = 0
```

Figure 4.8 Results of running the aoi_st2 test bench of Figure 4.7.

Figure 4.9(c) shows the formation of the 4-to-16 decoder in terms of two numbers of 3-to-8 decoders. The 3-to-8 decoders have an "Enable" input each (designated 'en' – one being of the active high and the other of the active low type); these are connected to the most significant bit of the 4-bit input to form the 4-to-16 decoder. The 3-to-8 decoder can again be formed in terms of two 2-to-4 decoders in the same manner as shown in Figure 4.9(b). The 2-to-4 decoder block used here is shown in Figure 4.9(a). The logic of building a complex circuit unit in terms of repeated use of smaller and smaller circuit units followed here is used in the design description as well. Figure 4.10 shows the design description of a 2-to-4 decoder module and a test bench for the same. The decoder module (dec2_4) accepts a 2-bit-wide vector input b and decodes it into a 4-bit-wide vector output a. It has an additional "Enable" input designated "en"; the outputs are enabled only if en = 1. The input en has been introduced to facilitate expansion of the decoder capacity by repeated instantiation as explained above. The test bench for the decoder is more illustrative than exhaustive; that is, it does not test the module for all possible input values. Results of the simulation run are shown in Figure 4.11.

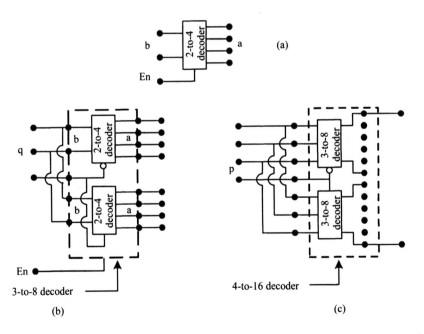

Figure 4.9 Formation of 4-to-16 decoder circuit in terms of smaller decoders: (a) 2-to-4 decoder, (b) 3-to- 8 decoder in terms of two 2-to-4 decoders, and (c) 4-to-16 decoder in terms of two 3-to-8 decoders.

```
module dec2_4 (a,b,en);
output [3:0] a;
input [1:0]b; input en;
wire [1:0]bb;
not(bb[1],b[1]),(bb[0],b[0]);
and(a[0],en, bb[1],bb[0]),(a[1],en, bb[1],b[0]),
(a[2],en, b[1],bb[0]),(a[3],en, b[1],b[0]);
endmodule
//test bench
module tst_dec2_4();
wire [3:0]a;
reg[1:0] b; reg en;
dec2_4 dec(a,b,en);
initial
begin
  {b,en} =3'b000;
#2{b,en} =3'b001;
#2{b,en} =3'b011;
#2{b,en} =3'b101;
#2{b,en} =3'b111;
end
initial
$monitor ($time  ,   "output a =  %b, input b  = %b ",
a, b);
endmodule
```

Figure 4.10 Design description of a 2-to-4 decoder circuit and its test bench.

Figure 4.12 shows a 3-to-8 decoder module formed by repeated instantiation of the 2-to-4 decoder of Figure 4.10. The eight AND gate instantiations ensure that the outputs are enabled only when enn — a separate "Enable" signal — goes active. Following the same logic, the module for the 4-to-16 decoder is described in Figure 4.13. A test bench to test the module through all the possible input states is also included in the figure. Figure 4.14 shows the results of running the test-bench.

```
//output
//#        0 output a =  0000, input b  = 00
//#        2 output a =  0001, input b  = 00
//#        4 output a =  0010, input b  = 01
//#        6 output a =  0100, input b  = 10
//#        8 output a =  1000, input b  = 11
```

Figure 4.11 Results of running the test bench of Figure 4.10.

```
module dec3_8(pp,q,enn);
output[7:0]pp;
input[2:0]q;
input enn;
wire qq;
wire[7:0]p;
not(qq,q[2]);
dec2_4 g1(.a(p[3:0]),.b(q[1:0]),.en(qq));
dec2_4 g2(.a(p[7:4]),.b(q[1:0]),.en(q[2]));
and g30(pp[0],p[0],enn);
and g31(pp[1],p[1],enn);
and g32(pp[2],p[2],enn);
and g33(pp[3],p[3],enn);
and g34(pp[4],p[4],enn);
and g35(pp[5],p[5],enn);
and g36(pp[6],p[6],enn);
and g37(pp[7],p[7],enn);
endmodule
```

Figure 4.12 A 3-to-8 decoder module formed by repeated instantiation of the 2-to-4 decoder module in Figure 4.10.

```
module dec4_16(m,n);
output[15:0]m;
input[3:0]n;
wire nn;
//wire en;
not(nn,n[3]);
dec3_8 g3(.pp(m[7:0]),.q(n[2:0]),.enn(nn));
dec3_8 g4(.pp(m[15:8]),.q(n[2:0]),.enn(n[3]));
endmodule

//test-bench
module dec4_16_stimulus;
wire[15:0]m;
//wire l,m,n;
reg[3:0]n;
dec4_16 gg(m,n);
initial
```

continued

continued

```
begin
  n=4'b0000;#2n=4'b0000;#2n=4'b0001;
#2n=4'b0010;#2n=4'b0011;#2n=4'b0100;
#2n=4'b0101;#2n=4'b0110;#2n=4'b0111;
#2n=4'b1000;#2n=4'b1001;#2n=4'b1010;
#2n=4'b1011;#2n=4'b1100;#2n=4'b1101;
#2n=4'b1110;#2n=4'b1111;#2n=4'b1111;
end
initial $monitor($time," m = %b ,n = %b , gg.g3.qq = %b
, gg.g4.g1.bb = %b " , m,n,gg.g3.qq,gg.g4.g1.bb);
//gg.g3.qq displays the enable line of dec3_8 called
g3-g1
//gg.g4.g1.bb displays the bb wire in dec2_4
initial #40 $stop ;
endmodule
```

Figure 4.13 A 4-to-16 decoder module formed by repeated instantiation of the 3-to-8 decoder module of Figure 4.12. A test bench for the same is also shown.

```
//output
//#         0 m = 0000000000000001 ,n = 0000 ,
gg.g3.qq = 1 , gg.g4.g1.bb = 11
//#         4 m = 0000000000000010 ,n = 0001 ,
gg.g3.qq = 1 , gg.g4.g1.bb = 10
//#         6 m = 0000000000000100 ,n = 0010 ,
gg.g3.qq = 1 , gg.g4.g1.bb = 01
//#         8 m = 0000000000001000 ,n = 0011 ,
gg.g3.qq = 1 , gg.g4.g1.bb = 00
//#        10 m = 0000000000010000 ,n = 0100 ,
gg.g3.qq = 0 , gg.g4.g1.bb = 11
//#        12 m = 0000000000100000 ,n = 0101 ,
gg.g3.qq = 0 , gg.g4.g1.bb = 10
//#        14 m = 0000000001000000 ,n = 0110 ,
gg.g3.qq = 0 , gg.g4.g1.bb = 01
//#        16 m = 0000000010000000 ,n = 0111 ,
gg.g3.qq = 0 , gg.g4.g1.bb = 00
//#        18 m = 0000000100000000 ,n = 1000 ,
gg.g3.qq = 1 , gg.g4.g1.bb = 11
//#        20 m = 0000001000000000 ,n = 1001 ,
gg.g3.qq = 1 , gg.g4.g1.bb = 10
//#        22 m = 0000010000000000 ,n = 1010 ,
```

continued

continued

```
gg.g3.qq = 1 , gg.g4.g1.bb = 01
//#      24 m = 0000100000000000 ,n = 1011 ,
gg.g3.qq = 1 , gg.g4.g1.bb = 00
//#      26 m = 0001000000000000 ,n = 1100 ,
gg.g3.qq = 0 , gg.g4.g1.bb = 11
//#      28 m = 0010000000000000 ,n = 1101 ,
gg.g3.qq = 0 , gg.g4.g1.bb = 10
//#      30 m = 0100000000000000 ,n = 1110 ,
gg.g3.qq = 0 , gg.g4.g1.bb = 01
//#      32 m = 1000000000000000 ,n = 1111 ,
gg.g3.qq = 0 , gg.g4.g1.bb = 00
```

Figure 4.14 Results of running the test bench of Figure 4.13 for the 4-to-16 decoder.

Observations:–

- The nested tree of modules with the inputs and outputs in each case are shown in Figure 4.15.

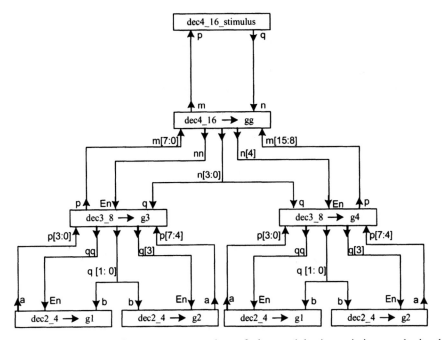

Figure 4.15 Block diagram representation of the module instantiations and signal assignments for the stimulus module of Figure 4.10.

- Two signals within the two nested modules are monitored in dec4_16_stimulus. Formation of their hierarchical addresses is also shown in Figure 4.15. (Hierarchical addressing is addressed in detail in Chapter 11.)
- The module dec3_8 is instantiated twice in the module dec4_16. Here the port declarations are done by declaring the port names on a one-to one basis. The order has not been maintained as in the defining module.

4.5.2.1   Decoder Synthesis

The synthesized circuit of the 2-to-4 decoder module of Figure 4.10 (dec2_4) is shown in Figure 4.16. The AND gate cells available in the library are all of the two-input type; hence six such cells (designated as ix5, ix7, ix11, ix13, ix15, and ix19) are utilized to realize the four numbers of three-input AND gates instantiated in the design module. The NOT gates are realized through two NOT gate cells in the library (designated as ix1 and ix3). The wider lines in the figure signify bus-type interconnections. The synthesized circuit of the 3-to-8 decoder module of Figure 4.12 (dec3_8) is shown in Figure 4.17. The two instantiations of the dec2_4 module (g1 and g2) are shown as black boxes. Similarly, Figure 4.18 shows the synthesized circuit of the 4-to-16 decoder module of Figure 4.13 (dec4_16). The two instantiations of the dec3_8 module (g3 and g4) appear as black boxes inside. Figure 4.19 shows the complete hierarchy of instantiations in the synthesized circuit. In the figure boxes g3 and g4 represent instantiations of the 3-to-8 decoders used in the module. Each of these has two numbers of the 2-to-4 decoders – designated as g1 and g2; these are shown enclosed inside boxes.

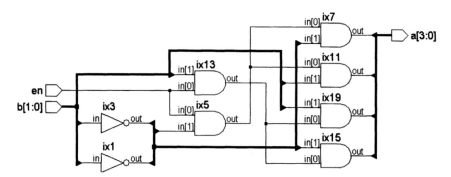

Figure 4. 16 The synthesized circuit of the 2-to-4 decoder of Figure 4.10.

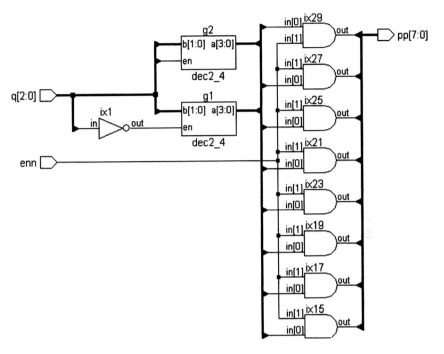

Figure 4.17 The synthesized circuit of the 3-to- 8 decoder of Figure 4.12.

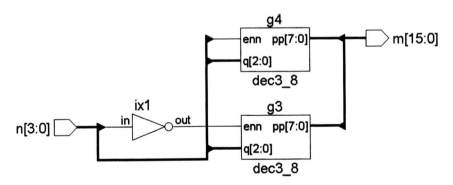

Figure 4.18 The synthesized circuit of the 4-to-16 decoder of Figure 4.13.

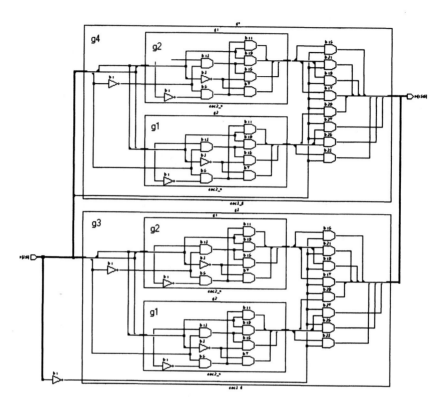

Figure 4.19 Four-to-sixteen decoder – hierarchy of instantiations.

4.6   TRI-STATE GATES

Four types of tri-state buffers are available in Verilog as primitives. Their outputs can be turned ON or OFF by a control signal. The direct buffer is instantiated as

Bufif1 nn (out, in, control);

The symbol of the buffer is shown in Figure 4.20. We have

- out as the single output variable
- in as the single input variable and
- control as the single control signal variable.

When
control = 1,
out = in.

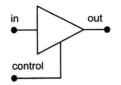

Figure 4.20 A tri-state buffer.

When
control = 0,
out is cut off from the input and tri-stated. The output, input and control signals should appear in the instantiation in the same order as above. Details of bufif1 as well as the other tri-state type primitives are shown in Table 4.4. In all the cases shown in Table 4.4, out is the output, in is the input, and control, the control variable.

Table 4.4 Instantiation and functional details of tri-state buffer primitives

Typical instantiation	Functional representation	Functional description
bufif1 (out, in, control);	in out control	Out = in if control = 1; else out = z
bufif0 (out, in, control);	in out control	Out = in if control = 0; else out = z
notif1 (out, in, control);	in out control	Out = complement of in if control = 1; else out = z
notif0 (out, in, control);	in out control	Out = complement of in if control = 0; else out = z

The truth tables of the tri-state buffers are given in Appendix B. The following observations are common to all the tri-state buffer primitives:

- If the control signal has a value that corresponds to the buffer being on, two possibilities exist:
  - The output has the same value as the input if the input is 0 or 1.
  - The output is at **x** otherwise (*i.e.*, if the input is **x** or **z**).
- If the control signal has a value that corresponds to the control signal being off, the output is at **z** state irrespective of the value of the input.
- If the control signal is at **x** or **z**, three possibilities arise:
  - If the input is at **x** or **z**, the output is at **x**.
  - If the input is at 0 state, the output is **L** for bufif1 and bufif0. It is at **H** for notif1 and notif0.
  - If the input is at 1 state, the output is **H** for bufif1 and bufif0. It is at **L** for notif1 and notif0.

Note that **H** corresponds to 1 or **z** state while **L** corresponds to 0 or **z** state.

4.7   ARRAY OF INSTANCES OF PRIMITIVES

The primitives available in Verilog can also be instantiated as arrays. A judicious use of such array instantiations often leads to compact design descriptions. A typical array instantiation has the form

and gate [7 : 4] (a, b, c);

where a, b, and c are to be 4 bit vectors. The above instantiation is equivalent to combining the following 4 instantiations:

and gate [7] (a[3], b[3], c[3]), gate [6] (a[2], b[2], c[2]), gate [5] (a[1], b[1], c[1]), gate [4] (a[0], b[0], c[0]);

The assignment of different bits of input vectors to respective gates is implicit in the basic declaration itself. A more general instantiation of array type has the form

and gate[mm : nn](a, b, c);

where mm and nn can be expressions involving previously defined parameters, integers and algebra with them. The range for the gate is 1+ (*mm-nn*); *mm* and *nn* do not have restrictions of sign; either can be larger than the other.

4.7.1   Example 4.4 A Byte Comparator

A circuit to compare two variables each of one byte is given in Figure 4.21. The circuit outputs a flag d; d is 1 if the two bytes are equal; else it is 0. The output is activated only if the enable signal en = 1. If en = 0, the circuit output is tri-stated. The module description is given in Figure 4.22 along with a test-bench. The simulated output is in Figure 4.23.

Observations:

- In all array-type instantiations, the array sizes are to be matched.
- The order of assignments to outputs, inputs, *etc.*, in the individual gates will be decided by the order of the bits. Thus the array instantiation

$$\text{or gg[3:1] (a[3:1], b[4:2], c);}$$

is equivalent to the combination of instantiations

$$\text{or gg[3] (a[3], b[4], c[2]), gg[2] (a[2], b[3], c[1]), gg[1] (a[1], b[2], c[0]);}$$

- If the vector sizes in the port list do not match the array size specified, assignments will be done starting from the right; that is, the rightmost instantiation will be assigned the rightmost inputs and outputs and the following instantiations will be made assignments in the order specified. However, it is desirable to avoid such ill-matched instantiations.

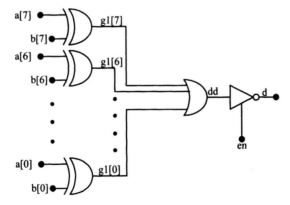

Figure 4.21 A byte comparator.

- In the general case the array size is specified in terms of two constant expressions. These can involve constants, previously defined parameters and algebraic operators: Such an instantiation can have a form as

and gate [offset*2+size-1: offset*2] (a, b, c);

where 'offset' and 'size' are parameters whose values should have been assigned earlier (operators are discussed in detail in Chapter 6).

```
module comp(d,a,b,en);
input en;
input[7:0]a,b;
output d;
wire [7:0]c;
wire dd;
xor g1[7:0] (c,b,a);
or(dd,c);
notif1(d,dd,en);
endmodule

module comp_tb;
reg[7:0]a,b;
reg en;
comp gg(d,a,b,en);
initial
      begin
      a   = 8'h00;
      b   = 8'h00;
      en = 1'b0;
      end
always
#2      en = 1'b1;
always
begin
 #2    a = a+1'b1;
 #2    b = b+2'd2;
end
initial $monitor($time," en = %b , a = %b ,b = %b ,d =
%b ",en,a,b,d);
initial #30 $stop;
endmodule
```

Figure 4.22 Module of an 8-bit comparator and its test bench.

```
# 0  en = 0,  a = 00000000,  b = 00000000,  d = z
# 2  en = 1,  a = 00000001,  b = 00000000,  d = 0
# 4  en = 1,  a = 00000001,  b = 00000010,  d = 0
# 6  en = 1,  a = 00000010,  b = 00000010,  d = 1
# 8  en = 1,  a = 00000010,  b = 00000100,  d = 1
#10  en = 1,  a = 00000011,  b = 00000100,  d = 0
#12  en = 1,  a = 00000011,  b = 00000110,  d = 0
#14  en = 1,  a = 00000100,  b = 00000110,  d = 1
#16  en = 1,  a = 00000100,  b = 00001000,  d = 1
#18  en = 1,  a = 00000101,  b = 00001000,  d = 0
#20  en = 1,  a = 00000101,  b = 00001010,  d = 0
#22  en = 1,  a = 00000110,  b = 00001010,  d = 1
#24  en = 1,  a = 00000110,  b = 00001100,  d = 1
#26  en = 1,  a = 00000111,  b = 00001100,  d = 0
#28  en = 1,  a = 00000111,  b = 00001110,  d = 0
```

Figure 4.23 Results of the simulation run of the test bench in Figure 4.22.

4.8   ADDITIONAL EXAMPLES

A set of representative examples is discussed here with the following aims:–

- Bring out the flexibility associated with the use of primitives and their instantiations.
- Illustrate the use of different features of Verilog discussed in the chapter.
- Focus attention on the fact that any combinational circuit can be designed at the gate level.

Details of the examples considered are summarized in Table 4.5

Table 4.5 Summary of the examples considered in Section 4.8

Circuit function	Figure numbers			Remarks
	Module & Test-bench	Simulation results	Synthesized circuit	
Half-adder	4.24	4.25	4.26	
Full-adder	4.27	4.28	4.29 & 4.30	Instantiates the half-adder twice as ha1 and ha2 in Figure 4.27
2-to-1 Mux	4.37	4.38	4.39	Realized with tri-state buffers
4-to-1 Mux	4.31	4.32	4.33	Simple & direct
	4.34	4.35	4.36	The above type with an additional tri-state output facility
	4.40	4.41	4.42	Realized with tri-state buffers

```
module ha(s,ca,a,b);
input a,b;
output s,ca;
xor(s,a,b);
and(ca,a,b);
endmodule

//test-bench
module tstha();
reg a,b;
wire s,ca;
ha hh(s,ca,a,b);
initial
begin
a=0;b=0;
end
always
begin
#2 a=1;b=0;
#2 a=0;b=1;
#2 a=1;b=1;
#2 a=0;b=0;
end
initial $monitor($time ,  "   a = %b , b = %b ,out carry
= %b , outsum = %b  " ,a,b,ca,s);
initial #24 $stop;
endmodule
```

Figure 4.24 Design module and a test bench for a half-adder.

```
   output
   #   0   a = 0 , b = 0 ,out carry = 0 , outsum = 0
   #   2   a = 1 , b = 0 ,out carry = 0 , outsum = 1
   #   4   a = 0 , b = 1 ,out carry = 0 , outsum = 1
   #   6   a = 1 , b = 1 ,out carry = 1 , outsum = 0
   #   8   a = 0 , b = 0 ,out carry = 0 , outsum = 0
   # 10   a = 1 , b = 0 ,out carry = 0 , outsum = 1
   # 12   a = 0 , b = 1 ,out carry = 0 , outsum = 1
   # 14   a = 1 , b = 1 ,out carry = 1 , outsum = 0
   # 16   a = 0 , b = 0 ,out carry = 0 , outsum = 0
   # 18   a = 1 , b = 0 ,out carry = 0 , outsum = 1
   # 20   a = 0 , b = 1 ,out carry = 0 , outsum = 1
   # 22   a = 1 , b = 1 ,out carry = 1 , outsum = 0
```

Figure 4.25 Results of running the test bench of the half-adder module in Figure 4.24.

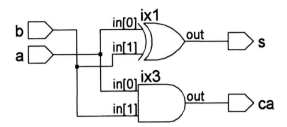

Figure 4.26 Synthesized output of the half-adder module of Figure 4.24.

```
module fa(sum,cout,a,b,cin);
input a,b,cin;
output sum,cout;
wire s,c1,c2;
ha ha1(s,c1,a,b), ha2(sum,c2,s,cin);
or(cout,c2,c1);
endmodule

//test-bench
module tst_fa();
reg a,b,cin;
fa ff(sum,cout,a,b,cin);
initial
begin
a =0;b=0;cin=0;
end
always
        begin
        #2 a=1;b=1;cin=0;#2 a=1;b=0;cin=1;
        #2 a=1;b=1;cin=1;#2 a=1;b=0;cin=0;
        #2 a=0;b=0;cin=0;#2 a=0;b=1;cin=0;
        #2 a=0;b=0;cin=1;#2 a=0;b=1;cin=1;
        #2 a=1;b=0;cin=0;#2 a=1;b=1;cin=0;
        #2 a=0;b=1;cin=0;#2 a=1;b=1;cin=1;
        end
initial $monitor($time ," a = %b, b = %b, cin = %b,
outsum = %b, outcar = %b ", a,b,cin,sum,cout);
initial #30 $stop ;
endmodule
```

Figure 4.27 Design module and a test bench for a full-adder.

```
//output
#0   a = 0, b = 0, cin = 0, outsum = 0, outcar = 0
#2   a = 1, b = 1, cin = 0, outsum = 0, outcar = 1
#4   a = 1, b = 0, cin = 1, outsum = 0, outcar = 1
#6   a = 1, b = 1, cin = 1, outsum = 1, outcar = 1
#8   a = 1, b = 0, cin = 0, outsum = 1, outcar = 0
#10  a = 0, b = 0, cin = 0, outsum = 0, outcar = 0
#12  a = 0, b = 1, cin = 0, outsum = 1, outcar = 0
#14  a = 0, b = 0, cin = 1, outsum = 1, outcar = 0
#16  a = 0, b = 1, cin = 1, outsum = 0, outcar = 1
#18  a = 1, b = 0, cin = 0, outsum = 1, outcar = 0
#20  a = 1, b = 1, cin = 0, outsum = 0, outcar = 1
#22  a = 0, b = 1, cin = 0, outsum = 1, outcar = 0
#24  a = 1, b = 1, cin = 1, outsum = 1, outcar = 1
#26  a = 1, b = 1, cin = 0, outsum = 0, outcar = 1
#28  a = 1, b = 0, cin = 1, outsum = 0, outcar = 1
```

Figure 4.28 Results of running the test bench of the full-adder module in Figure 4.27.

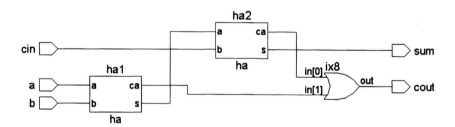

Figure 4.29 Synthesized output of the full-adder module of Figure 4.27.

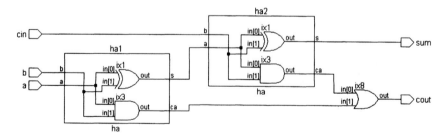

Figure 4.30 Synthesized circuit hierarchy of the full-adder module in Figure 4.27.

```
module mux4_1(y,i,s);
input [3:0] i;
input [1:0] s;
output y;
wire [1:0] ss;
wire [3:0]yy;
not (ss[0],s[0]),(ss[1],s[1]);
and (yy[0],i[0],ss[0],ss[1]);
and (yy[1],i[1],s[0],ss[1]);
and (yy[2],i[2],ss[0],s[1]);
and (yy[3],i[3],s[0],s[1]);
or (y,yy[3],yy[2],yy[1],yy[0]);
endmodule

//test-bench
module tst_mux4_1();
reg [3:0]i;
reg [1:0] s;
mux4_1 mm(y,i,s);
initial
    begin
    #2{i,s} = 6'b 0000_00;
    #2{i,s} = 6'b 0001_00;
    #2{i,s} = 6'b 0010_01;
    #2{i,s} = 6'b 0100_10;
    #2{i,s} = 6'b 1000_11;
    #2{i,s} = 6'b 0001_00;
    end
initial
$monitor($time," input s = %b,y = %b" ,s,y);
endmodule
```

Figure 4.31 Design module and a test bench for a 4-to-1 mux module.

```
//output
//#          0 input s = xx ,y = x
//#          2 input s = 00 ,y = 0
//#          4 input s = 00 ,y = 1
//#          6 input s = 01 ,y = 1
//#          8 input s = 10 ,y = 1
//#         10 input s = 11 ,y = 1
//#         12 input s = 00 ,y = 1
```

Figure 4.32 Results of running the test bench of the 4-to- mux module in Figure 4.31.

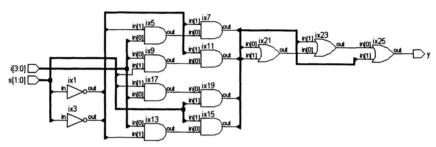

Figure 4.33 Synthesized output of the 4-to-1 Mux module of Figure 4.31.

```
module trimux4_1(o,e,i,s);
input e;
input [1:0]s;
input [3:0]i;
output o;
tri o;
wire y,y1,y2,y3,y4;
wire [1:0]ss;
not(ss[0],s[0]),(ss[1],s[1]);
and g1(y1,ss[0],ss[1],i[0]);
and g2(y2,ss[1],s[0],i[1]);
and g3(y3,ss[0],s[1],i[2]);
and g4(y4,s[1],s[0],i[3]);
or(y,y3,y2,y1,y2);
bufif1 buf2(o,y,e);
endmodule

//TESTBENCH
module tst_trimux4_1();
reg [1:0]s;
reg [3:0]i;
reg e;
wire o;
trimux4_1 tmx4_1(o,e,i,s);
initial
begin
e =0;i =2'b00;
end
always
    begin
    #6 e=0;s=2'b00;i=4'b0001;
    #6 e=1;s=2'b01;i=4'b0010;
```

continued

continued

```
    #6 e=1;s=2'b10;i=4'b0100;
    #6 e=1;s=2'b10;i=4'b1000;
    end
initial $monitor($time ," input e = %b , s= %b , i = %b
, output o = %b " ,e,s,i,o);
initial #48 $stop;
endmodule
```

Figure 4.34 Design module and a test bench for a 4-to-1 mux module with tri-state output.

```
    output
# 0 input e = 0 , s= xx , i = 0000 , output o = z
# 6 input e = 0 , s= 00 , i = 0001 , output o = z
#12 input e = 1 , s= 01 , i = 0010 , output o = 1
#18 input e = 1 , s= 10 , i = 0100 , output o = 1
#24 input e = 1 , s= 10 , i = 1000 , output o = 0
#30 input e = 0 , s= 00 , i = 0001 , output o = z
#36 input e = 1 , s= 01 , i = 0010 , output o = 1
#42 input e = 1 , s= 10 , i = 0100 , output o = 1
```

Figure 4.35 Results of running the test bench of the 4-to-1 mux module in Figure 4.34.

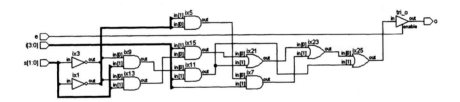

Figure 4.36 Synthesized output of the 4-to-1 mux module of Figure 4.34

```
module ttrimux2_1(out,e,i,s);
input[1:0]i;
input e;
input s;
output out;
wire o;
bufif0  g1(o,i[0],s);
bufif1  g2(o,i[1],s);
```

continued

continued

```
bufif1  g3(out,o,e);
endmodule

//testbench
module ttst_ttrimux2_1();
reg e;
reg [1:0]i;
reg s;
ttrimux2_1 mm(out,e,i,s);
initial
begin
e =0;  i = 2'b 00;end
always
    begin
    #4 e =0;{i,s} = 3'b 01_0;
    #4 e =1;{i,s} = 3'b 01_0;
    #4 e =1;{i,s} = 3'b 10_1;
    #4 e =1;{i,s} = 3'b 00_1;
    #4 e =1;{i,s} = 3'b 10_1;
    #4 e =1;{i,s} = 3'b 01_0;
    #4 e =1;{i,s} = 3'b 00_0;
    #4 e =1;{i,s} = 3'b 11_0;
    end
initial $monitor($time ," enable e = %b ,
s= %b , input i = %b ,output out = %b ",e ,s,i,out);
initial #48 $stop;
endmodule
```

Figure 4.37 Design module and a test bench for a 2-to-1 mux module formed with tri-state buffers.

```
output
# 0 enable e = 0,  s= x,  input i = 00,output out = z
# 4 enable e = 0,  s= 0,  input i = 01,output out = z
# 8 enable e = 1,  s= 0,  input i = 01,output out = 1
#12 enable e = 1,  s= 1,  input i = 10,output out = 1
#16 enable e = 1,  s= 1,  input i = 00,output out = 0
#20 enable e = 1,  s= 1,  input i = 10,output out = 1
#24 enable e = 1,  s= 0,  input i = 01,output out = 1
#28 enable e = 1,  s= 0,  input i = 00,output out = 0
#32 enable e = 1,  s= 0,  input i = 11,output out = 1
#36 enable e = 0,  s= 0,  input i = 01,output out = z
#40 enable e = 1,  s= 0,  input i = 01,output out = 1
#44 enable e = 1,  s= 1,  input i = 10,output out = 1
```

Figure 4.38 Results of running the test bench of the 2-to-1 mux module in Figure 4.37.

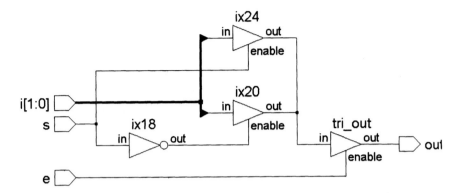

Figure 4.39 Synthesized output of the 2-to-1 mux module of Figure 4.37.

```
module ttrimux4_1(out,e,i,s);
input[3:0]i;
input e;
input[1:0]s;
output out;
tri o;
tri [1:0]o1;
bufif0  g1(o1[0],i[0],s[0]);
bufif1  g2(o1[0],i[1],s[0]);
bufif0  g3(o1[1],i[2],s[0]);
bufif1  g4(o1[1],i[3],s[0]);
bufif0  g5(o,o1[0],s[1]);
bufif1  g6(o,o1[1],s[1]);
bufif1  g7(out,o,e);
endmodule

//testbench
module ttst_ttrimux4_1();
reg e;
reg [3:0]i;
reg [1:0]s;
ttrimux4_1 mm(out,e,i,s);
initial
```

continued

continued

```
begin
        e = 0;
        i = 4'b 0000;
end
always
    begin
    #4 e =0;{i,s} = 6'b 0001_00;
    #4 e =1;{i,s} = 6'b 0001_00;
    #4 e =1;{i,s} = 6'b 0010_01;
    #4 e =1;{i,s} = 6'b 0000_01;
    #4 e =1;{i,s} = 6'b 0100_10;
    #4 e =1;{i,s} = 6'b 0101_10;
    #4 e =1;{i,s} = 6'b 1000_11;
    #4 e =1;{i,s} = 6'b 0000_11;
    end
initial $monitor($time ," enable e = %b , s= %b , input
i = %b ,output out = %b ",e ,s,i,out);
initial #48 $stop;
endmodule
```

Figure 4.40 Design module and a test bench for a 4-to-1 mux module formed with tri-state buffers.

```
output
# 0 enable e =0,s=xx, input i =0000, output out = z
# 4 enable e =0,s=00, input i =0001, output out = z
# 8 enable e =1, s=00,input i =0001 ,output out = 1
#12 enable e =1, s=01,input i =0010 ,output out = 1
#16 enable e =1, s=01,input i =0000 ,output out = 0
#20 enable e =1, s=10,input i =0100 ,output out = 0
#24 enable e =1, s=10,input i =0101 ,output out = 1
#28 enable e =1, s=11,input i =1000 ,output out = 1
#32 enable e =1, s=11,input i =0000 ,output out = 0
#36 enable e =0, s=00,input i =0001 ,output out = z
#40 enable e =1, s=00,input i =0001 ,output out = 1
#44 enable e =1, s=01,input i =0010 ,output out = 1
```

Figure 4.41 Results of running the test bench of the 4-to-1 mux module in Figure 4.40.

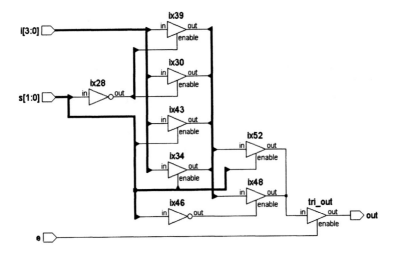

Figure 4.42 Synthesized output of the 4-to-1 mux module of Figure 4.40.

4.9   EXERCISES

1. Modify the test bench of Figure 4.1 and test the functionality of each of the basic gate primitives namely, OR, NOR, NAND, EXOR, EXNOR, NOT, and BUF.

For all the Exercises below prepare test benches and run the same.

2. Draw the half-adder circuit in terms of EX-OR gates and AND gate. Prepare a half-adder module in terms of EX-OR and AND gate primitive.

3. Prepare a full-adder module using half-adder module and OR gate Primitive.

4. Prepare a 4-bit adder module in terms of full-adder and half-adder modules. Treat the two 4-bit numbers as vectors for all input combination.

5. Prepare a module to generate a look-ahead-carry bit for the above problem.

6. Prepare modules for addition of 16 bit words and 32 bit words.

7. Prepare a module for conversion of an 8-bit number into its respective BCDs.

8. Prepare a module to add 2 BCDs

9. Prepare a module for the conversion of a pair of BCDs into the corresponding byte.

10. Prepare a module to generate Excess-3 code type of 4-bit output from a BCD.

11. Prepare a module to generate a BCD from an Excess-3 code digit.

12. Prepare an adder module to add Excess-3 coded digits.

13. Prepare a module to convert a set of 8 bits in gray code into an equivalent binary number.

14. Prepare an adder module to convert an 8-bit binary number into gray code.

15. Prepare a half-subtractor module and use it to form a 4-bit subtractor module.

16. Prepare a module to generate the 1's complement of a 4-bit number.

17. Prepare a module to generate 2's complement of a 4-bit number.

18. A set of 5-bit numbers is available as vectors – b [4:0]; b[4] is the sign bit. b [3:0] represent the number in 1's complement form.   Prepare

    a)   a module to add two such numbers

    b)   a module to subtract one such number from the other

19. Repeat the above problem when the numbers are in 2's complement form.

20. Prepare a module to multiplex two input bits into one output bit.

21. Prepare a module to demultiplex one bit into 2 bits.

22. Use the 2 to 4 decoder module and prepare

    a)   a 4 to 1 multiplexer module

    b)   a 1 to 4 demultiplexer module

23. A is an 8-bit vector. Prepare a module to form another 8-bit vector B with its bits forming the mirror image of A.

24. A 16-bit barcode driver output is available. Generate the corresponding 4 bit output from these (Priority Encoder)

25. Prepare a module to generate 16-bit barcode driver outputs from a 4-bit binary number.

26. Prepare a module to generate 7-segment driver outputs from a 4-bit number.

27. Two 4-bit binary numbers a and b are available. Prepare a comparator module. The comparator module will generate 2 output bits. One bit is 0 if a > b and 1 if a < b. The second bit is 1 if a = b and 0 otherwise.

28. Prepare a 2-bit ALU module and its test bench. Let the module inputs – A and B – be 2-bit wide. D is the 2-bit output. Ci is the carry input and Co is the carry output. F is the function select vector. If F = 1, D = A + B; if F = 2, D = A + B + Ci; if F = 3, D =A - B; if F = 4, D = A – B - Ci; if F = 5, D = A OR B; if F = 6, D = A AND B; if F = 7, D = A XOR B.

29. Prepare a module for addition of bytes, instantiating the nibble adder of Exercise 4.4 repeatedly. Use the look-ahead-carry output of Exercise 4.5 to generate the carry bit from bit position 3 to bit position 4.

30. Use arrays of instances and redo the 4-to-16 decoder module of Figure 4.13.

5

GATE LEVEL MODELING – 2

5.1   INTRODUCTION

Design of combinational circuits was discussed in detail in the last chapter. Flip-flops too can be designed in a similar manner - that is, in terms of gate primitives. The same can be extended to registers, register files, memory, and so on. These can be combined with combinational circuits to form designs at the MSI level. Design of different types of flip-flops is discussed here through a series of examples. Subsequently, constructs available to account for different types of propagation delays are discussed. Constructs to represent source and load impedances and their use along with propagation delays are dealt with subsequently [IEEE].

5.2   DESIGN OF FLIP-FLOPS WITH GATE PRIMITIVES

The basic RS latch can be designed using gate primitives. Two instantiations of NAND or NOR gates suffice here. More involved flip-flops, registers, *etc.*, can be built around these. Some of the level triggered versions of such flip-flops are taken up for design. Subsequently, the edge-triggered flip-flop of the 7474 type is developed in a skeletal form. More generalized versions are left as exercises.

Example 5.1 A Simple Latch

Figure 5.1 shows the design description of a simple latch formed with two NAND gates. A test bench for the same is shown in Figure 5.2 along with the results of the simulation run for 20 time steps. The test-bench has a block within a **begin-end** construct which reassigns values to rb and sb at two successive time step intervals. The whole sequence described within the block lasts for 10 ns. Defining the block within the **always** construct repeats the above assignment sequence cyclically until the simulation stops. The latch has been synthesized, and the synthesized circuit is shown in Figure 5.3.

```
module sbrbff(sb,rb,q,qb);
input sb,rb;
output q,qb;
nand(q,sb,qb);
nand(qb,rb,q);
endmodule
```

Figure 5.1 A module to instantiate the AND gate primitive and test it.

```
module tstsbrbff; //test-bench
reg sb,rb;
wire q,qb;
sbrbff ff(sb,rb,q,qb);
initial
begin
    sb =1'b1;
    rb =1'b0;
end
always
begin
    #2 sb =1'b1;rb =1'b1;
    #2 sb =1'b0;rb =1'b1;
    #2 sb =1'b1;rb =1'b1;
    #2 sb =1'b1;rb =1'b0;
    #2 sb =1'b1;rb =1'b1;
end
initial $monitor($time, " sb = %b,  rb = %b,
q = %b,  qb  = %b",sb,rb,q,qb);
initial #20 $stop;
endmodule
```

```
Simulation results
          #   0 sb = 1 ,  rb = 0 ,  q = 0 ,  qb  = 1
          #   2 sb = 1 ,  rb = 1 ,  q = 0 ,  qb  = 1
          #   4 sb = 0 ,  rb = 1 ,  q = 1 ,  qb  = 0
          #   6 sb = 1 ,  rb = 1 ,  q = 1 ,  qb  = 0
          #   8 sb = 1 ,  rb = 0 ,  q = 0 ,  qb  = 1
          # 10 sb = 1 ,  rb = 1 ,  q = 0 ,  qb  = 1
          # 14 sb = 0 ,  rb = 1 ,  q = 1 ,  qb  = 0
          # 16 sb = 1 ,  rb = 1 ,  q = 1 ,  qb  = 0
          # 18 sb = 1 ,  rb = 0 ,  q = 0 ,  qb  = 1
```

Figure 5.2 A test bench for the flip-flop of Figure 5.1 and results of running the test bench.

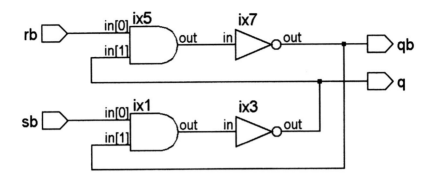

Figure 5.3 Synthesized circuit of the flip-flop module of Figure 5.1.

Example 5.2 An RS Flip-Flop

The design module of an RS flip-flop along with a test bench for the same is shown in Figure 5.4. The module is a slight modification of the flip-flop of Figure 5.1. The simulation results are shown in Figure 5.5. The synthesized circuit is shown in Figure 5.6. One can easily relate the difference between this circuit and that of Figure 5.3 to the corresponding difference between the respective design modules.

```
module srff(s,r,q,qb);
input s,r;
output q,qb;
wire ss,rr;
not(ss,s),(rr,r);
nand(q,ss,qb);
nand(qb,rr,q);
endmodule

module tstsrff; //test-bench
reg s,r;
wire q,qb;
srff ff(s,r,q,qb);
initial
```

continued

continued
```
begin
        s =1'b1;
        r =1'b0;
end
always
begin
        #2  s =1'b0;r =1'b0;
        #2  s =1'b0;r =1'b1;
        #2  s =1'b0;r =1'b0;
        #2  s =1'b1;r =1'b0;
        #2  s =1'b0;r =1'b0;
end
initial $monitor($time, " s = %b,  r = %b,  q = %b,  qb  =
%b ",s,r,q,qb);
initial #20 $stop;
endmodule
```

Figure 5.4 Module of an RS flip-flop with NAND gates and a test bench for the same.

```
#  0 s = 1 ,· r = 0 , q = 1 , qb  = 0
#  2 s = 0 , r = 0 , q = 1 , qb  = 0
#  4 s = 0 , r = 1 , q = 0 , qb  = 1
#  6 s = 0 , r = 0 , q = 0 , qb  = 1
#  8 s = 1 , r = 0 , q = 1 , qb  = 0
# 10 s = 0 , r = 0 , q = 1 , qb  = 0
# 14 s = 0 , r = 1 , q = 0 , qb  = 1
# 16 s = 0 , r = 0 , q = 0 , qb  = 1
# 18 s = 1 , r = 0 , q = 1 , qb  = 0
```

Figure 5.5 Results of running the test bench for the flip-flop of Figure 5.4.

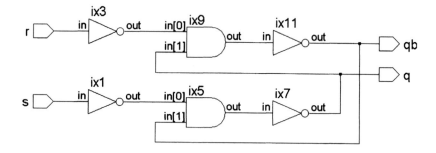

Figure 5.6 Synthesized circuit of the flip-flop module of Figure 5.4.

Example 5.3 A Clocked RS Flip-Flop

The module in Figure 5.7 is for a clocked RS flip-flop. It is the RS flip-flop of Figure 5.4 with the clock signal gating the R and S inputs. A test bench for the flip-flop is also shown in the figure. The clock waveform in the test bench is a square wave with a period of 4 ns [see Example 7.5 for details]. The simulation results are shown in Figure 5.8. Figure 5.9 shows the synthesized circuit of the flip-flop.

```
module srffcplev(cp,s,r,q,qb);
input cp,s,r;
output q,qb;
wire ss,rr;
nand(ss,s,cp),(rr,r,cp),(q,ss,qb),(qb,rr,q);
endmodule

module srffcplev_tst;// test-bench
reg cp,s,r;
wire q,qb;
srffcplev ff(cp,s,r,q,qb);
initial
begin
        cp=1'b0;
        s =1'b1;
        r =1'b0;
end
always #2cp=~cp;
always
begin
        #4  s =1'b0;r =1'b0;
        #4  s =1'b0;r =1'b1;
        #4  s =1'b0;r =1'b0;
        #4  s =1'b1;r =1'b0;
        #4  s =1'b0;r =1'b0;
end
initial $monitor($time,"cp = %b ,s = %b , r = %b , q =
%b , qb  = %b " ,cp,s,r,q,qb);
initial #20 $stop;
endmodule
```

Figure 5.7 Module of a clocked RS flip-flop with NAND gates and a test bench for the same.

```
#  0 cp = 0,  s = 1,  r = 0,  q = x,  qb  = x
#  2 cp = 1,  s = 1,  r = 0,  q = 1,  qb  = 0
#  4 cp = 0,  s = 0,  r = 0,  q = 1,  qb  = 0
#  6 cp = 1,  s = 0,  r = 0,  q = 1,  qb  = 0
#  8 cp = 0,  s = 0,  r = 1,  q = 1,  qb  = 0
# 10 cp = 1,  s = 0,  r = 1,  q = 0,  qb  = 1
# 12 cp = 0,  s = 0,  r = 0,  q = 0,  qb  = 1
# 14 cp = 1,  s = 0,  r = 0,  q = 0,  qb  = 1
# 16 cp = 0,  s = 1,  r = 0,  q = 0,  qb  = 1
# 18 cp = 1,  s = 1,  r = 0,  q = 1,  qb  = 0
```

Figure 5.8 Results of running the test bench for the flip-flop of Figure 5.7.

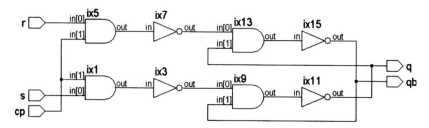

Figure 5.9 Synthesized circuit of the flip-flop module of Figure 5.7.

Example 5.4 A D-Latch

The design description of a D latch is given in Figure 5.10. It has one instantiation of the basic flip-flop of Figure 5.1. A test bench for the latch is also included in the figure. The simulation results are shown in Figure .5.11. Two versions of the synthesized circuit are shown in Figure 5.12 and Figure 5.13, respectively. The basic latch [sbrbff] — which was instantiated in the module of Figure 5.10 — is shown as a black box in Figure 5.12. The internals of the latch are shown in Figure 5.13, which brings out the hierarchy clearly.

```
module dlatch(en,d,q,qb);
input d,en;
output q,qb;
wire dd;
wire s,r;
not n1(dd,d);
nand (sb,d,en);
nand g2(rb,dd,en);
```

continued

continued   ⋮

```
sbrbff ff(sb,rb,q,qb);//Instantiation of the sbrbff
endmodule

module tstdlatch; //test-bench
reg d,en;
wire q,qb;
dlatch ff(en,d,q,qb);
initial
begin
        d  = 1'b0;
        en = 1'b0;
end
always #4 en =~en;
always #8 d=~d;
initial $monitor($time," en = %b , d = %b , q = %b , qb
= %b " , en,d,q,qb);
initial #40 $stop;
endmodule
```

Figure 5.10 Module of a D latch and a test bench for the same.

```
#   0 en = 0, d = 0, q = x, qb  = x
#   4 en = 1, d = 0, q = 0, qb  = 1
#   8 en = 0, d = 1, q = 0, qb  = 1
# 12 en = 1, d = 1, q = 1, qb  = 0
# 16 en = 0, d = 0, q = 1, qb  = 0
# 20 en = 1, d = 0, q = 0, qb  = 1
# 24 en = 0, d = 1, q = 0, qb  = 1
# 28 en = 1, d = 1, q = 1, qb  = 0
# 32 en = 0, d = 0, q = 1, qb  = 0
# 36 en = 1, d = 0, q = 0, qb  = 1
```

Figure 5.11 Results of running the test bench for the D latch of Figure 5.10.

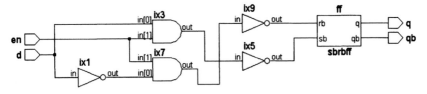

Figure 5.12 Synthesized circuit of the D latch module of Figure 5.10.

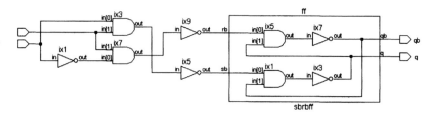

Figure 5.13 Synthesized circuit of the D latch module of Figure 5.10 showing hierarchy.

Example 5.5 An Edge-Triggered Flip-Flop

Figure 5.14 shows the circuit of an edge-triggered flip-flop. It is a simplified version of the 7474 IC. The circuit is a combination of three latches – designated as FF1, FF2, and FF3 in the figure. FF3 is similar to the latch considered in Example 5.1. FF1 and FF2 are minor modifications of FF3. The design modules for FF1 and FF2 are given in Figure 5.15. All three latches are instantiated to form the edge-triggered flip-flop. A test bench for the flip-flop is also included in the figure. With a square waveform for the clock – cp – the waveform for the d input is chosen to bring out the edge-triggered nature of operation of the flip-flop. The output obtained by running the test bench is shown in Figure 5.16; the respective waveforms are shown in Figure 5.17. One can see that the output changes only at the positive edges of the clock, and it assumes the value of the input at that instant of time.

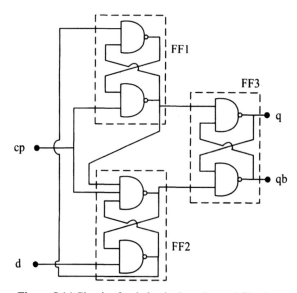

Figure 5.14 Circuit of a skeletal edge-triggered flip-flop.

```verilog
module dffgatnew1(cp,d,q,qb);
input d,cp;
output q,qb;
wire sb,rb;
wire s,r;
sbrbffdff ff1(rb,cp,s);
sbrbff1 ff2(s,d,cp,r,rb);
sbrbff ff3(s,r,q,qb);
endmodule

module tst_dffgatnew1; //test-bench
reg d,cp;
wire q,qb;
dffgatnew1 ff(cp,d,q,qb);
initial
begin
    d =1'b0;cp =1'b0;
    #2 cp =1'b1;#2 cp =1'b0;#2 cp =1'b1;#2 cp =1'b0;
    #2 cp =1'b1;#2 cp =1'b0;#2 cp =1'b1;#2 cp =1'b0;
end
initial
begin
    #3 d=1'b1;#2d=1'b1;#2d=1'b0;#3d=1'b0;#3d=1'b1;
end
initial $monitor($time," cp = %b , d = %b , q = %b , qb
= %b " , cp,d,q,qb);
initial #40 $stop;
endmodule

module sbrbffdff(sb,rb,qb);
input sb,rb;
output qb;
wire q;
nand(q,sb,qb);
nand(qb,rb,q);
endmodule

module sbrbff1(sb,rb,cp,q,qb); //test-bench
input sb,rb,cp;
output q,qb;
nand(q,sb,cp,qb);
nand(qb,rb,q);
endmodule
```

Figure 5.15 Module of a positive edge-triggered flip-flop and its test bench.

```
#  0 cp = 0 , d = 0 , q = x , qb  = x
#  2 cp = 1 , d = 0 , q = 0 , qb  = 1
#  3 cp = 1 , d = 1 , q = 0 , qb  = 1
#  4 cp = 0 , d = 1 , q = 0 , qb  = 1
#  6 cp = 1 , d = 1 , q = 1 , qb  = 0
#  7 cp = 1 , d = 0 , q = 1 , qb  = 0
#  8 cp = 0 , d = 0 , q = 1 , qb  = 0
# 10 cp = 1 , d = 0 , q = 0 , qb  = 1
# 12 cp = 0 , d = 0 , q = 0 , qb  = 1
# 13 cp = 0 , d = 1 , q = 0 , qb  = 1
# 14 cp = 1 , d = 1 , q = 1 , qb  = 0
# 16 cp = 0 , d = 1 , q = 1 , qb  = 0
```

Figure 5.16 Results of running the test bench for the flip-flop of Figure 5.15.

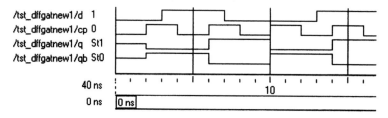

/tst_dffgatnew1/d   1
/tst_dffgatnew1/cp 0
/tst_dffgatnew1/q   St1
/tst_dffgatnew1/qb St0

40 ns   10

0 ns   0 ns

Figure 5.17 Clock (cp), data input (d), and output waveforms for the edge-triggered flip-flop with the test bench in Figure 5.15.

Synthesized circuits of the latches FF1 (sbrbffdff) and FF2 (sbrbff1) are shown in Figure 5.18 and Figure 5.19, respectively. The synthesized circuit for the overall flip-flop is shown in Figure 5.20. FF1, FF2, and FF3 are represented as boxes there; only their interconnections are shown. The comprehensive circuit in terms of the elementary gates is not shown.

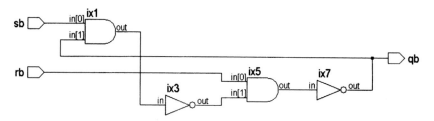

Figure 5.18 Synthesized circuit of the flip-flop sbrbffdff of Figure 5.15.

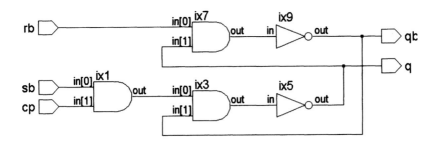

Figure 5.19 Synthesized circuit of the flip-flop sbrbff1 of Figure 5.15.

The flip-flop of Figure 5.14 can be made comprehensive with slight modifications. It can be replicated and with suitable additions, expanded substantially into register files and full-fledged memory [see the Exercises at the end of the chapter].

5.3   DELAYS

Verilog has the facility to account for different types of propagation delays of circuit elements. Any connection can cause a delay due to the distributed nature of its resistance and capacitance. Due to the manufacturing tolerances, these can vary over a range in any given circuit [Bignel, Sedra]. Similar delays are present in gates too. These manifest as propagation delays in the 0 to 1 transitions and 1 to 0 transitions from input to the output. Such propagation delays can differ for the two types of transitions. A variety of such delays can be accommodated in Verilog. Sometimes manufacturers adjust input and output impedances of circuit elements to specific levels and exploit them to reduce interface hardware. These too can be accommodated in Verilog design descriptions [Ciletti, Palnitkar].

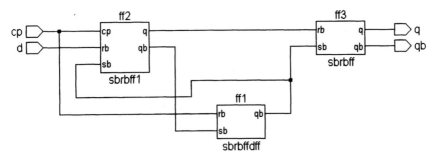

Figure 5.20 Synthesized circuit of the flip-flop dffgatnew1 in Figure 5.15.

5.3.1   Net Delay

One of the simplest delays is that of a direct connection – a net. It can be part of the declaration statement

   wire #2 nn; // nn is declared as a net with a propagation delay of 2 time steps

Here nn is declared as a net with an associated propagation delay of 2 time steps. The delay is the same for the positive as well as the negative transitions. The same is illustrated in Figure 5.21(a), which connects two circuit blocks through a net nn with a delay of 2 time steps associated with it. The module in Figure 5.22 is a simple realization of the same. A test bench for the module is also shown in the figure. The simulation results are shown in Figure 5.21(b), which bring out the effect of the net delay clearly.

Similar delays can be assigned to other types of nets as well. Whenever a variable or a signal is defined as a net and no delay is specified for it, the associated delay is taken as zero. This is true of instantiations of modules as well. The impedance connected as well as the type of loading can differ for the two transitions. The propagation delay too can differ accordingly. Two such delays can be specified as follows:

Wire   # (2, 1) nm;

Here nm is declared as a net with two distinct propagation delays; the positive (0 to 1) transition has a delay of 2 time steps associated with it. The negative

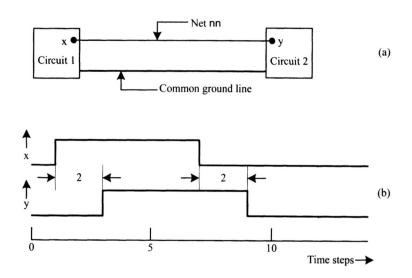

Figure 5.21 A net connecting two circuit blocks and the delay through it: (a) Connection diagram (b) Typical signal waveforms at the input and output ends of the net.

```
module netdelay(x,y);
input x;
output y;
wire #2 nn;
not (nn,x); //circuit1 in Figure 5.21
buf y = x; //circuit2 in Figure 5.21
endmodule

module tst_netdelay ; //test-bench
reg x;
wire y;
netdelay  nd(x,y);
initial
begin
        x =1'b0;
      #6 x =~x;
end
initial #20 $stop;
endmodule
```

Figure 5.22 A module to illustrate net delay and a test bench for the same.

(1 to 0) transition has a delay of 1 time step. The delays are explained in Figure 5.23. The module of Figure 5.22 has been modified and shown in Figure 5.24; the propagation delays are different for rise and fall here.

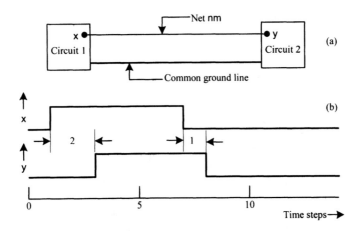

Figure 5.23 A net connecting two circuit blocks and the delays through it: (a) Connection diagram (b) Typical signal waveforms at the input and output ends of the net.

```
module netdelay1(x,y);
input x;
output y;
wire #(2,1) nn;
not (nn,x);
y=nn;
endmodule

module tst_netdelay1; //test-bench
reg x;
wire y;
netdelay1   nd(x,y);
initial
begin
        x =1'b0;
     #6 x =~x;
end
initial #20 $stop;
endmodule
```

Figure 5.24 A module to demonstrate different delays for rise and fall times on a net.

5.3.2   Gate Delay

Gates too can have delays associated with them. These can be specified as part of the instantiation itself.

and #3 g (a, b, c);

The above represents an AND gate description with a uniform delay of 3 ns for all transitions from input to output. A more detailed description can be as follows:

and #(2, 1) (a, b, c);

With the above statement the positive (0 to 1) transition at the output has a delay of 2 time steps while the negative (1 to 0) transition has a delay of 1 time step. Figure 5.25 shows a module to illustrate the delays associated with gate primitives. A test bench for the same is also shown in the figure. The results of running the test bench are shown in Figure 5.27. The AND gate instantiation in Figure 5.25 has different delays for the output transitions; respective waveforms are shown in Figure 5.26.

```
module gade(a,a1,b,c,b1,c1);
input b,c,b1,c1;
output a,a1;
or #3gg1(a1,c1,b1);
and #(2,1)gg2(a,c,b);
endmodule

module tst_gade();//test-bench
reg b,c,b1,c1;
wire c,c1;
gade ggde(a,a1,b,c,b1,c1);
initial
begin
b =1'b0;c =1'b0;b1 =1'b0;c1=1'b0;
end
always
begin
    #5 b =1'b0;c =1'b0;b1 =1'b1;c1=1'b1;
    #5 b =1'b1;c =1'b1;b1 =1'b0;c1=1'b0;
    #5 b =1'b1;c =1'b0;b1 =1'b1;c1=1'b0;
    #5 b =1'b0;c =1'b1;b1 =1'b0;c1=1'b1;
    #5 b =1'b1;c =1'b1;b1 =1'b1;c1=1'b1;
    #5 b =1'b1;c =1'b1;b1 =1'b1;c1=1'b1;
end
initial $monitor($time  ,   "   b= %b , c = %b , b1 = %b
,c1 = %b , a = %b ,a1 = %b" ,b,c,b1,c1,a,a1);
initial #30 $stop;
endmodule
```

Figure 5.25 Module to demonstrate the delays with gates.

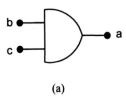

(a)

Figure 5.26 AND gate instantiation with different delays for the positive and negative transitions and associated waveforms: (a) Gate instantiated.

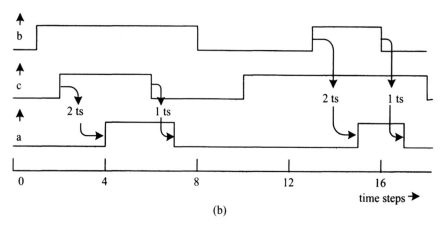

(b)

Figure 5.26 (cont'd) **(b)** associated waveforms (time step has been abbreviated to "ts" in the diagram).

In a more detailed design description, delays can be associated with nets as well as gates. Consider the design description shown in Figure 5.28(a). It has a total of 8 different time delay values specified. All these are hypothetical and different from each other. It is done intentionally to bring out the effect of each of them on the concerned gates and signals. The circuit for this design description is shown in Figure 5.28(b). Typical waveforms of input signals as well as other signals are shown in Figure 5.29, to illustrate the different delays in the design description. Figures 5.29(a) and 5.29(b) illustrate how changes in one of the inputs – b1 – affect the other signals; the signals and gates affected are shown

```
#  0 b= 0, c = 0 , b1 = 0 ,c1 = 0 , a = x ,a1 = x
#  1 b= 0, c = 0 , b1 = 0 ,c1 = 0 , a = x ,a1 = 0
#  3 b= 0, c = 0 , b1 = 0 ,c1 = 0 , a = 0 ,a1 = 0
#  5 b= 0, c = 0 , b1 = 1 ,b1 = 1 , a = 0 ,a1 = 0
#  7 b= 0, c = 0 , b1 = 1 ,c1 = 1 , a = 0 ,a1 = 1
# 10 b= 1, c = 1 , b1 = 0 ,c1 = 0 , a = 0 ,a1 = 1
# 11 b= 1, c = 1 , b1 = 0 ,c1 = 0 , a = 0 ,a1 = 0
# 13 b= 1, c = 1 , b1 = 0 ,c1 = 0 , a = 1 ,a1 = 0
# 15 b= 1, c = 0 , b1 = 1 ,c1 = 0 , a = 1 ,a1 = 0
# 17 b= 1, c = 0 , b1 = 1 ,c1 = 0 , a = 1 ,c1 = 1
# 18 b= 1, c = 0 , b1 = 1 ,c1 = 0 , a = 0 ,c1 = 1
# 20 b= 0, c = 1 , b1 = 0 ,c1 = 1 , a = 0 ,a1 = 1
# 25 b= 1, c = 1 , b1 = 1 ,c1 = 1 , a = 0 ,a1 = 1
# 28 b= 1, c = 1 , b1 = 1 ,c1 = 1 , a = 1 ,a1 = 1
```

Figure 5.27 Results of running the test bench of above module in Figure 5.25.

highlighted in Figure 5.29(a). Throughout this period, input c1 is taken as at 1 state while inputs b2 and c2 remain at 0 state. The propagation delays of signals at point P and Q and that for the signal a are shown in Figure 5.29(b). These conform to the delays specified in the design segment of Figure 5.28(a). Subsequently, input c1 goes down to 0 state and input b1 remains at 0 state itself. Only signal b2 changes. The affected signals and gates are shown highlighted in Figure 5.29(c). The waveforms of signals affected and the associated propagation designs are shown in Figure 5.29(d). These too conform to the program segment of Figure 5.28(a).

```
module gates(b1,b2,c1,c2,a);
input b1,b2,c1,c2;
wire #(2,1)a1,a2;
output a;
and #(3,4)g1(a1,b1,c1);
and #(5,6)g2(a2,b2,c2);
or  #(8,7)g3(a,a1,a2);
endmodule

module tst_gates;//test-bench
reg b1,b2,c1,c2;
gates gg(b1,b2,c1,c2,a);
initial
begin
     b1=1'b0;c1=1'b0;b2=1'b0;c2=1'b0;
end
initial #100 $stop;

always
begin
    #2b1=1'b0;c1=1'b0;b2=1'b1;c2=1'b1;
    #2b1=1'b1;c1=1'b1;b2=1'b0;c2=1'b0;
    #2b1=1'b0;c1=1'b1;b2=1'b0;c2=1'b0;
    #2b1=1'b0;c1=1'b0;b2=1'b1;c2=1'b0;
    #2b1=1'b1;c1=1'b0;b2=1'b1;c2=1'b1;
    #2b1=1'b1;c1=1'b1;b2=1'b0;c2=1'b0;
    #2b1=1'b1;c1=1'b1;b2=1'b1;c2=1'b0;
    #2b1=1'b0;c1=1'b0;b2=1'b1;c2=1'b1;
end
initial $monitor($time," b1= %b , c1 = %b ,b2 = %b , c2
= %b ,   a = %b ",b1,c1,b2,c2,a);
endmodule
```

Figure 5.28(a) A design having eight different time delay values.

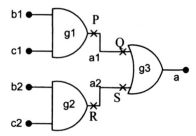

Figure 5.28(b) The circuit for the module considered in Figure 5.28(a).

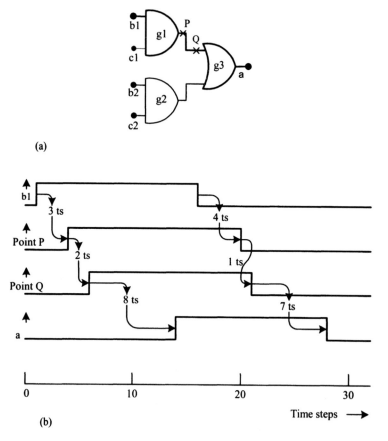

Figure 5.29 Illustration of signal delays in the design description segment in Figure 5.28: (a) The circuit portion active during changes to signal b1. (b) Signal waveforms following changes to signal b1 (time step has been abbreviated as ts in the diagram).

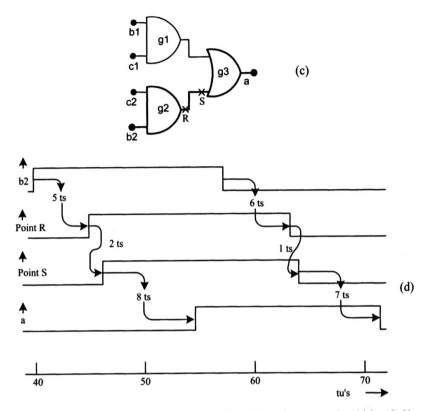

Figure 5.29 (cont'd) (c) The circuit portion active during changes to signal b2. (d) Signal waveforms following changes to signal b2 (time step has been abbreviated as ts in the diagrams).

5.3.3   Delays with Tri-state Gates

For tri-state gates the delays associated with the control signals can be different from those of the input as well as the output. The instantiation inclusive of this is shown in Figure 5.30 for a tri-state buffer of the **bufif1** type. Three time delay values are specified:

1. The first number represents the delay associated with the positive (0 to 1) transition of the output.
2. The second number represents the delay associated with the negative (1 to 0) transition of the output.
3. The third number represents the delay for the output to go to the hi-Z state as the control signal changes from 1 to 0 (*i.e.*, ON to OFF command).

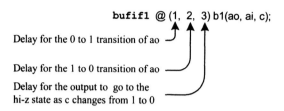

Figure 5.30 Delays associated with a typical tri-state gate.

Delays for the other tri-state buffers – namely **bufif0**, **notif1** and **notif0** – may be specified in a similar manner.

The turn-off time — 2 time steps here — represents the time for which the charge will be stored in the output line after the control line turns off. Values of delay time and storage time can be specified in this manner for all the types of tri-state type gates. The following are noteworthy here:

- Delays and storage times can be specified on the gate primitives and the nets but not on regs.
- All three time values are separately specified in the most versatile case.
- If only two time-values are specified, these are interpreted by Verilog as the rise (0 to 1) and fall (1 to 0) time, respectively. The turn-off time (delay) is taken as the smaller of these two.
- If only one time value is specified, it is taken as the rise time, the fall time, and the turn-off time.
- If no time value is specified, the rise and fall times at the output are taken as zero. The turn-off is taken as instantaneous.

Normally the delay time of any IC varies over a range for ICs from different production batches (as well as in any one batch). It is customary for manufacturers to specify delays and their range in the following manner:

- *Max delay*: The maximum value of the delay in a batch; that is, the delay encountered in practice is guaranteed to be less than this in the worst case.
- *Min. delay*: Minimum value of delay in a batch; that is, the specified signal is guaranteed to be available only after a minimum of time specified.
- *Typ. delay*: Typical or representative value of the delay.

Each of the delays in a gate primitive or for a net can be specified in terms of these three values. For example

and #(2:3:4) g1(a0, a1, a2);

can instantiate an AND gate with the following time delay specifications:

- The 0 to 1 rise time and the 1 to 0 fall time are equal.
- The minimum value of either is 2 time steps. Typical value is 3 time steps and the maximum value is 4 time steps.
- Note that the colon that separates the numbers signifies that the timings specified are the minimum, typical, and maximum values. At the time of simulation, one can specify the simulation to be carried out with any of these three delay values. If the same is not specified, the simulation is carried out with the typical delay value.

The group of minimum, typical, and maximum delay values for the propagation delays can be specified separately for any gate primitive. Thus an AND gate primitive can be specified as

and #(1:2:3, 2:4:6) g2(b0, b1, b2);

Here for the 0 to 1 transition of the output (rise time) the gate has a minimum delay value of 1 ns, a typical value of 2 ns, and a maximum value of 3 ns. Similarly, for the 1 to 0 transition (fall time) the gate has a minimum delay value of 2 ns, a typical delay value of 4 ns, and a maximum delay value of 6 ns. Such delay specifications can be associated with nets as well as tri-state type gates also.

Examples

wire #(1:2:3) a; /* The net a has a propagation delay whose minimum, typical and maximum values are 1 ns, 2 ns, and 3 ns, respectively*/

bufif1 #(1:2:3, 2:4:6, 3:6:9) g3 (a0, b0, c0);

/* The different delay values for the buffer are as follows:
- The output rise time (0 to 1 transition) has a minimum value of 1 ns, a typical value of 2 ns and a maximum value of 3 ns.
- The output fall time (1 to 0 transition) has a minimum value of 2 ns, a typical value of 4 ns and a maximum value of 6 ns.
- The output turn-off time (1 to 0) has a minimum value of 3 ns, a typical value of 6 ns, and a maximum value of 9 ns. */

A typical design can have a number of circuit blocks like gates, flip-flops, *etc.*, with associated interconnections. The individual nets and gates may have their own separate delays. The following general observations are in order regarding the overall delays through the circuit:

- A normal design can have many gates and nets in its signal paths. The delay through any path for a signal depends on the path and the type of transitions at each stage.

- The cumulative delay for a signal in a path puts an upper limit on the maximum operating frequency *vis-à-vis* the signal.
- A signal may go through multiple paths in a design to arrive at one gate. It is necessary to match the delays within specified tolerances for reliable operation of the device.
- In larger designs, one has to identify the longest signal path (critical path). This puts an upper limit on the operating frequency apart from causing mal-operation in a worst-case scenario. One of the practices in design is to re-route selected signals or redo selected design segments to reduce critical path delays.

5.3.4   General Definitions for Delays

Specific numerical values have been used for all the delays in the examples so far. However, Verilog LRM allows constant expressions to be used for any of the delay values. The expressions used may involve simple algebra in terms of integers and known quantities (but not variables).

5.4   STRENGTHS AND CONTENTION RESOLUTION

In practical situations, outputs of logic gates and signals on nets in a circuit have associated source impedances. When the outputs of two gates are joined together, the signal level is decided by the relative magnitudes of the source impedances. Sometimes a disparity between the impedances is intentionally introduced to minimize circuit hardware. Effects of such differences in the impedances are indirectly introduced in design descriptions by assigning "strengths" to specific signals (see also Section 3.9). Signal strength declarations are of two types – those associated with outputs of gate primitives and those with nets.

5.4.1   Strengths of Gate Primitives

Gate output strengths can be specified separately. Table 5.1 gives the names associated with strengths, respective abbreviations, and their order by weight. These hold good for logic 1 state as well as the 0 state.

Table 5.1 Strength levels associated with outputs of gate primitives

Name	supply	strong	pull	weak	High impedance	
Abbreviations	su1	st1	pu1	we1	HiZ1	
	su0	st0	pu0	we0		HiZ0
Strength	Strongest			Weakest		

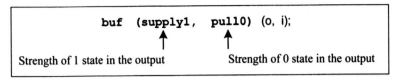

Figure 5.31 Format for specifying strengths in the instantiation of a gate primitive.

The strengths associated with the output of a gate primitive can be specified separately for the two logic levels. The format for the same is shown in Figure 5.31 for a specific case; the format remains the same for all types of gate primitives.

5.4.2   Strength Contention in Gate Primitives

When two signals of opposite polarity and differing strengths drive a line, the output status is decided by the stronger signal. However, if the signals are of equal strength, the output is indeterminate. Different contention possibilities arise here. The variety is brought out through examples.

Example 5.6 Strength Contention

Consider the module in Figure 5.32. The logic levels taken by the signal o for different combinations of inputs to the two buffers g1 and g2 are shown in Table 5.2. Contentions of signals with other combinations of levels can be resolved in the same manner.

Table 5.2 Outputs for different inputs for the example of Figure 5.32

Logic value of input i1	Logic value of input i2	Logic value of output o	Remarks
0	0	0	No contention
0	1	1	Contention; the stronger signal – i2 – prevails
1	0	1	Contention; the stronger signal – i1 – prevails
1	1	1	No contention

```
module contres(o,i1,i2);
input i1,i2;
output o;
buf(supply1,pull0)g1(o,i1), g2(o,i2);//note that the
endmodule// same net is driven by both the gates.

module tst_contres; //TEST BENCH
reg i1,i2;
contres cc(o,i1,i2);
initial
begin
     i1 =0;
     i2 =0;
end    //no contention
always
begin
     #4 i1 =0; i2 = 1;// contention; the stronger
     #4 i1 =1; i2 = 0;// signal prevails.
     #4 i1 =1; i2 = 1;//no contention.
end
initial $monitor($time,"i1=%b,i2=%b,o=%b",i1,i2,o);
initial #40$stop;
endmodule
```

```
output
#                           0 i1 = 0 , i2 = 0 , o = 0
#                           4 i1 = 0 , i2 = 1 , o = 1
#                           8 i1 = 1 , i2 = 0 , o = 1
#                          12 i1 = 1 , i2 = 1 , o = 1
#                          16 i1 = 0 , i2 = 1 , o = 1
#                          20 i1 = 1 , i2 = 0 , o = 1
#                          24 i1 = 1 , i2 = 1 , o = 1
#                          28 i1 = 0 , i2 = 1 , o = 1
#                          32 i1 = 1 , i2 = 0 , o = 1
#                          36 i1 = 1 , i2 = 1 , o = 1
#                          40 i1 = 0 , i2 = 1 , o = 1
```

Figure 5.32 A module to illustrate strength contention; the test bench and simulation results are also shown in the figure.

The outputs for the four input combinations are given in the table. Whenever there is a contention, the logic value of the output is decided by the stronger signal. In fact the design description here realizes an OR gate at the output side without additional hardware. It does not lead to any ambiguity.

Consider the Example in Figure 5.33, which is a slightly modified version of that in Figure 5.32. The output logic values for different input combinations are given in Table 5.3. The gate outputs are decided by following the same logic as in the last case. However, in one case — when both gates "drag" the output with equal strength in opposite directions — the output logic level is indeterminate — that is, **x**.

```
module contres1(o,i1,i2);
input i1,i2;
output o;
buf(strong1 ,pull0)g1(o,i1); buf(pull1,pull0)g2(o,i2);
endmodule

module tst_contres1; //TEST BENCH
reg i1,i2;
contres1 cc(o,i1,i2);
initial
begin
i1 =0;i2 =0;end       //no contention
always
begin
#4 i1 = 0; i2 = 1; //contention between pull0 due to
//i1 and pull1 due to i2; output is x
#4 i1 =1; i2 =0;   //contention; output is 1 since
//strong1 of i1 prevails.
#4 i1 =1 ;i2 = 1; //no contention.
end
initial $monitor($time   ," i1 = %b , i2 = %b ,o = %b "
,i1,i2,o);
initial #40 $stop;
endmodule
```

```
output
  #                            0 i1 = 0, i2= 0 ,o = 0
  #                            4 i1 = 0, i2= 1 ,o = x
  #                            8 i1 = 1, i2= 0 ,o = 1
  #                           12 i1 = 1, i2= 1 ,o = 1
  #                           16 i1 = 0, i2= 1 ,o = x
  #                           20 i1 = 1, i2= 0 ,o = 1
  #                           24 i1 = 1, i2= 1 ,o = 1
  #                           28 i1 = 0, i2= 1 ,o = x
  #                           32 i1 = 1, i2= 0 ,o = 1
  #                           36 i1 = 1, i2= 1 ,o = 1
```

Figure 5.33 Illustration of strength contention resulting in **x**-type output; the test bench and simulation results are also shown in the figure.

Table 5.3 Outputs for different inputs in the example of Figure 5.33

Logic value of input i1	Logic value of input i2	Logic value of output o	Remarks
0	0	0	No contention
0	1	x	Contention; both signals being of equal strength, the output is indeterminate
1	0	1	Contention; the stronger signal i1 prevails and forces the output to logic state 1
1	1	1	No contention

5.4.3   Net Charges

Whenever a net is driven by a signal, it takes the logic value of the signal. When the signal source is tri-stated, the net too gets tri-stated. In practice the net can have a capacitor associated with it, which can store the signal level even after the signal source dries up (*i.e.*, tri-stated). To account for this situation, a charge storage capacity is associated with the net. Such nets are declared with the keyword **trireg**. By virtue of the inherent capacitance associated with them, trireg nets can never be in the high impedance state – that is, they can assume 0, 1, or **x** value only. A **trireg** net can be in one of two possible states only:

- *Driven state*: When driven by a source or multiple sources, the net assumes the strength of the source. It can be any of the strengths specified in Table 5.1 except the high impedance value.
- *Capacitive state*: When the driven source (sources) is (are) tri-stated, the net retains the last value it was in – by virtue of the capacitance associated with it. The value can be 0, 1 or **x** (but not the high impedance value).

When in the capacitive state, a net can have a storage strength associated with it. Three such storage strengths are possible – namely **large, medium,** and **small**. Their details are shown in Table 5.4.   When a storage strength is not specified, it is assigned the default value – **medium**. For a **trireg** net one cannot assign storage strength capacity separately for the 0 and the 1 states.

A **trireg** net can be driven with possibilities of contention from two or more sources; such cases are considered in Chapter 10.

Table 5.4 Capacitive storage strengths on nets

Name	**large**	**medium**	**small**
Strength	Strongest		Weakest

Example 5.7 Net Storage

Consider the design in Figure 5.34. As long as the signal control = 1, the signal out follows the signal in. When control goes to 0, out is disconnected from the input and it "floats." It retains the last value due to the capacitance storage capacity. The storage strength is **medium**, signifying a medium value of capacitance.

```
module charge(out,in,control);
output out;
trireg(medium)out;
input in,control;
bufif1 g1(out,in,control);
endmodule

module tst_charge;  //TESTBENCH
reg in, control;
charge c1(out,in,control);
initial
    begin
    in =0;control =0;//when control=0 output is x
    #2 control =0;in =0;
    #2 control =1;in =0;
    #2 control =1;in =1;
    #2 control =0;in =0; // output is retained at
    end // the last value
initial $monitor($time ," in= %b ,control = %b , out=
%b " ,in,control,out);
initial #40$stop;
endmodule
```

```
output
        #      0 in = 0 , control = x , out=x
        #      2 in = 0 , control = 0 , out=x
        #      4 in = 0 , control = 1 , out=0
        #      6 in = 1 , control = 1 , out=1
        #      8 in = 0 , control = 0 , out=1
```

Figure 5.34 Illustration of net storage; the test bench and simulation results are also shown in the figure.

5.4.4   Contention Between Net and Gate Primitive Outputs

In case of a contention between a signal output from a gate and the charge on a net, the contention is decided by the relative strengths of the signals on each. Table 5.5 combines all the strengths – those of the gate outputs as well as those of tri-stated nets and – lists them in the order of their relative strengths. The abbreviations associated with the strengths are not repeated here.

5.4.5   Net Types and Port Assignments

All input ports of modules have to accept inputs from outside when instantiated and respond to changes in them. Hence they have to be of net type. Note that input ports cannot be of reg type since their values cannot be changed from outside. The output ports of instantiated modules can be of net or reg types. **Inout** ports have to function as input or output ports; hence they too have to be of net types.

The port assignments in an instantiation can be to scalars, vectors, part vectors, or concatenated vectors. However, their sizes should match those of the ports in the module definitions. Further, the type restrictions mentioned above have to be complied with.

In many situations the net types in the module definition and its instantiation may differ in the case of input and **inout** ports. In such cases the resulting net type can be of only one type. Either the net type declared in the module definition or that in the instantiation (external type) dominates. The choice is decided by a specific protocol in the LRM. Table 5.6 gives details. As can be seen from the table, whenever the two net types lead to a logical clash, the external data type prevails (identified by an asterisk in the table).

Table 5.5 Signal strength names and their relative weights

Signal strength name	Strength level
Supply (drive)	Strongest 7
Strong (drive)	6
Pull (drive)	5
Large (capacitance)	4
Weak (drive)	3
Medium (capacitance)	2
Small (capacitance)	Weakest 1
High impedance	0

Table 5.6 Net assignments with port connections

Internal net	External net							
	Wire & tri	Wand & triand	Wor & trior	Trireg	Tri0	Tri1	Su0	Su1
Wire & tri	E	E	E	E	E	E	E	E
Wand & triand	I	E	*	*	*	*	E	E
Wor & trior	I	I	E	*	*	*	E	E
Trireg	I	I	*	E	E	E	E	E
Tri0	I	I	*	I	E	*	E	E
Tri1	I	I	*	I	*	E	E	E
Su0	I	I	I	I	I	I	E	*
Su1	I	I	I	I	I	I	*	E

Notes   "E" signifies that the external net prevails, and "I" that the internal net prevails.
               "*" signifies a logical clash; the external net prevails.

5.5   NET TYPES

wire is possibly the simplest type of net declaration. **trireg** considered in the last section is another. A variety of other net types are possible. Most of them are provided for specific types of contention resolution.

5.5.1   wand and wor Types of Nets

Strengths on nets can be decided in ways other than a direct declaration also. These offer additional flexibility to the circuit designer. Consider the example of Figure 5.33 for which the input–output values are shown in Table 5.3. For the signal input combination i1 = 0 and i2 = 1, signal o is indeterminate. However, it may be made specific in two alternate ways: '**wand**' and **wor** are two types of net declarations for such contention resolution. **wand** is a wire declaration, which resolves to AND logic in case of contention. **wor** is a wire declaration, which resolves to OR logic in case of a contention. Use of **wand** and **wor** nets is illustrated here through two simple examples crafted for the purpose.

Example 5.8 Illustration of wand type net

Figure 5.35 shows a design module where the outputs of two buffers drive the same net; the net has been declared to be a **wand** type, and any contention with the possibility of indeterminate output is resolved according to AND logic. A test bench and simulation results are also shown in the figure. The input and output logic values and the nature of contention resolutions wherever it occurs are listed out in Table 5.7 also. Contention can be seen to be resolved in two possible ways:

1.   When i1 = 1 and i2 = 0, the stronger signal i1 at the 1 level prevails and o = 1. The contention is resolved according to the strengths.
2.   When i1 = 0 and i2 = 1, both signals being equally strong, the value of o is decided according to AND logic.

The synthesized version of the circuit is shown in Figure 5.36; the circuit translates into an AND gate which is erroneous (this is not consistent with the desired input–output relationship shown in Table 5.7).

```
module wand1(i1,i2,o);
input i1,i2;
output o;
wand o;
buf(strong1,pull0)g1(o,i1);
buf(pull1,pull0)g2(o,i2);
endmodule

module tst_wand1; //testbench
reg i1,i2;
wand1 ww(i1,i2,o);
initial
begin
        i1=0;i2=0;//o =0; no contention
    #2i1=0;i2=1;//o =0; contention resolved
    //according to wand declaration
    #2i1 =1;i2 =0;//out=1; contention resolved by
    //stronger signal
    #2i1 =1;i2=1;//out =1; no contention.
end
initial $monitor($time,"i1=%b,i2=%b,o=%b",i1,i2,o);
endmodule
```

```
output
        #                           0i1=0,i2=0,o=0
        #                           2i1=0,i2=1,o=0
        #                           4i1=1,i2=0,o=1
        #                           6i1=1,i2=1,o=1
```

Figure 5.35 A design module to illustrate use of the wand-type net; a test bench and the results of simulation are also shown.

Table 5.7 Output values for different inputs of the design in Figure 5.35

Logic value of i1	Logic value of i2	Logic value of o	Remarks
0	0	0	No contention
0	1	0	Contention resolved according to **wand** declaration
1	0	1	Contention resolved by the stronger signal
1	1	1	No contention

Example 5.9 Illustration of wor-type net

Consider the design segment in Figure 5.35 with o being declared as a **wor** type of net instead of a **wand** type. The corresponding design module is shown in Figure 5.37. A test bench and simulation results are also shown in the figure. The outputs for all possible combinations of inputs are given in Table 5.8. Contention can be seen to be resolved in two possible ways:

1. When i1 = 1 and i2 = 0, the stronger signal i1 at the 1 level prevails and o = 1. The contention is resolved according to the strengths.
2. When i1 = 0 and i2 = 1, both signals being equally strong, the value of o is decided according to OR logic.

The synthesized version of the circuit is shown in Figure 5.38; the circuit translates into an OR gate; this is consistent with the desired input–output relationship shown in Table 5.8.

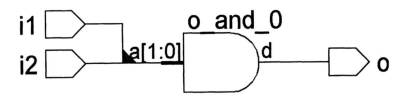

Figure 5.36 Synthesized version of the module with the wand-type net in Figure 5.35 above.

```
module wor1(i1,i2,o);
input i1,i2;
output o;
wor o;
buf(strong1,pull0)g1(o,i1);
buf(pull1,pull0)g2(o,i2);
endmodule

module tst_wor1;//testbench
reg i1,i2;
wor1 ww(i1,i2,o);
initial
begin
    i1=0;i2=0;//out =0 no contention
#2 i1=0;i2=1;//out =1 contention resolved according
//to wor declaration
#2 i1 =1;i2 =0;//out=1 contention resolved by
//stronger signal
#2 i1 =1;i2=1;//out =1 no contention.
end
initial $monitor($time,"i1=%b,i2=%b,o=%b",i1,i2,o);
endmodule
```

```
Output
          #                   0 i1=0,  i2=0,  o=0
          #                   2 i1=0,  i2=1,  o=1
          #                   4 i1=1,  i2=0,  o=1
          #                   6 i1=1,  i2=1,  o=1
```

Figure 5.37 A design module to illustrate use of the **wor**-type net; a test bench and the results of simulation are also shown.

Table 5.8 Output values for different inputs of the design in Figure 5.37

Logic value of i1	Logic value of i2	Logic value of o	Remarks
0	0	0	No contention
0	1	1	Contention resolved according to **wor** declaration
1	0	1	Contention resolved by the stronger signal
1	1	1	No contention

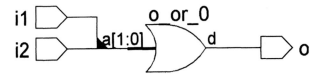

Figure 5.38 Synthesized version of the module with the **wor**-type net in Figure 5.37.

One can see that **wand** and **wor** are keywords to implement wired-or type logic. Observations:

- Many synthesizers do not support wired-or logic. **wand** and **wor** may be used to advantage when supported by the synthesizer.
- The net **triand** is functionally identical to the net **wireand**. Similarly, the net **trior** is functionally identical to the net **wireor**.
- All synthesizers support **wire**. **Triand, trior, tri0,** and **tri1** (discussed below) may not be supported by some.

5.5.2   Tri

The keyword **tri** has a function identical to that of **wire**. When a net is driven by more than one tri-state gate, it is declared as **tri** rather than as **wire**. The distinction is for better clarity. Similarly, **Triand** and **trior** are the counterparts of **wand** and **wor**, respectively.

Example 5.10 Illustration of tri-type net

Consider the design segment in Figure 5.39. Here the signal on net out is controlled by the control signal En. If En = 1, signal a is steered to the net out and the output of gate g2 is tri-stated. On the other hand, if En = 0, signal b is steered to the net out and the gate g1 is tri-stated. If the buffers are controlled by independent Enable signals, the output is resolved according to the respective strengths.

```
. . .
tri out;
wire a, b, En;
bufif1 g1(out, a, En);
bufif0 g2(out, b, En);
. . .
```

Figure 5.39 A segment of a design to illustrate **tri** type of net.

5.5.3   Tri0 and tri1

If the output of a tri-state buffer is to be pulled up to the 1 state when tri-stated, it is declared as net **tri1**. Similarly, it is declared as **tri0** if it is to be pulled down to 0 state when tri-stated. **Tri0** and **tri1** provide respective default outputs and avoid any following circuit having a tri-stated input. In turn, it may manifest as an added load at the concerned gate output. The example in Figure 5.40, which shows a design segment, illustrates an application. Table 5.9 lists the output values of signals considered in the design segment of Figure 5.40.

Referring to the figure (and the table), one can see that when En = 0, all three buffers g0, g1, and g2 are off. Net o3, being a wire is tri-stated and is in **z** state. However, net o1, being of **tri0** type, is pulled down to 0 state irrespective of the input value. Net 02, being of **tri1** type, is pulled up to 1 state. When En = 1, all three buffers are ON and the respective outputs follow the input. Thus though g0, g1, and g2 are functionally identical, they behave differently due to the difference in the type of the respective output nets.

Reset, Chip Enable and similar signals can be pulled up or down as required with **tri0** or **tri1**; this signifies the normal status –that is, the chip is disabled or the reset is disabled. As and when the chip is to be enabled, the same is done by enabling the buffer for the required period. Similarly, the *reset* can be activated for a specified period to reset the chip; subsequently, the reset can be deactivated to restore normal operation of the chip.

```
. . .
tri0 o1;
tri1 o2;
wire o3;
bufif1 g0 (o1, I, En), g3 (o2, I, En);
buif1 g1(o3, I, En);
. . .
```

Figure 5.40 A segment of a design to illustrate **tri0** and **tri1** types of net.

Table 5.9 Output values for different inputs of the segment in Figure 5.40

Logic value of I	Logic value of En	Logic value of o1	Logic value of o2	Logic value of o3
0	0	0	1	Z
0	1	0	0	0
1	0	0	1	Z
1	1	1	1	1

5.5.4   supply0 and supply1

supply0 and **supply1** are the keywords signifying the high- and low-side supplies. Nets to be connected to the Vcc supply are declared as **supply1**, and those to be grounded are declared as **supply0**. Their use is illustrated in Chapter 10.

5.5.5   Ambiguous Strengths

Certain **x** or **z** type of input port values of gate primitives can lead to outputs of apparently ambiguous strengths. A number of such situations can arise. Such cases are brought out and illustrated in the LRM. Nevertheless, such ambiguous situations may be avoided in practice.

5.5.6   Combining Delays & Strengths

So far we have discussed incorporation of strengths in net declarations and instantiations of primitives. Incorporation of a variety of delays and specifying tolerances on them were dealt with in the previous sections. One can combine delays and strengths in net declarations as well as in instantiation of gate primitives. The formats for the same are illustrated below

Wire (drive_strength_1, drive_strength_0) # (delay_0_1, delay_1_0, turn_off_delay) signal1, signal2;

Gate_type (drive_strength_1, drive_strength_0) # (delay_0_1, delay_1_0, turn_off_delay) instance_1(signal1, signal2);

For each of the delays above, one can also specify the minimum, typical, and maximum values. Such values can be specified in terms of constant expressions also. All these have been dealt with separately in detail earlier. Hence combining them and illustrating through examples is not done again here.

5.6   DESIGN OF BASIC CIRCUITS

Elementary gates are the basic building blocks of all digital circuits – whether combinational, sequential, or involved versions combining both. Conversely, any digital circuit can be split up into constituent elementary gates. The variety of examples of combinational circuits considered in the last chapter, and the sequential circuit examples at the beginning of this chapter are testimony to this. Any digital circuit however involved it may be, can be realized in terms of gate primitives. The step-by-step procedure to be adopted may be summarized as follows:

1. Draw the circuit in terms of the gates.
2. Name gates and signals.
3. Using the same nomenclature as above, do the design description.
4. As the functional blocks like encoder, decoder, half-adder, full-adder, *etc.*, get more and more involved, treat each as a building block with corresponding inputs and outputs.
5. Make more involved circuits in terms of the building blocks – as far as possible. Each block within another block manifests as an instantiation of one module within another.

Example 5.11 ALU

We consider the design of an ALU as an example of a relatively complex design. The ALU considered carries out four functions:

- Addition of two 4-bit numbers.
- Complementing all the bits of a 4-bit vector.
- Bit-by-bit AND operation on two nibbles.
- Bit-by-bit XOR operation on two nibbles.

A set of 2 mode select bits selects the function to be carried out from amongst the above four. The design has been evolved in a step-by-step manner. Figure 5.41 shows a 4-bit adder module and a test-bench for it. The simulation results are given in Figure 5.42. The adder module is built up by repeated instantiation of the full-adder module considered in Section 4.8. The synthesized version of the adder is shown in Figure 5.43. The full-adder module instantiations appear here as black boxes with respective inputs and outputs.

```
module add4g(sum,carry,a,b,cin);
input[3:0]a,b;
input cin;
output[3:0]sum;
output carry;
wire [2:0]cc;
fa a0(sum[0],cc[0],a[0],b[0],cin);
fa a1(sum[1],cc[1],a[1],b[1],cc[0]);
fa a2(sum[2],cc[2],a[2],b[2],cc[1]);
fa a3(sum[3],carry,a[3],b[3],cc[2]);
endmodule

module tstadd4g; //Test bench
reg[3:0]a,b;
reg cin;
wire[3:0]sum;
```

continued

continued
```
wire carry;
add4g gg(sum,carry,a,b,cin);
initial
begin
     a =4'h0;b=4'h0;cin=0;
end
always
begin
     #2 a=4'h0;b=4'h0;cin=1'b0;
     #2 a=4'h1;b=4'h0;cin=1'b1;
     #2 a=4'h1;b=4'h0;cin=1'b1;
     #2 a=4'h5;b=4'h3;cin=1'b0;
     #2 a=4'h7;b=4'h0;cin=1'b1;
     #2 a=4'h8;b=4'h9;cin=1'b1;
     #2 a=4'h0;b=4'h0;cin=1'b0;
     #2 a=4'hb;b=4'h7;cin=1'b0;
     #2 a=4'h0;b=4'h0;cin=1'b0;
     #2 a=4'hf;b=4'hf;cin=1'b0;
     #2 a=4'hf;b=4'hf;cin=1'b1;
end
initial $monitor($time," a = %b, b = %b, cin = %b,
outsum = %b, outcar = %b ", a, b, cin, sum, carry);
initial #30 $stop ;
endmodule
```

Figure 5.41 A 4-bit adder module and its test bench

```
output
# 0 a =0000,b =0000,cin = 0,outsum =0000,outcar =0
# 2 a =0001,b =0000,cin = 0,outsum =0001,outcar =0
# 4 a =0001,b =0000,cin = 1,outsum =0010,outcar =0
# 6 a =0001,b =0001,cin = 1,outsum =0011,outcar =0
# 8 a =0101,b =0011,cin = 0,outsum =1000,outcar =0
#10 a =0111,b =0110,cin = 1,outsum =1110,outcar =0
#12 a =1000,b =1001,cin = 1,outsum =0010,outcar =1
#14 a =1010,b =0001,cin = 1,outsum =1100,outcar =0
#16 a =1011,b =0111,cin = 0,outsum =0010,outcar =1
#18 a =1000,b =1000,cin = 0,outsum =0000,outcar =1
#20 a =1111,b =1111,cin = 0,outsum =1110,outcar =1
#22 a =1111,b =1111,cin = 1,outsum =1111,outcar =1
#24 a =0001,b =0000,cin = 0,outsum =0001,outcar =0
#26 a =0001,b =0000,cin = 1,outsum =0010,outcar =0
#28 a =0001,b =0001,cin = 1,outsum =0011,outcar =0
```

Figure 5.42 Simulation results of running the test bench in Figure 5.41.

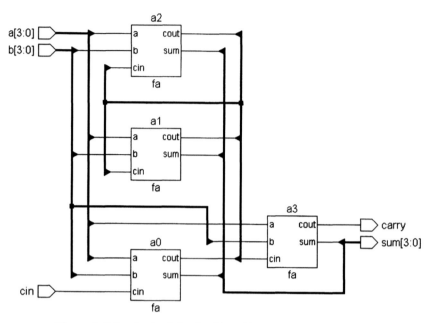

Figure 5.43 Synthesized circuit of the adder module of Figure 5.41.

Figure 5.44 shows a module to AND two nibbles. It is done through direct instantiation of AND gate primitives for two inputs. The corresponding synthesized circuit is shown in Figure 5.45.

```
module andg4(c,a,b);
input[3:0]a,b;

output[3:0]c;
and(c[0],a[0],b[0]);
and(c[1],a[1],b[1]);
and(c[2],a[2],b[2]);
and(c[3],a[3],b[3]);
endmodule
```

Figure 5.44 A 4-bit adder module.

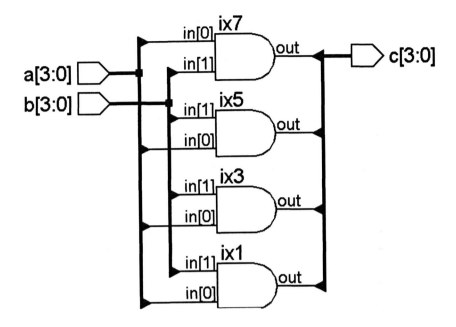

Figure 5.45 Synthesized circuit of the AND module of Figure 5.44 andg4.

The module in Figure 5.46 carries out the bit-wise XOR operation on 2 nibbles. Its synthesized circuit is shown in Figure 5.47. Similarly, the module in Figure 5.48 complements 2 nibbles in a bit-wise manner. The corresponding synthesized circuit is shown in Figure 5.49.

```
module xorg(c,a,b);
input[3:0]a,b;
//input cen;
output[3:0]c;
wire [3:0]cc;
xor x0(c[0],a[0],b[0]);
xor x1(c[1],a[1],b[1]);
xor x2(c[2],a[2],b[2]);
xor x3(c[3],a[3],b[3]);
endmodule
```

Figure 5.46 A 4-bit XOR module.

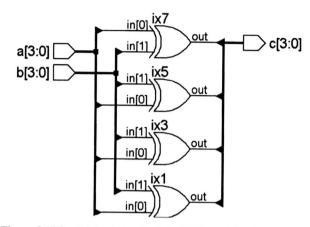

Figure 5.47 Synthesized circuit of the XOR module of Figure 5.46.

```
module compl(c,a);
input[3:0]a;
output[3:0]c;
not(c[0],a[0]);
not(c[1],a[1]);
not(c[2],a[2]);
not(c[3],a[3]);
endmodule
```

Figure 5.48 A module to complement a 4-bit vector.

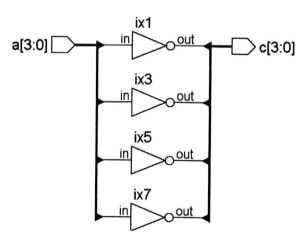

Figure 5.49 Synthesized circuit of the module in Figure 5.48.

```
module dec2_4 (a,b,en);
output [3:0] a;
input [1:0]b;
input en;
wire [1:0]bb;
not(bb[1],b[1]),(bb[0],b[0]);
and(a[0],en,bb[1],bb[0]),(a[1],en,bb[1],b[0]),
(a[2],en,b[1],bb[0]),(a[3],en,b[1],b[0]);
endmodule
```

Figure 5.50 A 2-to-4 decoder module.

A 2-bit binary number with its 4 distinct states can be used to select any one of the 4 desired functions; it calls for the use of a 2-to-4 decoder. Such a module is shown in Figure 5.50, and its synthesized circuit is shown in Figure 5.51.

As explained above, the decoder outputs can be used to select anyone of the 4 functional outputs and steer it to the final output; a 4-to-1 mux serves this purpose. The mux module is shown in Figure 5.52; its synthesized circuit is in Figure 5.53.

The overall ALU module is shown in Figure 5.54. It instantiates all the above modules. Depending on the mode specified, one of the four functions is selected by the 2-to-4 decoder; its output is multiplexed on to the output by the 4-to-1 mux. The ALU module here has been synthesized and shown in Figure 5.55. Each functional block instantiated in Figure 5.54 appears here as a corresponding distinct black box.

More functions can be added, if desired, to make the ALU more comprehensive. The ALU size can be increased to 16 or 32 bits by repeated instantiation (after some minor modifications) of the 4-bit module in a more comprehensive module.

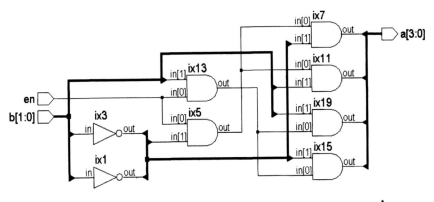

Figure 5.51 Synthesized circuit of the decoder module of Figure 5.50.

```
module mux4_1alu(y,i,e);
input [3:0] i;
input e;
output [3:0]y;
bufif1 g1(y[3],i[3],e);
bufif1 g2(y[2],i[2],e);
bufif1 g3(y[1],i[1],e);
bufif1 g4(y[0],i[0],e);
endmodule
```

Figure 5.52 A 4-to-1 mux module.

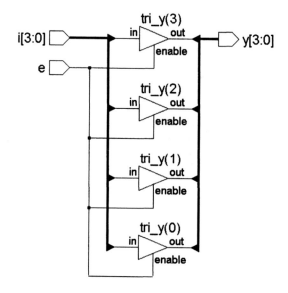

Figure 5.53 Synthesized circuit of the mux module of Figure 5.52.

```
module alu_4g(a,b,c,carry,cin,cen,s);
input [3:0]a,b;
input[1:0]s;
input cen,cin;
output [3:0]c;
output carry;
wire [3:0] data0,data1,data2,data3,e;
wire carry1 ;
dec2_4 •m5(e,s,cen);
add4g m1(data0,carry1,a,b,cin);
```

continued

continued

```
compl m2(data1,a);
xorg  m3(data2,a,b);
andg4 m4(data3,a,b);
bufif1 g5(carry,carry1,cen);
mux4_1alu m6(c,data0,e[0]);
mux4_1alu m7(c,data1,e[1]);
mux4_1alu m8(c,data2,e[2]);
mux4_1alu m9(c,data3,e[3]);
endmodule
```

Figure 5.54 A 4-bit ALU module.

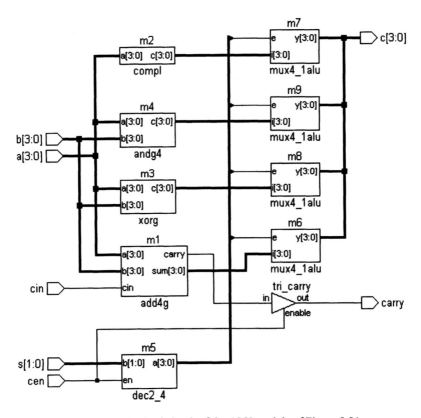

Figure 5.55 Synthesized circuit of the ALU module of Figure 5.54.

5.7   EXERCISES

In each of the cases below, prepare the test bench and test the design

1. Realize each of the flip-flops below using NOR gates.

    RS flip-flop; D-latch; Clocked RS flip-flop; Edge-triggered D flip-flop; Master-slave flip-flop.

2. Figure 5.56 shows the circuit of a flip-flop. Prepare the design module and test it.   Explain why it does not work.

3. Modify the flip-flop in Figure 5.56 above with 2 ns delay for sb. Test the flip-flop with different waveforms for d and clk; in each case ensure that the clock does not remain high continuously for more than 1 ns. Explain the need for this restriction.

4. Figure 5.57 shows the basic memory cell built around a d-latch. One can write data into it or read data from it.

    a. When rd/wrb input is low, the flip-flop is in write mode; data are an input line; data on data line are written into the latch, when clk is given a positive pulse.

    b. When rd/wrb input is high, the flip-flop is in read mode; data stored in the latch are made available on the data line.

    Build a module around the d-latch to realize the memory cell.

5. Expand the above to form a byte-wide memory cell.

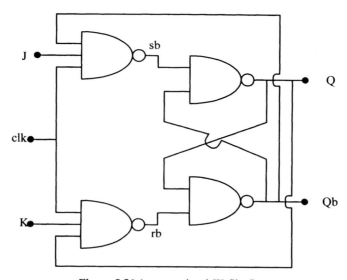

Figure 5.56 A conventional JK flip-flop.

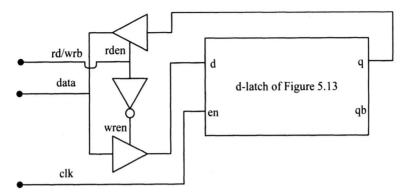

Figure 5.57 A d-latch with necessary additional circuitry to form a memory cell.

6. Replicating the memory element above, one can form a memory. Consider a memory of 16 locations addressed by a 4-bit-wide address bus. The memory will have a 4-to-16 address decoder. It will have an rd/wrb input for reading from it and writing data into it. The decoded address can be used to gate the rden and wren inputs to the respective tri-state buffers. Prepare the design module for the memory.

7. Consider the 4-to-1 mux module in Figure 4.31 and its synthesized circuit in Figure 4.33. Identify the signal paths in which maximum number of gates is involved. What is the number of gates in the path here?

    Identify the signal paths in which the number of gates involved is a minimum. What is the number of gates here and which are these?

8. For each of the gate primitives in Exercise 7 above, take the minimum, typical, and maximum delays to be 1 ns, 2 ns and 3 ns respectively. With the typical delay values, estimate the minimum and maximum delays of transmission. Verify by simulation. Repeat the exercise with minimum and minimum delay values.

9. In Exercise 8 above, assign the minimum delay values for the shortest paths and maximum for the longest paths. Using these, estimate the minimum and maximum time delays for the mux (see also the pin-to-pin delay specifications in Chapter 11).

10. Identify the ALU functions in the 8085 processor. Design an ALU module to carry out these.

11. Identify the ALU functions in the 8088 processor. Design an ALU module to carry out these (ignore the instructions for multiplication and division).

12. a[1:0] and b[1:0] are two 2-bit numbers. Their product – designated as m[3:0] – is in general a 4-bit number; it is formed as follows:

    Form m[0] by AND operation on a[0] and b[0].

Through a half-adder add the bits a[1]&b[0] and a[0]&b[1]. The sum bit is m[1]. Let the carry bt be c.

Through a half-adder add the bits a[1]&b[1] and c obtained above. The sum bit is m[2] and the carry bit m[3].

Design a 2-bit multiplier following the above steps and test it for all possible input value combinations.

13. Let abcd and efgh be two 4-bit numbers where a, b,...., g, h represent the respective bit values. The 4-bit numbers are multiplied as follows:

Form the four 4-bit numbers 00cd, 00gh, ab00, and ef00.

Form the following four intermediate products using 2-bit multipliers:

        00cd with 00gh
        00cd with ef00
        ab00 with 00gh
        ab00 with ef00

Add all the above four intermediate products to get the final 7-bit result.

Design a 4-bit multiplier module following the above steps. Instiantiate 2-bit multiplier module, half- and full-adder modules, *etc.*, wherever necessary.

14. Following steps analogous to the above, design an 8-bit multiplier.

15. Write down the Boolean logic expressions for all the product bits of a 4-bit multiplier; using these, design an 8-bit multiplier.

6

MODELING AT DATA FLOW LEVEL

6.1   INTRODUCTION

Gate level design description makes use of the gate primitives available in Verilog. These are repeatedly and judiciously instantiated to achieve the full design description. Digital designers familiar with the basic logic gates and SSI / MSI circuits can describe the desired target circuit in terms of them on paper and proceed with the design description based on them. This was the approach followed in the last two chapters; it is practical for comparatively smaller designs – say those involving tens of gates. One can define modules in terms of primitives involving tens of gates and instantiate them in macro-modules. This increases the complexity of designs that can be handled by one order. Beyond that the gate level design description becomes too complicated to be practical.

Data flow level description of a digital circuit is at a higher level. It makes the circuit description more compact as compared to design through gate primitives. We have a number of operands and operations representing the simulations directly or indirectly. The operations are carried out on the operand(s) in singles or in combinations [IEEE]. The results are assigned to nets. The operand-operation-assignments representing data flow are carried out repeatedly to complete the design description [Thomas & Morby]. Further, these can be combined judiciously with the gate instantiations wherever necessary. With such combinations, design description of a comprehensive nature can be accommodated.

6.2   CONTINUOUS ASSIGNMENT STRUCTURES

A simple two input AND gate in data flow format has the form

`assign` c = a && b;

Here

- "`assign`" is the keyword carrying out the assignment operation. This type of assignment is called a continuous assignment.

- **a** and **b** are operands – typically single-bit logic variables.
- "**&&**" is a logic operator. It does the bit-wise AND operation on the two operands **a** and **b**.
- "**=**" is an assignment activity carried out.
- **c** is a net representing the signal which is the result of the assignment.

In general, an operand can be of any one of the following types:

- A constant number [including real].
- Net of a scalar or vector type including part of a vector.
- Register variable of a scalar or vector type including part of a vector.
- Memory element.
- A call to a function that returns any of the above. The function itself can be a user-defined or of a system type [see Chapter 9].

There are other types of operators as well [see Section 6.5]. All types of combinational circuits can be modeled using continuous assignments. One need not necessarily resort to instantiation of gate primitives.

An AND gate module which uses the above assignment is shown in Figure 6.1. The test bench for the same is shown in Figure 6.2, and the waveforms of nets a, b, and c obtained with the simulation are shown in Figure 6.3. [The simulation software used has the facility to capture the waveforms of selected signals in the "run" phase; this has been invoked to get the waveforms in Figure 6.3. No separate **$monitor** command is included in the test bench of Figure 6.2. The same approach has been adopted with many of the test benches elsewhere in the book.]

Multiple assignments can be carried out through a direct extension of the structure adopted in the above case. Consider the AOI gate in Figure 6.4. A few patterns of the assignments for the circuit are given in Figure 6.5 to Figure 6.7.

```
module andgdf(c,a,b);
output c;
input a,b;
wire c;
assign c = a&&b;
endmodule
```

Figure 6.1 A module with an AND gate at the data flow level.

```
module tst_andgdf; //TESTBENCH
reg a,b;
wire c;
initial
begin
```

continued

continued

```
        a = 1'b0;
        b = 1'b0;
  #4    a = 1'b1;
  #4    b = 1'b1;
  #4    a = 1'b0;
  #4    b = 1'b0;
  #4    a = 1'b1;
end
andgdf g1(c,a,b);
initial #20 $stop;
endmodule
```

Figure 6.2 A test bench for the module in Figure 6.1.

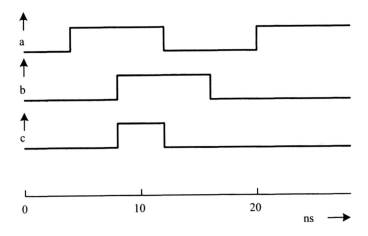

Figure 6.3 Waveforms of nets a, b, and c obtained with the simulation of the module in Figure 6.1 with the test bench in Figure 6.2.

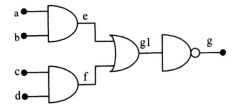

Figure 6.4 An A-O-I gate circuit.

```
assign e = a&&b, f = c&&d, g1 = e|f, g = ~g1;
```

Figure 6.5 A data flow level assignment statement to realize the A-O-I gate in Figure 6.4.

```
assign e = a & b, f = c & d;
assign g1 = e|f, g = ~g1;
```

Figure 6.6 Another set of data flow level assignment statements to realize the A-O-I gate in Figure 6.4.

```
assign e  = a & b;
assign f  = c & d;
assign g1 = e ! f;
assign g  = ~g1;
```

Figure 6.7 Yet another set of data flow level assignment statements to realize the A-O-I gate in Figure 6.4.

Observations:

- The semicolon terminates an assignment statement. Commas separate different assignments in an assignment statement.
- "|" is the bit-wise OR operator and "~" the bit-wise negation operator in Verilog.
- All the quantities in the left-hand side of a continuous assignment have to be of net type. Thus e, f, g, and g1 have to be declared as nets.
- All the operations in an assignment are evaluated whenever any of the operands in the assignment changes value. Further, all the assignments are carried out concurrently. Hence the order of the assignments or the statements is immaterial.
- The right-hand sides of assignment statements can be nets, regs, or function calls. Here a, b, c, and d can be nets or regs. All other variables have to be nets.

The module for the A-O-I gate of Figure 6.4 is given in Figure 6.8 – it is formed around the assignment statement of Figure 6.5. The same can be tested through a test bench.

6.2.1   Combining Assignment and Net Declarations

The assignment statement can be combined with the net declaration itself making the assignment implicit in the net declaration itself. Thus the two statements

wire c;
assign c = a & b;

can be combined as

wire c = a & b;

The above simplification cannot be carried over to multiple declarations. With this proviso, the module of Figure 6.8 can be modified as shown in Figure 6.9. In the modules of Figures 6.8 and 6.9, a, b, c, and d are declared as **input** and g as **output**. As was explained in Section 4.2, these would be taken as nets if there are no separate declarations concerning their types. However, the intermediate quantities – e, f, and g1– should be declared as **wire**. Synthesized version of the A-O-I circuit is shown in Figure 6.10.

```
module aoi2(g,a,b,c,d);
output g;
input a,b,c,d;
wire e,f,g1,g;
assign e = a && b,f = c && d, g1 = e||f, g=~g1;
endmodule
```

Figure 6.8 A compact description of the AOI module at the data flow level.

```
module aoi3(g,a,b,c,d);
output g;
input a,b,c,d;
wire g;
wire   e  = a && b;
wire   f  = c && d;
wire   g1 = e||f;
assign g  = ~g1;
endmodule
```

Figure 6.9 Alternate design module to realize the A-O-I gate in Figure 6.4.

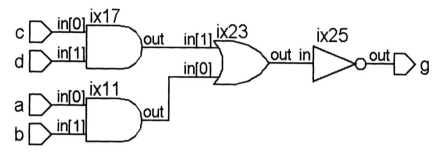

Figure 6.10 Synthesized circuit of the A-O-I gate module of Figure 6.9.

6.2.2   Continuous Assignments and Strengths

A net to which a continuous assignment is being made can be assigned strengths for its logic levels. The procedure is akin to the strength allocation to the outputs of primitives. The AOI gate of Figure 6.9 is modified with strength allocations to the output and is shown in Figure 6.11. The assignment to g can be combined with the wire declaration into a single statement as

```
wire (pull1, strong0) g = ~g1;
```

As mentioned earlier, one can have only one assignment in the statement here. In a bigger design, g in Figure 6.11 can be assigned to other expressions or primitives also. Any resulting contention in the output values will be resolved on the lines discussed in Chapter 4.

```
module aoi4 (g, a, b, c, d);
output g;
input a, b, c, d;
wire g;
wire e  = a &&b;
wire f  = c & &d;
wire g1 = e || f;
assign (pull1, strong0) g = ~g1;
endmodule
```

Figure 6.11 The module of Figure 6.9 modified with strength allocation to the output.

6.3   DELAYS AND CONTINUOUS ASSIGNMENTS

Delays can be incorporated at the data flow level in different ways [Ciletti]. Consider the combination of statements in Figure 6.12. The assignment takes effect with a time delay of 2 time steps. If a or b changes in value, the program waits for 2 time steps, computes the value of c based on the values of a and b at the time of computation, and assigns it to c. In the interim period, a or b may change further, but c changes and takes the new value only 2 time steps after the change in a or b initiates it. Typical waveforms for a, b, and c are shown in Figure 6.13. Note that the changes in a and b of duration less than 2 time steps are ignored *vis-à-vis* assignment to the net c.   The following may be noted with respect to the waveforms:

- a changes at 0 ns, 2 ns, 5 ns, 8 ns, 9 ns, 12 ns and 13 ns; b changes at 0 ns, 2 ns, 6 ns, 8 ns and 13 ns. All these trigger changes to c.
- In every case change to c comes into effect with a time delay of 2 time steps – that is, at the 2nd, 4th, 7th, 8th, 10th, 11th, 14th and 15th ns, respectively.
- Whenever c changes, its new value is decided by the values of a and b at that instant of time. In effect, c changes at 2nd, 4th and 7th ns only.

```
wire c, a, b;
assign #2    c = a & b;
```

Figure 6.12 Illustration of combining delays with assignments.

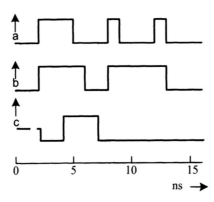

Figure 6.13 Waveforms of signals a, b, and c for the design segment of Figure 6.12.

The program segment in Figure 6.14 also gives the same output as shown in Figure 6.13. If the time delay is in the net and not in the assignment proper, its effect is not any different. Consider the program segment in Figure 6.15. Here the changes in the values of d are computed immediately following those in a and b. The assignment takes effect immediately. The delay in the net c causes a delay of 2 time steps in the assignment to c. Such a delay is not present for d. Typical waveforms for the program segment are shown in Figure 6.16. Note the following:

- a changes at 2 ns, 5 ns, 8 ns, 9 ns, 12 ns and 13 ns; b changes at 2 ns, 6 ns, 8 ns and 13 ns. All these trigger changes to c and d also.
- In every case, change to c comes into effect with a time delay of 2 time steps – that is, in effect, c changes at 2nd, 4th and 7th ns only.
- Whenever c changes, its new value is decided by the values of a and b at that instant of time.
- In every case, changes to d come into effect immediately.

```
wire a, b;
wire  #2 c = a & b;
```

Figure 6.14 Alternate description for the program segment of Figure 6.10.

```
wire a, b, d;
wire #2      c;
assign c = a & b;
assign d = a & b;
```

Figure 6.15 Illustration of combining delays with assignments.

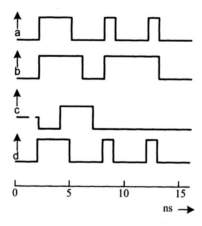

Figure 6.16 Waveforms of Signals a, b, c, and d for the design segment of Figure 6.15.

6.4   ASSIGNMENT TO VECTORS

The continuous assignments are equally applicable to vectors. A single statement can describe operations involving vectors wherever possible. This is illustrated in the adder module in Figure 6.17, which adds two 8-bit numbers. Here it is assumed that the sum is also of 8 bits. However to account for the possibility of a carry bit being generated in the course of the addition process, it is desirable to increase the vector size of c by one bit.

6.4.1   Concatenation of Vectors

One can concatenate vectors, scalars, and part vectors to form other vectors. The concatenated vector is enclosed within braces. Commas separate the components –scalars, vectors, and part vectors. If a and b are 8- and 4-bit wide vectors, respectively and c is a scalar

{a, b, c}

stands for a concatenated vector of 13 bits width. The vector components are formed in the order shown – c is the least significant bit and a[7] the most significant bit and the other bits are in between in the order specified. The concatenation can be with selected segments of vectors also. For example,

{a(7:4), b(2:0)}

represents a 7-bit vector formed by combining the 4 most significant bits of vector a with the 3 least significant bits of vector b. The size of each operand within the braces has to be specified fully to form the concatenated vector. Hence unsized constant numbers cannot be used as operands here.

Example 6.1 Eight-Bit Adder

Figure 6.18 shows the design description of an 8-bit adder, where the output vector is formed directly by concatenation. The adder takes a carry input and gives out a carry output. The adder module here can form the "seed" adder block in a multi-byte adder chain.

```
module add_8(a,b,c);
input[7:0]a,b;
output[7:0]c;
assign c = a + b ;
endmodule
```

Figure 6.17 An adder module at data flow level where the nets are vectors.

```
module add_8_c(c,cco,a,b,cci);
input[7:0]a,b;
output[7:0]c;
input cci;
output cco;
assign {cco,c} = (a + b + cci);
endmodule
```

Figure 6.18 A complete 8-bit adder module at data flow level.

When it is necessary to replicate vectors, scalars, *etc.,* to form other vectors, the same can be arrived at in a compact manner using the repetition multiplier again through concatenation. Thus,

{2{p}}

represents the concatenated vector

{p, p}

and

{2{p}, q}

represents the concatenated vector

{p, p, q}.

The two statements

assign GND=supply0;
p={8{GND}};

together ground the 8 bits of the vector p.

Concatenation operation can be nested to form bigger vectors when component combinations are repeated. For example,

{a, 3 {2{b , c}, d}}

is equivalent to the vector

{a, b, c, b, c, d, b, c, b, c, d, b, c, b, c, d }

6.5   OPERATORS

A set of operators is available in Verilog. The operator symbols are similar to those in C language [Gottfried]. With these operators we can carry out specified operations on the operands and assign the results to a net or a vector set of nets as the case may be. A few such operands have already been used in the examples so far. We discuss here the different operators, their types, and the operations carried out by each. Subsequently the use of operators is illustrated through a set of examples.

6.5.1   Unary Operators

Unary operators do an operation on a single operand and assign the result to the specified net. The unary operators in Verilog are given in Table 6.1. All unary operators get precedence over binary and ternary operators. The operators "+" and "–" preceding an integer or a real number change its sign. These are also unary operators, though not separately listed in Table 6.1.

6.5.2   Binary Operators

Most operators available are of the binary type. A binary operator takes on two operands; the operator comes in between the two operands in the assignment. The binary operators are grouped into type categories and discussed separately. The following are to be noted:

- The arithmetic operators treat both the operands as numbers and return the result as a number.
- All net and **reg** operand values are treated as unsigned numbers.
- Real and integer operands may be signed quantities.
- If either of the operand values has a zero value, the entire result has a zero value (?).

The result of any arithmetic operation — with the "+" or "–" or with any of the other arithmetic operators discussed later — will have an **x** value if any of the operand bits has an **x** or a **z** value.

6.5.2.1   Arithmetic Operators
The arithmetic operators of the binary type are given in Table 6.2. The modulus operand is similar to that in C language – It provides the remainder of the division

Table 6.1 Unary operators and their symbols

Operator type	Symbol	Remarks
Logical negation	!	Only for scalars
Bit-wise negation	~	For scalars and vectors
Reduction AND	&	For vectors – yields a single bit output
Reduction NAND	~&	
Reduction OR	|	
Reduction NOR	~|	
Reduction XOR	^	
Reduction XNOR	~^ or ^~	

Table 6.2 Arithmetic operators and their symbols

Operand type	Symbol	Remarks
Multiplication	*	
Division	/	The result is x if the denominator is zero
Modulus	%	
Addition	+	
Subtraction	−	

of two numbers. The module in Figure 6.17 is an example of the illustration of the use of the arithmetic binary operator "+" (for addition). Other arithmetic operators are also used in a similar manner.

Observations:

- In integer division the fractional part of the result is truncated and ignored.
- If any bit of an operand is **x** or **z** in an arithmetic operation, the result takes the **x** value.
- If the first operand of a modulus operator is negative, the result is also a negative number.

Depending on the type of definition of a number, a modulus operation can lead to different results. Typical examples are given in Table 6.3.

6.5.2.2   Logical Operators

There are two logical operators involving two operands. The operands concerned can be variables or expressions involving variables. In both cases the result of the operation is a single bit of value 1 (true) or 0 (false). If a bit in one of the operands is x or z, the result of evaluation of the expression has an **x** value. The operator details are shown in Table 6.4. The modules in Figure 6.8 and Figure 6.9 are examples of the illustration of the use of logical binary operators.

6.5.2.3   Relational Operators

There are four relational operators; their details are shown in Table 6.5.   A relational operator treats both the operands as binary numbers and compares them. The result is a 1 (true) bit or a 0 (false) bit. If a bit in either operand is **x** or **z**, the result has **x** (unknown) value.   The operands can be variables or expressions involving variables. Operands of net or **reg** type are treated as unsigned numbers. Real and integers can be positive or negative (*i.e.*, signed) numbers.

Table 6.3 Typical modulus operations and their results

Expressions involving modulus operator	Result of the operation	Remarks
15 % 5	0	Results are obvious
14 % 5	4	
4'hf % 5	0	The numbers 4'hf and 4'he are in hex format with decimal values of 15 and 14, respectively. But the denominator 5 is in decimal form.
4'he % 5	4	
6'o15 % 5	3	6'o15 is an octal number with a decimal value of 13.
–4 % 3	–1	
4 % –3		Illegal form

Table 6.4 Binary logical operators and their symbols

Operator type	Symbol	Possible output value
AND	&&	Single-bit output
OR	‖	

Table 6.5 Relational operators and their symbols

Operator type	Symbol	Possible output value
Greater than	>	Single-bit output
Less than	<	
Greater than or equal to	>=	
Less than or equal to	<=	

6.5.2.4   Equality Operators

The equality operator makes a bit-by-bit comparison of the two operands and produces a result bit. The result bit is a 1 (true) if the operand condition is satisfied; otherwise it is 0 (false). The operands can be variables or expressions involving variables. If the operands are of unequal length, the shorter one is zero filled to match the larger operand. The operators in this category are only of two types – those to test the equality and those to test inequality. The four operators in this category are given in Table 6.6.

6.5.2.5   Bit-wise Logical Operators

The operator does a specified bit-by-bit operation on the two operands and produces a set of result bits. The result is (bit-wise) as wide as the wider operand.

Table 6.6 Equality operators and their symbols

Operand symbol	Description of operand	Possible logical value of result
==	(The symbol comprises two consecutive equal signs.) If the two operands are equal bit by bit, the result is 1 (true); else the result is 0 (false). If either operand has a **x** or **z** bit, the result is **x**.	0, 1, or **x**
!=	(The symbol comprises of an exclamation mark followed by an equal sign.) A bit-by-bit comparison of the two operands is made. The result is a 1 if there is a mismatch for at least one bit position.	0, 1, or **x**
===	(The symbol comprises of three consecutive equal signs.) The operand bits can be 0, 1, **x**, or **z**. If the two operands match on a bit by bit basis, the result is a 1 (true) bit; else it is 0 (false). Note that the result is never **x** here.	0 or 1
!==	(The symbol comprises an exclamation mark followed by 2 consecutive equal signs). The operand bits can be 0, 1, **x**, or **z**. If the two operands do not match on a bit by bit basis, the result is a 1 (true) bit; else it is 0 (false). Note that the result is never **x** here.	0 or 1

If the width of one of the operands is less than that of the other, it is bit-extended by filling zero bits and the widths are matched. Subsequently, the specified operation is carried out. If one of the operands has an **x** or **z** bit in it, the corresponding result bit is **x**. Either operand can be a single variable or an expression involving variables. Table 6.7 gives the four operators of this category.

6.5.2.6   Operator Truth Table

The truth tables for different types of bit-wise operators are given in Table 6.8. Note that an **z** input is treated as an **x** value (Compare these with their counterparts for respective gate primitives in Chapter 4.)

Table 6.7 Bit-wise logical operators and their symbols

Operator type	Symbol	Possible result
AND	&	0, 1, or **x**
OR	|	
XOR	^	
XNOR	~^ or ^~	

Table 6.8 Truth tables for bit-wise operators

AND

		Input 2		
		0	1	**x**
Input 1	0	0	0	0
	1	1	0	**x**
	x	0	**x**	**x**

OR

		Input 2		
		0	1	**x**
Input 1	0	0	1	**x**
	1	1	1	1
	x	**x**	1	**x**

XOR

		Input 2		
		0	1	**x**
Input 1	1	1	0	**x**
	x	**x**	**x**	**x**

XNOR

		Input 2		
		0	1	**x**
Input 1	0	1	1	**x**
	x	**x**	**x**	**x**

Negation

Input	0	1	**x**
Output	1	0	**x**

6.5.2.7   Shift Operators

Table 6.9 shows the two operators of this category. The << operator executes left shift operation, while the >> operator executes the right shift operation. In either case the operand specified on the left is shifted by the number of bits specified on the right. The shifting is done irrespective of whether the bits are 0, 1, **x**, or **z**. The bits shifted out are lost. The vacated positions created as a result of the shifting are filled with zeroes. If the right operand is **x** or **z**, the result has an x value. If the right operand is negative, the left operand remains unchanged.

6.5.3   Ternary Operator

Verilog has only one ternary operator – the conditional operator. It checks a condition and does a branching. It is a versatile and powerful operator. It enhances the potential of design description substantially (as can be seen through the examples below). The general form is

A?b:c

The conditional operation is made up of two operators – "?" and ":" – and three operands. The two operands separate the three operators in the order shown. The operational sequence of the operation is as follows:

Table 6.9 Shift-type operators and their symbols

Operand	Typical usage	Operation
>>	A >> b	The set of bits representing A are shifted right repeatedly b times.
<<	A<< b	The set of bits representing A are shifted left repeatedly b times.

- "A" is evaluated first.
- If A is true, b is evaluated.
- If A is false, c is evaluated.

If A evaluates to an ambiguous result, both b and c are evaluated. Then they are combined on a bit-by-bit basis to form the resultant bit stream. The result bit can have the following three possible values:

- 0 if the corresponding bits of b and c are 0.
- 1 if the corresponding bits of b and c are 1.
- x otherwise.

As an example, consider the assignment statement

assign y = w ? x : z;

where w, x, y and z are binary bits. If the bit w is true (1), y is assigned the value of x: otherwise – that is, if w is false (0) – y is assigned the value of z. The assignment statement here multiplexes x and z onto y; w is the control signal here. Consider the assignment

assign flag = (adr1 == adr2)?1'b1 : 1'b0;

Here adr1 and adr2 are two multibit vectors representing two addresses. If the two are identical, the flag bit is set to zero; else it is reset.

assgn zero_flag = (|byte)? 0:1;

All the bits of the byte are ORed together here. The zero_flag is set if the result is zero.

assign c = s ? a: b; //The net c is connected to a if s=1; else it is connected to b

The statement realizes a 2 to 1 mux. b and c have to be scalars or vectors of the same width. The assignment can be expanded to realize larger muxes.

The conditional operator can be nested [see Figure 6.19]. Nesting gives rise to a variety of uses of the operator. As an example, consider the formation of an ALU. ALU can be defined in a compact manner using the ternary operator.

assign d = (f==add)?(a+b): ((f==subtract)?(a-b): ((f==compl)?~a: ~b));

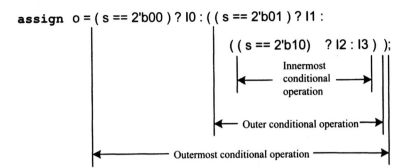

Figure 6.19 Illustration of nested conditional operations.

In the example here, f is taken as a control word. If it is equal to the number add, d is to be equal to the sum of a and b. If f is equal to the number subtract, d is to be equal to the difference between a and b. If it is equal to the number compl, d is to be the complement of a. Otherwise (*i.e.*, f = 3) d is taken as the complement of b. As another example consider a mux; the assignment statement in Figure 6.18 represents a 4-to-1 mux formed with a nested set of ternary operators. The construct in the figure can be judiciously used to form muxes of larger sizes.

Example 6.2 ALU

Figure 6.20 shows an ALU module. It is built around a single executable statement present as a continuous assignment. A test bench for the ALU is also shown in the figure. The synthesized circuit is shown in Figure 6.21. Results of running the test bench are shown in Figure 6.22. Some of the combinational circuit operations required are realized inside the "modgen" blocks of the FPGA used. The nature of the ALU description in the module decides the translation into circuit. Contrast this with the ALU considered at the gate level of design in Section 5.7 where each functional block is instantiated separately and the selected set of outputs steered to the final output. Each such instantiated module translates into a separate circuit block. Their outputs are mux'ed into the final output vector. There is a one-to-one correspondence between the elements of the design description and their respective realizations.

```
module alu_df1 (d, co, a, b, f,cci);
//a SIMPLE ALU FOR ILLUSTRATION PURPOSES
output [3:0] d;
output co;
wire[3:0]d;
```

continued

continued

```
wire co;
input cci;
input [3 : 0 ] a, b;
input [1 : 0] f;//f is a two-bit function select input;
assign {co,d}=(f==2'b00)?(a+b+cci):((f==2'b01)?(a-b)
         :((f==2'b10)? {1'bz,a^b}:{1'bz,~a}));
/*co is the carry bit in case of addition;it is the
borrow bit in case of subtraction. In the other two
cases, co is not required. Hence it is assigned z
value.*/
endmodule

module tst_aludf1; //test-bench
reg [3:0]a,b;
reg[1:0] f;
reg cci;
wire[3:0]d;
wire co;
alu_df1 aa(d,co,a,b,f,cci);
initial
begin
    cci= 1'b0;
    f   = 2'b00;
    a   = 4'b0;
    b   = 4'h0;
end
always
    begin
    #2 cci = 1'b0;f = 2'b00;a = 4'h1;b = 4'h0;
    #2 cci = 1'b1;f = 2'b00;a = 4'h8;b = 4'hf;
    #2 cci = 1'b1;f = 2'b01;a = 4'h2;b = 4'h1;
    #2 cci = 1'b0;f = 2'b01;a = 4'h3;b = 4'h7;
    #2 cci = 1'b1;f = 2'b10;a = 4'h3;b = 4'h3;
    #2 cci = 1'b1;f = 2'b11;a = 4'hf;b = 4'hc;
    end
initial $monitor($time, " cci = %b , a= %b ,b = %b ,
f = %b ,d =%b ,co= %b ",cci ,a,b,f,d,co);
initial #30 $stop;
endmodule
```

Figure 6.20 A 4-bit 4-function ALU and a test bench for the same.

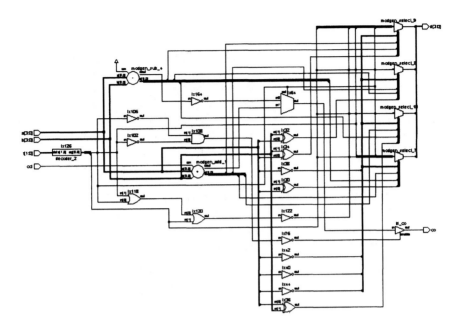

Figure 6.21 Synthesized circuit of the ALU in Example 6.18.

```
output listing
#  0  cci = 0 ,  a= 0000 ,b = 0000 ,f = 00 ,d =0000 ,co= 0
#  2  cci = 0 ,  a= 0001 ,b = 0000 ,f = 00 ,d =0001 ,co= 0
#  4  cci = 1 ,  a= 1000 ,b = 1111 ,f = 00 ,d =1000 ,co= 1
#  6  cci = 1 ,  a= 0010 ,b = 0001 ,f = 01 ,d =0001 ,co= 0
#  8  cci = 0 ,  a= 0011 ,b = 0111 ,f = 01 ,d =1100 ,co= 1
#10  cci = 1 ,  a= 0011 ,b = 0011 ,f = 10 ,d =0000 ,co= z
#12  cci = 1 ,  a= 1111 ,b = 1100 ,f = 11 ,d =0000 ,co= z
#14  cci = 0 ,  a= 0001 ,b = 0000 ,f = 00 ,d =0001 ,co= 0
#16  cci = 1 ,  a= 1000 ,b = 1111 ,f = 00 ,d =1000 ,co= 1
#18  cci = 1 ,  a= 0010 ,b = 0001 ,f = 01 ,d =0001 ,co= 0
#20  cci = 0 ,  a= 0011 ,b = 0111 ,f = 01 ,d =1100 ,co= 1
#22  cci = 1 ,  a= 0011 ,b = 0011 ,f = 10 ,d =0000 ,co= z
#24  cci = 1 ,  a= 1111 ,b = 1100 ,f = 11 ,d =0000 ,co= z
#26  cci = 0 ,  a= 0001 ,b = 0000 ,f = 00 ,d =0001 ,co= 0
#28  cci = 1 ,  a= 1000 ,b = 1111 ,f = 00 ,d =1000 ,co= 1
```

Figure 6.22 Results of running the test bench for the ALU module in Figure 6.20.

Example 6.3 Four-to-One *Mux*

Figure 6.23 shows a 4-to-1 mux module realized through repeated similar
assignments. It multiplexes one out of four 4-bit-wide buses to the output side.
The assignments are done through 4-bit-wide switches. (The mux can be built up
in other ways too; for example, it can be built around the compact assignment
statement in Figure 6.20.) The synthesized version of the mux is shown in
Figure 6.24; it is essentially the vector counterpart of the 4-to-1 mux of
Figure 4.40.

```
module mux_df1(ao, a1, a2, a3, a4, f, en);
//f is a 2 bit selector input & en is an enable input
output [3:0] ao;
input[3:0] a1, a2, a3, a4;
input en;
input [1:0]f;
trireg [3:0]aa0;
parameter d=4'hz;
assign aa0=(f==2'b00)?a1:d;
assign aa0=(f==2'b01)?a2:d;
assign aa0=(f==2'b10)?a3:d;
assign aa0=(f==2'b11)?a4:d;
assign ao =(en)?aa0:d;
endmodule
```

Figure 6.23 A 4 to 1 vector multiplexor module at the data flow level.

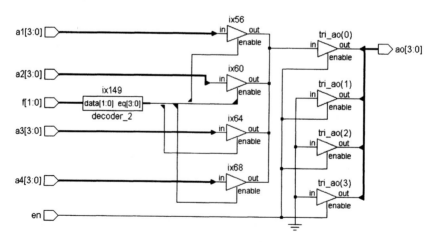

Figure 6.24 Synthesized circuit of the mux in Figure 6.21.

Example 6.4 BCD Adder

A BCD adder can be formed through a compact assignment using a ternary operator. The assignment statement has the form

assign {co, sumd} = (sumb<=4'b1001)?{1'b0,sumb} : (sumb + 4'b0110;

The adder module using the above assignment and a test-bench for the same are shown in Figure 6.25. The synthesized version of the circuit is shown in Figure 6.26. The results of running the test bench are given in Figure 6.27.

```
module bcd(co,sumd,a,b);
input [3:0]a,b;
output [3:0]sumd;
output co;
wire [3:0]sumb;
assign sumb = a + b;
assign{co,sumd}=(sumb<=4'b1001)?{1'b0,sumb}:(sumb+4'b01
10);
endmodule

module tst_bcd;//Test bench
reg [3:0]a,b;
wire co;
wire [3:0]sumd;
bcd bcc(co,sumd,a,b);
initial
    begin
            a = 4'h0 ; b = 4'h0;
    #2  a = 4'h1 ; b = 4'h0;
    #2  a = 4'h2 ; b = 4'h1;
    #2  a = 4'h4 ; b = 4'h5;
    #2  a = 4'h6 ; b = 4'h6;
    #2  a = 4'hd ; b = 4'h1;
    #2  a = 4'hf ; b = 4'h0;
    end
initial $monitor($time,"a = %b, b = %b, co = %b, sumd =
%b",a,b,co,sumd);
initial #16 $stop;
endmodule
```

Figure 6.25 A BCD adder module at the data flow level.

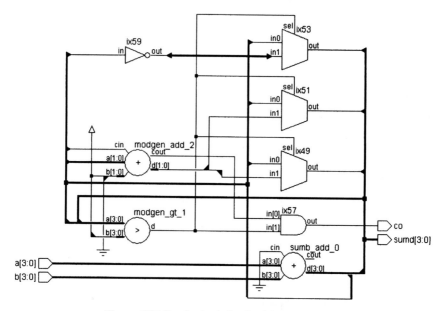

Figure 6.26 Synthesized circuit of the BCD adder.

```
# 0  a = 0000 ,  b = 0000 ,   co = 0 ,  sumd = 0000
# 2  a = 0001 ,  b = 0000 ,   co = 0 ,  sumd = 0001
# 4  a = 0010 ,  b = 0001 ,   co = 0 ,  sumd = 0011
# 6  a = 0100 ,  b = 0101 ,   co = 0 ,  sumd = 1001
# 8  a = 0110 ,  b = 0110 ,   co = 1 ,  sumd = 0010
#10  a = 1101 ,  b = 0001 ,   co = 1 ,  sumd = 0100
#12  a = 1111 ,  b = 0000 ,   co = 1 ,  sumd = 0101
```

Figure 6.27 Results of running the test bench for the BCD adder in Figure 6.24.

6.5.4   Operator Priority

A clear understanding of the operator precedence makes room for a compact design description. But it may lead to ambiguity and to inadvertent errors. Whenever one is not sure of the operator priorities, it is better to resort to the use of parentheses and ensure clarity and accuracy of expressions. Further, some synthesizers may not interpret the operator precedence properly. These too call for the apt use of parentheses.

The operators are arranged in tabular form and shown in Table 6.10. The table brings out the order of precedence. The order of precedence decides the priority for sequence of execution and circuit realization in any assignment statement. The following form the basic rules for the same:

Table 6.10 Operator precedence details

Unary operators	! & ~& | ~| ^ ~^ + −	Highest precedence
Binary operator	* ? /	
	+ -	
	<< >>	
	< <= > >=	
	== != === !==	
	& ^ ~^	
	|	
	&&	
	||	
Ternary operators	? :	Lowest precedence

- Unary operators have the highest priority and execute first.
- Subsequently the binary operators execute. Amongst these the algebraic operators have the highest precedence. Amongst the algebraic operators *, / and % have precedence over + and − operators.
- Subsequent precedence amongst the binary operators is as shown in the table.
- Conditional operator has the lowest precedence and hence is executed last.
- In any expression, operators associate from left to right. Ternary operator is the only exception to this; it associates from right to left.

6.5.4.1   Examples

P = Q − R + S;

Here R is subtracted from Q and then S is added to the result. However, operator precedence does not cause any ambiguity or change the result here.

P = Q − R / S;

In the above case the "divide" operator "/" has precedence over the "subtract" operator "−". Hence R will be divided by S, and the result will be subtracted from Q. If division of (Q − R) is desired, the expression has to be recast as

P = (Q − R) / S;

In a lengthier expression such as

P = a1 − a2 / a3 + a4 * a5;

the operation is equivalent to

P = a1 – (a2 / a3) + (a4 * a5);

Use of parentheses adds to clarity especially in operations involving more than two operators. The operation

P > Q – R

is the same as

P > (Q – R)

since the relational operator ">" has a lower precedence than the algebraic operator "–". Similarly, the expression

P + Q <= R

is the same as

(P + Q) <= R.

6.5.5   Bit Widths of Expressions

When expressions are evaluated or continuous assignments are made, the bit width of the result is decided by different factors. Three cases arise here:

- The operators decide the bit width of the result; logical operators like '&&' and "||" are examples.
- Widths of all operands are specified and they are consistent in all the expressions used. Bit-wise logic with all the operands having the same width are examples of this.
- Widths of all operands are not specified or do not match. The result of expression evaluation and assignments can lead to ambiguity here. However, the rules to resolve these lead mostly to a natural solution.

Bit widths of results of evaluating expressions are given in Table 6.11 for various cases.

6.6   ADDITIONAL EXAMPLES

The use of operands and their combinations are illustrated through a set of two examples here. They also illustrate how data flow level statements can be combined with instantiation of primitives in defining the modules. The results of running the test-benches are shown as waveforms of selected signals. **$monitor** or **$display** commands are not inserted in the test benches.

Table 6.11 Bit widths of expressions: A, B and C represent operands in the table; *opr* represents an operator

Expression	Bit width
Integer, unsized constant number	Compiler-specific
Sized constant number	Decided by the specified size
Opr A where opr is an unary operator out of +, - or ~	Same as that of A
Opr A where opr is an unary operator of Table 6.1	1
A opr B where opr is a logical operator of Table 6.4, a relational operator of Table 6.5 or an equality operator of Table 6.6	
A opr B where opr is an algebraic operator from Table 6.2 or a bit-wise logical operator from Table 6.7.	Width of A or B, whichever is higher
A opr B where opr is a shift operator from Table 6.8	Same as that of A
C ? A : B	Width of A or B, whichever is higher
{A, .., B}	The sum of the bit widths of all the operands
{N*{A, . . . , B}}	N times the sum of the bit widths of all the operands

Example 6.5 Bus Switcher

Figure 6.28 shows the module of a 4-bit bus switcher. A is a 4-bit input bus that is switched on to a selected 4-bit bus. The selection is done by a 2-bit select vector and carried out through a set of simple ternary operator-based assignments. The synthesized circuit of the switcher is shown in Figure 6.29. It decodes the 2-bit select vector into 4 lines that form the control lines for switching. The switching is done through a 4 × 4 tri-state buffer bank. The bus switcher can be easily scaled up to form switches of 8- or 16-bit widths.

```
module demux_df1(a1, a2, a3, a4, a, f);
//A 1 to 4 demux module at data flow level:
output[3:0]a1, a2, a3, a4; // output vectors
input[3:0]a; //a 4 bit input vector
input[1:0]f; //f is the select vector
parameter d = 4'hz;
```

continued

continued

```
assign a1=(f==2'b00)?a:d;
assign a2=(f==2'b01)?a:d;
assign a3=(f==2'b10)?a:d;
assign a4=(f==2'b11)?a:d;
endmodule
```

Figure 6.28 A 4-bit switcher module at the data flow level.

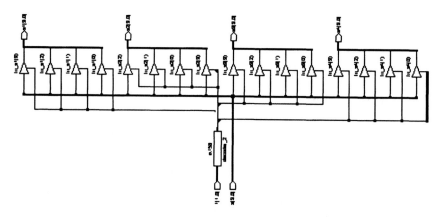

Figure 6.29 Synthesized circuit of the 4-bit switcher.

Example 6.6 Ring Counter

A ring counter is built here in a step-by-step manner. Firstly the simple latch of Example 5.1 has been modified to form another latch shown in Figure 6.30. It has two sets of inputs – sb, rb; and d, db – in place of the single set of sb and rb in Example 5.1. The synthesized circuit is shown in Figure 6.31. The basic cell in the design library being a 2-input AND gate, the NAND function is realized with 2 AND gates followed by a NOT gate. With the additional set of inputs here – d and db – set and reset operations can be carried out independently of the data input.

```
module srff7474(sb, d, rb, db, q, qb);
input sb, rb, d, db;
output q, qb;
nand(q,  sb, d ,qb);
nand(qb, rb, db, q);
endmodule
```

Figure 6.30 A basic latch module.

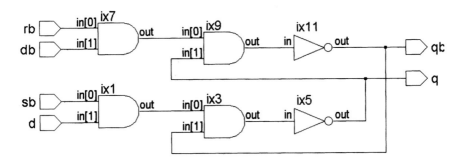

Figure 6.31 Synthesized circuit of the basic latch in Figure 6.30.

A positive edge-triggered flip-flop of the 7474 type is formed by repeated instantiation of the latch in the module of Figure 6.30. Such a flip-flop module is shown in Figure 6.32; it is an enhanced version of the edge-triggered flip-flop in Example 5.5 and in Figure 5.20. The synthesized circuit is shown in Figure 6.33. The srff7474 instantiations are represented there as black boxes.

Figure 6.34 shows a module, which has 4 instantiations of the above edge-triggered flip-flop. This cluster of 4 flip-flops can form the "seed module" of a wide variety of sequential circuits. Figure 6.35 shows the corresponding synthesized circuit.

```
module dff7474new(cp,d,sd,rd,q,qb);
input d,cp,sd,rd;
output q,qb;
wire sdd,rdd;
not(sdd,sd);
not (rdd,rd);
wire n1,n2,n1b,n2b;
  srff7474 ff1(sdd,n2b,rdd,cp,n1,n1b);
  srff7474 ff2(n1b,cp,rdd,d,n2,n2b);
  srff7474 ff3(sdd,n1b,rdd,n2,q,qb);
endmodule
```

Figure 6.32 An edge-triggered flip-flop built with the latch in Figure 6.30 and a test bench for the same.

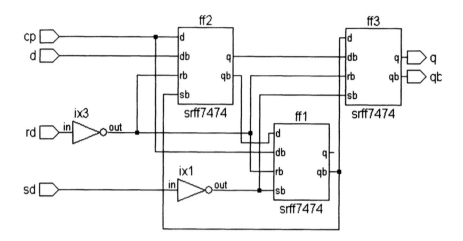

Figure 6.33 Synthesized circuit of the edge-triggered flip-flop in Figure 6.32.

The 4 flip-flops in Figure 6.34 and Figure 6.35 have been connected to form a simple 4-bit ring counter in Figure 6.36. Cen is the overall enabling signal for the ring counter connection. The connection is defined through a set of direct continuous assignments. A test-bench for the ring counter is also included in Figure 6.36. Initially the binary number 1000 is loaded into the set of the 4 flip-flops. Subsequently the flip-flops are connected in a ring counter fashion by enabling Cen. At every positive edge of the clock the data in the ring counter is shifted right by one bit and it circulates. Waveforms of the 4 flip-flops of the ring counter obtained when running the test bench are shown in Figure 6.37. The synthesized circuit of the ring counter is shown in Figure 6.38.

```
module unishrg(clk,d,sd,rd,q,qb);
input clk;
input[3:0]d,sd,rd;
output[3:0]q,qb;
dff7474new ff1(clk,d[0],sd[0],rd[0],q[0],qb[0]);
dff7474new ff2(clk,d[1],sd[1],rd[1],q[1],qb[1]);
dff7474new ff3(clk,d[2],sd[2],rd[2],q[2],qb[2]);
dff7474new ff4(clk,d[3],sd[3],rd[3],q[3],qb[3]);
endmodule
```

Figure 6.34 A module for a general set of 4 edge-triggered flip-flops.

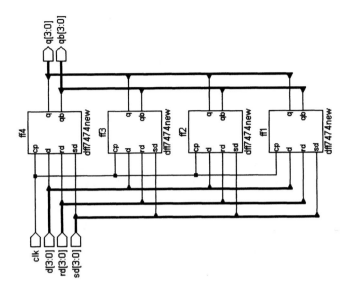

Figure 6.35 Synthesized circuit of the module for a general set of 4 edge-triggered flip-flops in Figure 6.34.

```
module rng_ctr(cen,clk,sd,rd,q,qb);
input clk,cen;
input[3:0]sd,rd;
output [3:0]q,qb;
wire [3:0]d;
unishrg uu(clk,d,sd,rd,q,qb);
assign d[1]=(cen)? q[0]:1'b0;
assign d[2]=(cen)? q[1]:1'b0;
assign d[3]=(cen)? q[2]:1'b0;
assign d[0]=(cen)? q[3]:1'b0;
endmodule

module tst_rng_ctr;//test-bench
reg clk,cen;
reg[3:0]sd,rd;
wire [3:0]q,qb;
rng_ctr rsh(cen,clk,sd,rd,q,qb);
initial
begin
        clk=0;sd=4'b1000;rd=4'b0111;
```

continued

continued

```
        #3sd=4'b0000;rd=4'b0000;
        #2cen=1'b1;
end
always
begin
#2clk =~clk;
end
initial #50 $stop;
endmodule
```

Figure 6.36 A module for a ring counter and a test bench for the same.

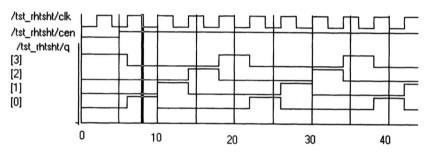

Figure 6.37 Waveforms of a selected set of signals obtained with the test bench in Figure 6.36. The numbers below indicate the time steps

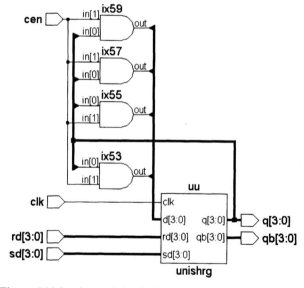

Figure 6.38 Synthesized circuit of the ring counter in Figure 6.36.

6.7   EXERCISES

1. Use continuous assignment statements to design circuits for the following:
    Byte comparator, Parity generator for one data byte, Binary byte to BCD
    code, a pair of BCD digits to binary, BCD to Ex-3 code, Ex-3 to BCD, byte
    multiplier, BCD nibble to 7-segment decoder [Bignel, Sedra, Tocci].

2. What is the result vector in each of the following concatenation operations?

    {3{a},b,c}; {3{a},2{b},c};{3{{a},2{b ,c}}};{3{3{a},2{b}},c};
    {{3{a},b},c};{3{a},b,2{c,1'b0}};};{3{{a, 2'b10, b},2{c,1'b0}}}.

3. Consider the program segment in Figure 6.39; test the segment through a
    test-bench with values of p and q ranging from 0 to 10. Explain why only
    r3 is correct. Declare r1, r2, and r3 to be 5 bits wide: Repeat the test run and
    comment on the results.

    .....
    reg[3:0] p, q, r1, r2, r3;
    ....
    ....
    ....
    assign r1 = p + q;
    assign r1 = p + q + 3'b0;
    assign r1 = p + q + 0;
    .....
    ....

Figure 6.39 Segment of a module for Exercise No. 3 above.

4. Realize the edge-triggered flip-flop of Figure 5.14 through continuous
    assignments for the gates. Test it through a test bench.

5. Form the NOR gate counterpart of the edge-triggered flip-flop of Figure
    5.14; realize it through continuous assignments. Test it through a test bench.

6. Use the set of 4 edge-triggered flip-flops of Figure 6.34 as the basis and
    form the following. In each case, form a test-bench and test the design.

    • A left-shift-type shift register.
    • An 8-bit shift register of the left shift type.
    • A 4-bit Johnson counter.
    • Have a select line sl. If sl = 1, q[0], q[1], and q[2] are to be
      connected to the data inputs d[1], d[2], and d[3], respectively and the
      set of flip-flops should function as a right-shift-type shift register. If
      sl = 0, q[3], q[2], and q[1] are to be connected to d[2], d[1], and
      d[0], respectively, to form a left-shift-type shift register.

7.   Use the edge-triggered flip-flop of Figure 6.32 as the basis and form (a) a
        ripple counter of the count up type, (b) a ripple counter of the count-down
        type, (c) an up down counter. In the case of the up down counter, U_Db is
        the mode signal. If it is high, the counter will count up. If it is low, the
        counter will count down.

8.   Maximum length sequences (Pseudo-random sequences)[Proakis]: Consider
        a set of r flip-flops connected in a shift register fashion. D[k] and q[k]
        represent the data input and output of the kth flip-flop, respectively. The
        flip-flops are clocked at regular intervals by the clock signal clk. D[1], the
        data input to the first flip-flop, is formed as the XOR function of a select set
        of flip-flop outputs; if these are selected suitably, the binary vector
        representing the outputs of all the flip-flops together takes all possible states
        in a "pseudo-random" fashion and repeats the sequence cyclically.
        Specifically, N – the total number of states the sequence passes through – is
        given by

$$N = 2^r - 1$$

        Table 6.12 gives the flip-flop numbers whose outputs are to be XOR'ed to
        form $d[1]$ to yield the maximum length sequence. Design the Maximum
        length sequence generator for different values of r. Give the clock input and
        obtain the output waveform.

Table 6.12 Details for Exercise 8 above

r	2	3	4	5	6	7	8	9	10	11	12
N	3	7	15	31	63	127	255	511	1023	2047	4095
FF numbers to be XOR'ed	2,1	3,1	4,1	5,2	6,2	7,1	8,5,3,1	9,4	10,3	11,1	12,6,4,1

7

BEHAVIORAL MODELING — 1

7.1   INTRODUCTION

Design descriptions at data flow level and gate level are close to the circuit. At every stage of the design description process, one can relate the modules and the instantiations with the corresponding logic or sequential blocks and their interconnections. The approach is practical and effective as long as the gate count remains within a few hundred. An increase in gate count may still be accommodated, if it is due to an increase in vector size –for example, when a system designed and tested at the 8-bit level is being scaled up to a 16- or 32-bit level. But with many of the VLSI's of today, one has to work at a different dimension – the circuit can have a million gates. The increase in vector size may still be accommodated at the data flow level (*e.g.*, 32- or 64-bit systems), since it calls only for scaling of a smaller design. But increase in terms of functional complexity makes the approach almost intractable for many designs.

Behavioral level modeling constitutes design description at an abstract level. One can visualize the circuit in terms of its key modular functions and their behavior; it can be described at a functional level itself instead of getting bogged down with implementation details. The description is carried out essentially with constructs similar to those in "C" language; the design itself is similar to programming in "C" [Gottfried]. For example, one can describe an FFT or a digital filter routine in terms of these constructs. The design can be simulated, debugged, and finalized. This completes the system level structure for the design. Subsequently, one can expand the design by describing the modules in terms of components closer to the data flow and gate level models. One can simulate and debug each such component module, check it for its functionality, integrate it with the main design and test conformity. Constructs for such layered expansion of design are available in behavioral modeling. Proceeding with the layered expansion of design, one can have the final design description at the RTL level itself. However, we may add here that such a top-down activity is more in the realm of design.

The constructs available in behavioral modeling aim at the system level description. Here direct description of the design is not a primary consideration in the Verilog standard. Rather, flexibility and versatility in describing the design are in focus [IEEE]. One should be able to describe the design and simulate it for its functionality. Hence the constructs aim essentially at these two aspects of the design. Synthesis tools available from different vendors can synthesize most of the constructs at the data flow as well as the gate levels, but not all constructs or combinations possible at the behavioral level can be synthesized. The extent to which the constructs at the behavioral level are accommodated in synthesis varies with vendors. The synthesized circuit need not guarantee optimum or near-optimum realization either. These limitations are in line with the basic purpose of behavioral level modeling mentioned above – that is, to complete an error or bug-free description and identify the functional modules required. Their synthesis is more often done following a more detailed design description at the RTL level.

7.2   OPERATIONS AND ASSIGNMENTS

The design description at the behavioral level is done through a sequence of assignments. These are called 'procedural assignments' – in contrast to the continuous assignments at the data flow level. Though it appears similar to the assignments at the data flow level discussed in the last chapter, the two are different. The procedure assignment is characterized by the following:

- The assignment is done through the "=" symbol (or the "<=" symbol) as was the case with the continuous assignment earlier.
- An operation is carried out and the result assigned through the "=" operator to an operand specified on the left side of the "=" sign – for example,
  $N = \sim N$;
  Here the content of **reg** N is complemented and assigned to the reg N itself. The assignment is essentially an updating activity.
- The operation on the right can involve operands and operators. The operands can be of different types – logical variables, numbers – real or integer and so on.
- All the operands are given in Tables 6.1 to 6.9. The format of using them and the rules of precedence remain the same.
- The operands on the right side can be of the net or variable type. They can be scalars or vectors.
- It is necessary to maintain consistency of the operands in the operation expression – *e.g.*,
  $N = m / l$;
  Here m and l have to be same types of quantities – specifically a **reg**, **integer**, **time**, **real**, **realtime**, or memory type of data – declared in advance.
- The operand to the left of the "=" operator has to be of the variable (*e.g.*, **reg**) type. It has to be specifically declared accordingly. It can be a scalar, a vector, a part vector, or a concatenated vector.

- Procedural assignments are very much like sequential statements in C. Normally they are carried out one at a time sequentially. As soon as a specified operation on the right is carried out, the result is assigned to the quantity on the left – for example

$N = m + l$;

$N1 = N * N$;

The above form a set of two procedures placed within an **always** block. Generally they are carried out sequentially in the order specified; that is, first m and l are added and the result assigned to N. Then the square of N is assigned to $N1$. Subsequently the following assignment, if any, is carried out. However, there can be exceptions to this which will be discussed later. The sequential nature of the assignments requires the operands on the left of the assignment to be of **reg** (variable) type. The basic sequential nature of assignments here is in direct contrast to the concurrent nature of assignments at the data flow level.

Procedural assignments within a process are of a variety of types. These are discussed later.

7.3   FUNCTIONAL BIFURCATION

Design description at the behavioral level is done in terms of procedures of two types; one involves functional description and interlinks of functional units. It is carried out through a series of blocks under an "**always**" banner – discussed later. The second concerns simulation – its starting point, steering the simulation flow, observing the process variables, and stopping of the simulation process; all these can be carried out under the "**always**" banner, an "**initial**" banner, or their combinations. However, each **always** and each **initial** block initiates an activity flow during simulation. In general the activity with all such blocks starts at the simulation time and flows concurrently during the whole simulation process. The concurrent flow of activity with all processes is characteristic of any behavioral level module. A procedure-block of either type – **initial** or **always** – can have a structure shown in Figure 7.1. A block starts with the declaration of the type of block – that is, **initial** or **always**. It may be followed by the definition of a triggering activity and then the body of the block. The body may be a single procedural assignment or a group of procedural assignments. In the latter case the block appears within a "**begin – end**" or similar blocks. The **initial** and **always** blocks have distinct characteristics. The two are discussed separately.

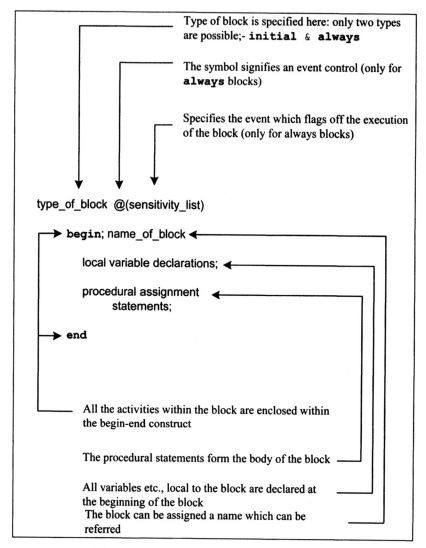

Figure 7.1 Structure of a typical procedural block.

7.3.1   begin – end Construct

If a procedural block has only one assignment to be carried out, it can be specified as below:

initial #2 a=0;

   The above statement assigns the value 0 to variable a at the simulation time of 2 ns. It is possibly the simplest initial block. More often more than one procedural assignment is to be carried out in an **initial** block. All such assignments are grouped together between "**begin**" and "**end**" declarations. Functionally, the construct is similar to the **begin–end** construct in Pascal or the { } construct in C language. The following are to be noted here:

- Every **begin** declaration must have its associated **end** declaration.
- **begin – end** constructs can be nested as many times as desired.
- For clarity in description and to avoid mistakes, nested **begin – end** blocks are separated suitably (see Figure 7.2).

7.3.2   Name of the Block

Any block can be assigned a name, but it is not mandatory. Only the blocks which are to be identified and referred by the simulator need be named. Needless to say the names assigned to different blocks have to be different. Names chosen should conform to the rules for the selection of names to variables [see Section 3.4]. Assigning names to blocks serves different purposes:

- Registers declared within a block are local to it and are not available outside. However, during simulation they can be accessed for simulation, *etc.*, by proper dereferencing [see Section 11.4].
- Named blocks can be disabled selectively when desired [see Section 8.6].

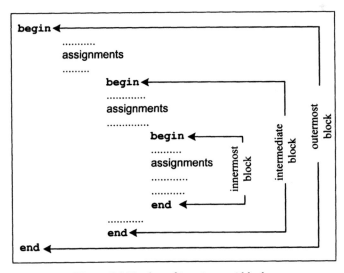

Figure 7.2 Nesting of **begin-end** blocks.

7.3.3   Local Variables

Variables used exclusively within a block can be declared within it. Such a variable need not be declared outside, in the module encompassing the block. Such local declarations conserve memory and offer other benefits too. Regs declared and used within a block are static by nature. They retain their values at the time of leaving the block. The values are modified only at the next entry to the block.

7.4   INITIAL CONSTRUCT

A set of procedural assignments within an **initial** construct are executed only once – and, that too, at the times specified for the respective assignments. Consider the **initial** process shown in Figure 7.3. It is characterized by the following:

- In any assignment statement the left-hand side has to be a storage type of element (and not a net). It can be a **reg**, **integer**, or **real** type of variable. The right-hand side can be a storage type of variable (**reg**, **integer**, or **real** type of variable) or a net.
- As already mentioned in Section 7.2, all the operations described in Tables 6.1 to 6.9 for continuous assignment can be used for procedural assignments as well. The context decides whether the assignment is of a continuous type or procedural type. In the latter case it is present within an **always** or an **initial** construct.
- All the procedural assignments appear within a **begin–end** block explained earlier.
- All the procedural assignments are executed sequentially – in the same order as they appear in the design description. The waveforms of a and b conforming to the assignments in the block are shown in Figure 7.4.
- Initially (at time $t = 0$ ns), a and b are set equal to zero.

```
reg a,b;
initial
    begin
            a = 1'b0;
            b = 1'b0;
    #2      a = 1'b1;
    #3      b = 1'b1;
    #1      a = 1'b0;
    #100$stop;
    end
```

Figure 7.3 A typical initial block.

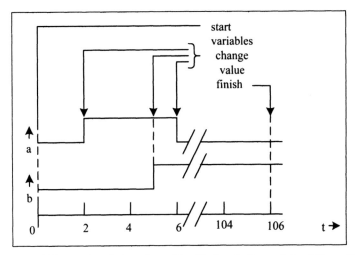

Figure 7.4 Nature of variation of a and b with time in the module of Figure 7.3.

- At time 2 ns **a** is made equal to 1. After 3 more nanoseconds – that is, at the 5th ns – **b** is made equal to 1.
- After one more ns – that is, at the 6th ns – **a** is made equal to 0.
- **$stop** is a system task. 100 ns later – that is, at the 106th ns – the simulation comes to an end (see Figure 7.4).

Integer values have been used here to decide time delay values. In a more general case the delay value can be a constant expression. It is evaluated and decided dynamically as the simulation proceeds.

The **initial** block above does three controlling activities during the simulation run.

- Initialize the selected set of **reg**'s at the start.
- Change values of **reg**'s at predetermined instances of time. These form the inputs to the module(s) under test and test it for a desired test sequence.
- Stop simulation at the specified time.

Figure 7.4 depicts the events for the above case; *t* is the time axis here.

Specific system tasks available in Verilog can be used to tabulate the values of selected variables. Providing such output display in a desired or preferred format is the activity of the simulation run. Two system tasks are useful here – **$display** & **$monitor** [see Section 3.15 and Chapter 11]. By way of illustration consider the simulation routine in Figure 7.5. It incorporates the block

```
module nil;
reg a, b;
initial
begin
        a = 1'b0;
        b = 1'b0;
        $display ("display: a = %b, b = %b", a, b);
    #2     a = 1'b1;
    #3     b = 1'b1;
    #1     a = 1'b0;
    #100 $stop;
end
initial
$monitor("monitor: a = %b, b = %b", a, b);
endmodule
```

Figure 7.5 A typical module with an **initial** block.

Figure 7.3 and two system tasks. The result of the simulation is shown in Figure 7.6. The $display task is a one-time activity. It is executed when encountered. At that instant in simulation the values of *a* and *b* are zero and the same are displayed. In contrast, $monitor is a repeated activity. It need be present only once in a simulation routine – all the specified variables will be monitored. If multiple $monitor tasks are present in the routine, only the last one will be active. All others will be ignored. In contrast, the $display task may appear any number of times in a module. It is executed every time it is encountered.

Simulators have the facility to observe the waveforms and changes in the magnitudes of different variables with simulation time. The necessary facility is provided with the help of user-friendly menus and icons. Waveforms of *a* and *b* obtained with the test bench of Figure 7.5 are shown in Figure 7.7; they can be seen to be consistent with their values shown in Figure 7.6.

```
output
        # display : a = 0 ,b = 0
        # monitor : a = 0 ,b = 0
        # monitor : a = 1 ,b = 0
        # monitor : a = 1 ,b = 1
        # monitor : a = 0 ,b = 1
```

Figure 7.6 Results of running the test bench in Figure 7.5.

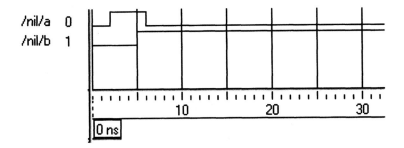

Figure 7.7 Results of running the test bench in Figure 7.5 shown as waveforms.

7.4.1   Multiple Initial Blocks

A module can have as many **initial** blocks as desired. All of them are activated at the start of simulation. The time delays specified in one **initial** block are exclusive of those in any other block. Consider the module in Figure 7.8 which is a modified version of that in Figure 7.5. It has four **initial** blocks. The $monitor task is declared separately (a healthy practice). The simulated results are shown in Figure 7.9. The following observations are in order here:

```
module nil1;
initial
reg a, b;
begin
        a = 1'b0;
        b = 1'b0;
        $display ($time,"display:  a = %b,  b = %b", a, b);
  #2    a = 1'b1;
  #3    b = 1'b1;
  #1    a = 1'b0;
end
initial #100$stop;
initial $monitor ($time, "monitor:  a = %b,  b = %b", a, b);
initial
        begin
  #2    b = 1'b1;
        end
endmodule
```

Figure 7.8 A typical module with multiple initial blocks.

```
output
        # display : a = 0 , b = 0
        # monitor : a = 0 , b = 0
        # monitor : a = 0 , b = 1
        # monitor : a = 1 , b = 1
        # monitor : a = 1 , b = 0
        # monitor : a = 1 , b = 1
        # monitor : a = 0 , b = 1
```

Figure 7.9 Results of running the test bench in Figure 7.8.

- All changes in *a* are brought about in one initial block.
- Changes to *b* are specified in two blocks, and both these blocks are executed concurrently.
- The progress of simulation time in different blocks is concurrent. However, those in one block are sequential. Changes in *b* are consistent with this.
- The $stop task is in an independent **initial** block. Hence simulation is terminated at 100 ns. Contrast this with the previous case (Figure 7.4), where sequential execution results in finish of simulation after 106 ns (even though in both the cases the statement "#100 $stop" remains the same).
- More than one activity may be scheduled for execution at one time instant. Those in one **initial** block are executed in the same order   as they appear – that is, sequentially.
  Thus, the two events
  a = 1'b0;
  b = 1'b0;
  are executed in the same sequential order – that is, *b* is set to 0 after *a* is set to 0, although both the activities are scheduled for execution at the same time.
- At 2 ns *a* changes to 1 and *b* changes to 0. These two activities are to be done concurrently. They are in different **initial** blocks. The order of their execution depends upon the implementation. This does not cause any anomaly in the present case. But it can be a potential source of problem in more involved designs and their simulation.

7.5   ALWAYS CONSTRUCT

The **always** process signifies activities to be executed on an "always basis." Its essential characteristics are:

- Any behavioral level design description is done using an always block.
- The process has to be flagged off by an event or a change in a net or a reg. Otherwise it ends in a stalemate.

- The process can have one assignment statement or multiple assignment statements. In the latter case all the assignments are grouped together within a "**begin – end**" construct.
- Normally the statements are executed sequentially in the order they appear.

7.5.1 Event Control

The **always** block is executed repeatedly and endlessly. It is necessary to specify a condition or a set of conditions, which will steer the system to the execution of the block. Alternately such a flagging-off can be done by specifying an event preceded by the symbol "**@**". The event can be a change in the variable specified in either direction or a change in a specified direction. For example,

- @(**negedge** clk) :
  executes the following block at the negative edge of the **reg** (variable) clk.
- @(**posedge** clk) :
  executes the following block at the positive edge of the **reg** (variable) clk.
- @clk :
  executes the following block at both the edges of clk.

The event can be a combination as well.

- @(prt or clr) :
  With the above event the block is executed whenever either of the variables prt or clr undergoes a change.
- @(**posedge** clk1 or **negedge** clk2) :
  With the above event the block is executed in two cases – whenever the clock clk1 changes from 0 to 1 state or the clock clk2 changes from 1 to 0. One can specify more elaborate events by OR'ing individual ones. The following are to be noted:
- The events can be changes in **reg**, **integer**, **real** or a signal on a net. These should be declared beforehand.
- No algebra or logic operation is permitted as an event. The OR'ing signifies "execute the block if any one of the events takes place."
- The edge transition on each event is to be specified separately
- Note the difference between the following:
    - (**posedge** clk1 or clk2): means "execute the block following if clk1 goes to 1 state or clk2 changes state (whether 0 to 1 or 1 to 0)."
    - (**posedge** clk1 or **posedge** clk2): means "execute the block following if clk1 goes to 1 state or clk2 goes to 1 state."

- The positive transition for a reg type single bit variable is a change from 0 to1. For a logic variable it is a transition from false to true.
- The **"posedge"** transition for a signal on a net can be of three different types:

  - 0 to1
  - 0 to **x** or **z**
  - **x** or **z** to 1

- The **"negedge"** transition for a signal on a net can be of three different types:-

  - 1 to 0
  - 1 to **x** or **z**
  - **x** or **z** to 0

- If the event specified is in terms of a multibit **reg**, only its least significant bit is considered for the transition. Changes in the other bits are ignored.
- The event-based flagging-off of a block is applicable only to the **always** block.
- According to the recent version of the LRM, the comma operator (,) plays the same role as the keyword **or**. The two can be used interchangeably or in a mixed form. Thus the following are identical:

  @ (a **or** b **or** c)
  @ (a **or** b, c)
  @ (a, b, c)
  @ (a, b **or** c)

7.6   EXAMPLES

A few simple design examples are considered here [Arnold, Bogart, Navabi]]; they are aimed at bringing out the potential flexibility at the behavioral level, despite the compactness in the module descriptions. Some of these examples have already been discussed in earlier chapters at the data flow as well as the gate levels.

Example 7.1 A Versatile Counter

We consider a versatile up-down counter module with the following facilities:

- *Clear input*: If it goes high, the counter is cleared and reset to zero.
- *U/D input*: If it goes high, the counter counts up; if it goes down, the counter counts down.
- The counter counts at the negative edge of the clock.
- The counter counts up or down between 0 and N where N is any 4-bit hex number.

The above counter design specifications are implemented in stages. The module in Figure 7.10 is an up counter which counts up repeatedly from 0 to a preset number N. A test-bench for the counter is also shown in the figure. N is an input to the module. The count advances at every negative edge of the clock. When the count reaches the value N, the count value a is reset to 0. The simulation results are shown as waveforms in Figure 7.11 (only partially shown). The periodic clock waveform (with a period of 4 ns), the incrementing of a at every negative edge of the clock and counting of a from 0 to the set value of N (=1011 in this specific

```
module counterup(a,clk,N);
input clk;
input[3:0]N;
output[3:0]a;
reg[3:0]a;
initial a=4'b0000;
always@(negedge clk) a=(a==N)?4'b0000:a+1'b1;
endmodule

module tst_counterup;//TEST_BENCH
reg clk;
reg[3:0]N;
wire[3:0]a;
counterup c1(a,clk,N);
initial
begin
    clk = 0;
    N   = 4'b1011;
end
always #2 clk=~clk;
initial $monitor($time,"a=%b,clk=%b,N=%b",a,clk,N);
endmodule
```

Figure 7.10 An up counter module.

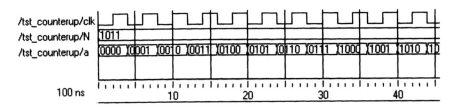

Figure 7.11 Partial results of running the test bench in Figure 7.10.

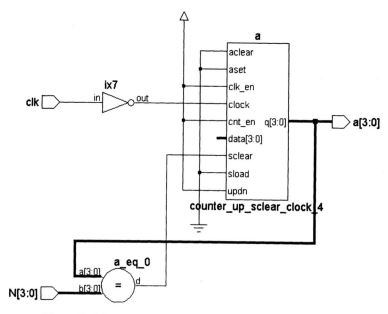

Figure 7.12 Synthesized circuit of the up counter in Figure 7.10.

case) can be seen from the figure. The synthesized circuit of the counter is shown in Figure 7.12. It has a versatile counter block and a comparator. The comparator compares the value of *a* with the set value of *N* and resets the counter when the two are equal – as specified in the design module.

The module of Figure 7.13 is a down counter. The count a decrements at the negative edge of the clock – clk. The counter counts down from *N* to zero. As soon as the count reaches the value 0, it is set back to *N*. The simulation results are shown tabulated in Figure 7.14 and as waveforms in Figure 7.15; these can be seen to be consistent with the design module. The synthesized circuit is shown in Figure 7.16. The basic blocks – namely versatile counter, comparator and buffer for the clock – are the same as those for the up counter of Figure 7.12. The comparator output loads the value of *N* back into the counter every time a reaches the set value of *N* (In contrast, in the case of the up counter above, the comparator resets the counter back to zero, whenever a reaches the set value of *N*.)

```
module counterdn(a,clk,N);
input clk;
input[3:0]N;
output[3:0]a;
reg[3:0]a;
initial a =4'b0000;
```

continued

continued

```
always@(negedge clk) a=(a==4'b0000)?N:a-1'b1;
endmodule

module tst_counterdn();//TEST_BENCH
reg clk;
reg[3:0]N;
wire[3:0]a;
counterdn cc(a,clk,N);
initial
begin
    N   = 4'b1010;
    Clk = 0;
end
always #2 clk=~clk;
initial $monitor($time,"a=%b,clk=%b,N=%b",a,clk,N);
initial #55 $stop;
endmodule
```

Figure 7.13 Design module of a down counter and a test bench for the same.

```
Output
   #   0a=1010,clk=0,N=1010
   #   2a=1010,clk=1,N=1010
   #   4a=1001,clk=0,N=1010
   #   6a=1001,clk=1,N=1010
   #   8a=1000,clk=0,N=1010
   #  10a=1000,clk=1,N=1010
   #  12a=0111,clk=0,N=1010
   #  14a=0111,clk=1,N=1010
   #  16a=0110,clk=0,N=1010
   #  18a=0110,clk=1,N=1010
   #  20a=0101,clk=0,N=1010
   #  22a=0101,clk=1,N=1010
   #  24a=0100,clk=0,N=1010
   #  26a=0100,clk=1,N=1010
   #  28a=0011,clk=0,N=1010
   #  30a=0011,clk=1,N=1010
   #  32a=0010,clk=0,N=1010
   #  34a=0010,clk=1,N=1010
   #  36a=0001,clk=0,N=1010
```

continued

continued

```
#  38a=0001,clk=1,N=1010
#  40a=0000,clk=0,N=1010
#  42a=0000,clk=1,N=1010
#  44a=1010,clk=0,N=1010
#  46a=1010,clk=1,N=1010
#  48a=1001,clk=0,N=1010
#  50a=1001,clk=1,N=1010
#  52a=1000,clk=0,N=1010
#  54a=1000,clk=1,N=1010
```

Figure 7.14 Results of running the test bench in Figure 7.13.

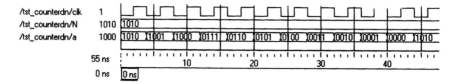

Figure 7.15 Results of running the test bench in Figure 7.13 – shown partly as waveform.

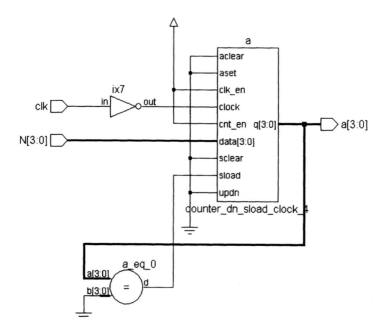

Figure 7.16 Synthesized circuit of the down counter in Figure 7.13.

The up and down modes of counting have been combined in the up down counter of Figure 7.17. A test bench is also shown in the figure. The test results are tabulated in Figure 7.18 and also shown as waveforms in Figure 7.19. Figure 7.20 shows the synthesized circuit; the counter block remains the same as in the last two cases; the mode control part of the circuit has been changed to meet the enhanced needs. The counting can be seen to be changing from "up" to the "down" type, when the mode control input u_d changes.

```
module updcounter(a,clk,N,u_d);
input clk,u_d;
input[3:0]N;
output[3:0]a;
reg[3:0]a;
initial a =4'b0000;
always@(negedge clk)
a=(u_d)?((a==N)?4'b0000:a+1'b1):((a==4'b0000)?N:a-
1'b1);
endmodule

module tst_updcounter();//TEST_BENCH
reg clk,u_d;
reg[3:0]N;
wire[3:0]a;
updcounter c2(a,clk,N,u_d);
initial
begin
     N   = 4'b0111;
     u_d = 1'b0;
     clk = 0;
end
always #2 clk=~clk;
always #34u_d=~u_d;
initial $monitor
($time,"clk=%b,N=%b,u_d=%b,a=%b",clk,N,u_d,a);
initial #64 $stop;
endmodule
```

Figure 7.17 Design module of an up down counter and a test bench for the same.

```
#                    0clk=0,N=0111,u_d=0,a=0111
#                    2clk=1,N=0111,u_d=0,a=0111
#                    4clk=0,N=0111,u_d=0,a=0110
#                    6clk=1,N=0111,u_d=0,a=0110
#                    8clk=0,N=0111,u_d=0,a=0101
#                   10clk=1,N=0111,u_d=0,a=0101
#                   12clk=0,N=0111,u_d=0,a=0100
#                   14clk=1,N=0111,u_d=0,a=0100
#                   16clk=0,N=0111,u_d=0,a=0011
#                   18clk=1,N=0111,u_d=0,a=0011
#                   20clk=0,N=0111,u_d=0,a=0010
#                   22clk=1,N=0111,u_d=0,a=0010
#                   24clk=0,N=0111,u_d=0,a=0001
#                   26clk=1,N=0111,u_d=0,a=0001
#                   28clk=0,N=0111,u_d=0,a=0000
#                   30clk=1,N=0111,u_d=0,a=0000
#                   32clk=0,N=0111,u_d=0,a=0111
#                   34clk=1,N=0111,u_d=1,a=0111
#                   36clk=0,N=0111,u_d=1,a=0000
#                   38clk=1,N=0111,u_d=1,a=0000
#                   40clk=0,N=0111,u_d=1,a=0001
#                   42clk=1,N=0111,u_d=1,a=0001
#                   44clk=0,N=0111,u_d=1,a=0010
#                   46clk=1,N=0111,u_d=1,a=0010
#                   48clk=0,N=0111,u_d=1,a=0011
#                   50clk=1,N=0111,u_d=1,a=0011
#                   52clk=0,N=0111,u_d=1,a=0100
#                   54clk=1,N=0111,u_d=1,a=0100
#                   56clk=0,N=0111,u_d=1,a=0101
#                   58clk=1,N=0111,u_d=1,a=0101
#                   60clk=0,N=0111,u_d=1,a=0110
#                   62clk=1,N=0111,u_d=1,a=0110
```

Figure 7.18 Results of running the test bench in Figure 7.17.

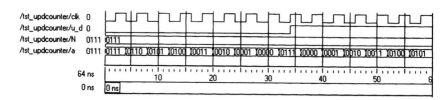

Figure 7.19 Results of running the test bench in Figure 7.17 – shown partly as waveforms.

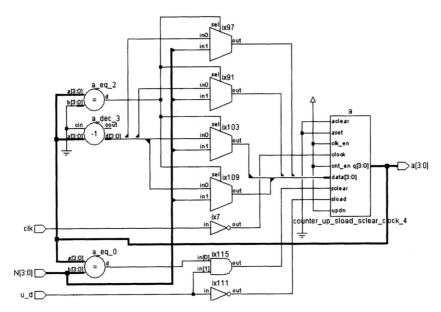

Figure 7.20 Synthesized circuit of the up down counter in Figure 7.17.

The counter as described in Figure 7.21 has an additional "clear" input. With this enhancement, it has become versatile (compare with 74196 or 74197). Note that despite the versatility offered by the design, the full counter has been described in the single line of executable statement reproduced below:

always@ (**negedge** clk or **posedge** clr)
a=(clr)?4'h0:((u_d)?((a==N)?4'b0000:a+1'b1):((a==4'b0000)?N:a-1'b1));

```
module clrupdcou(a,clr,clk,N,u_d);
input clr,clk,u_d;
input[3:0]N;
output[3:0]a;
reg[3:0]a;
initial a =4'b0000;
always@(negedge clk or posedge clr)
      a=(clr)?4'h0:((u_d)?((a==N)?4'b0000:a+1'b1):((a==
4'b0000)?N:a-1'b1));
 /*signals having priority over clk have to be included
in the sensitivity list*/
endmodule
```

continued

continued

```
module tst_clrupdcou;//TEST_BENCH
reg clr,clk,u_d;
reg[3:0]N;
wire [3:0]a;
clrupdcou cc11(a,clr,clk,N,u_d);
initial
begin
     N   = 4'b0111;
     Clr = 1'b1;u_d=1'b1;
     Clk = 0;
end
always
begin
     #2  clk = ~clk;
          clr = 1'b0;
end
always  #34 u_d<=~u_d;
initial $monitor($time
,"clk=%b,clr=%b,u_d=%b,N=%b,a=%b",clk,clr,u_d,N,a);
initial #60 $stop;
endmodule
```

Figure 7.21 Design module of an up down counter with clear facility and a test bench for the same.

The test bench for the counter is also shown in the figure. The test results are reproduced in Figure 7.22 as waveforms; the synthesized circuit is shown in Figure 7.23.

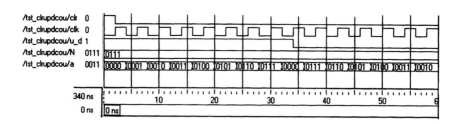

Figure 7.22 Results of running the test bench in Figure 7.21 – shown partly as waveforms.

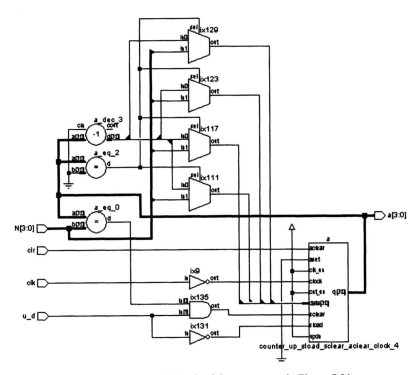

Figure 7.23 Synthesized circuit of the up counter in Figure 7.21.

Example 7.2 Shift Register

Figure 7.24 shows an 8-bit shift register module along with a test bench for the same. The register shifts by one bit to the right if r_l = 1 and to the left by one bit otherwise (*i.e.*, if r_l = 0). The whole shift register is described in a single line of procedural assignment, namely

always@(negedge clk) a=(r_l)?(a>>1'b1):(a<<1'b1);

The simulation results are given in tabular form in Figure 7.25 and as waveforms in Figure 7.26.

```
module shifrlter(a,clk,r_l);
input clk,r_l;
output [7:0]a;
reg[7:0]a;
initial a= 8'h01;
always@(negedge clk)
```

continued

continued

```
begin
     a=(r_l)?(a>>1'b1):(a<<1'b1);
end
endmodule

module tst_shifrlter;//test-bench
reg clk,r_l;
wire [7:0]a;
shifrlter shrr(a,clk,r_l);
initial
begin
     clk =1'b1;
     r_l = 0;
end
always #2 clk =~clk;
initial #16 r_l =~r_l;
initial
$monitor($time,"clk=%b,r_l = %b,a =%b ",clk,r_l,a);
initial #30 $stop;
endmodule
```

Figure 7.24 Design module of a shift register with facility for right or left shift and a test bench for the same.

```
Output
    #     0 clk=1, r_l = 0 , a = 00000001
    #     2 clk=0, r_l = 0 , a = 00000010
    #     4 clk=1, r_l = 0 , a = 00000010
    #     6 clk=0, r_l = 0 , a = 00000100
    #     8 clk=1, r_l = 0 , a = 00000100
    #    10 clk=0, r_l = 0 , a = 00001000
    #    12 clk=1, r_l = 0 , a = 00001000
    #    14 clk=0, r_l = 0 , a = 00010000
    #    16 clk=1, r_l = 1 , a = 00010000
    #    18 clk=0, r_l = 1 , a = 00001000
    #    20 clk=1, r_l = 1 , a = 00001000
    #    22 clk=0, r_l = 1 , a = 00000100
    #    24 clk=1, r_l = 1 , a = 00000100
    #    26 clk=0, r_l = 1 , a = 00000010
    #    28 clk=1, r_l = 1 , a = 00000010
```

Figure 7.25 Results of running the test bench in Figure 7.24.

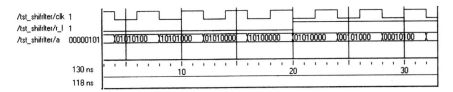

Figure 7.26 Results of running the test bench in Figure 7.24 – shown partly as waveforms.

Example 7.3   Clocked Flip-Flop

The module for a clocked flip-flop is shown in Figure 7.27. A test bench for the flip-flop is also included in the figure. The test results are shown in Figure 7.28 and Figure 7.29 in tabular form and as waveforms, respectively. The input can be seen to be sensed, latched, and presented as output at every negative edge of the clock. Otherwise the output remains frozen at the last latched value. The synthesized circuit of the flip-flop is shown in Figure 7.30.

```
module dff(do,di,clk);
output do;
input di,clk;
reg do;
initial
do=1'b0;
always@(negedge clk) do=di;
endmodule

module tst_dffbeh();//test-bench
reg di,clk;
wire do;
dff d1(do,di,clk);
initial
begin
    clk=0;
    di=1'b0;
end
always #3clk=~clk;
always #5 di=~di;
initial
$monitor($time,"clk=%b,di=%b,do=%b",clk,di,do);
initial #35 $stop;
endmodule
```

Figure 7.27 Design module of a D-flip-flop and a test bench for the same.

```
Output
    #                    0clk=0,di=0,do=0
    #                    3clk=1,di=0,do=0
    #                    5clk=1,di=1,do=0
    #                    6clk=0,di=1,do=1
    #                    9clk=1,di=1,do=1
    #                   10clk=1,di=0,do=1
    #                   12clk=0,di=0,do=0
    #                   15clk=1,di=1,do=0
    #                   18clk=0,di=1,do=1
    #                   20clk=0,di=0,do=1
    #                   21clk=1,di=0,do=1
    #                   24clk=0,di=0,do=0
    #                   25clk=0,di=1,do=0
    #                   27clk=1,di=1,do=0
    #                   30clk=0,di=0,do=0
    #                   33clk=1,di=0,do=0
```

Figure 7.28 Results of running the test bench in Figure 7.27.

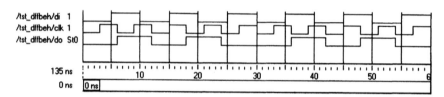

Figure 7.29 Results of running the test bench in Figure 7.27– shown partly as waveforms.

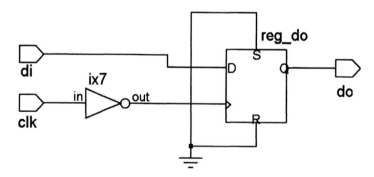

Figure 7.30 Synthesized circuit of the D-flip-flop in Figure 7.27.

Example 7.4 D Latch

Figure 7.31shows the module of a D latch along with its test bench. Whenever en is high, the output follows the input; the latch is transparent. When en goes low the output remains frozen at the last value. The simulation results are shown as waveforms in Figure 7.32.

```
module dffen(do,di,en); // d-latch
output do;
input di,en;
reg do;
initial
do=1'b0;
always@(di or en)
if(en)
do=di;
endmodule

module tst_dffbehen;//test-bench
reg di,en;
wire do;
dffen d1(do,di,en);
initial
begin
    en=0;
    di=1'b0;
end
always#7 en =~en;
always#4 di=~di;
initial
$monitor($time,"en=%b,di=%b,do=%b",en,di,do);
initial #50 $stop;
endmodule
```

Figure 7.31 Design module of a D-latch and a test bench for the same.

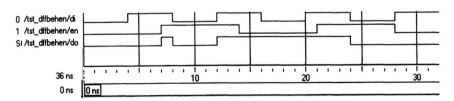

Figure 7.32 Results of running the test bench in Figure 7.31– shown partly as waveforms.

Example 7.5 Clock Waveform

Consider the design description line

always #3 clk = ~clk;

The sequence of operation taking place within this line segment is as follows:

- When the system comes across the statement, it schedules an activity 3 ns later.
- At the end of the 3 ns, the value of clk is sensed; the sensed value is complemented and then stored temporarily.
- Then the stored value is assigned to the clock, which completes the activity of the always block; once again, execution resumes at step 1.

The clock waveform is shown in Figure 7.33.

7.7   ASSIGNMENTS WITH DELAYS

Specific delays can be associated with procedural assignments. The delay refers to the specific activity it qualifies. A variety of possibilities of specifying delays to assignments exist. A clear understanding makes room for flexibility through their judicious use; the absence of a clear understanding can be disastrous! The variety and flexibility are brought here through simple illustrations.
Consider the assignment

always #3 b = a;

simulator encounters this at zero time and posts the entire activity to be done 3 ns later. Further, by virtue of the always nature of the activity, the assignment is scheduled to be repeated every 3 ns, irrespective of whether a changes in the meantime. Values of a at the 3rd, 6th, 9th, *etc.*, ns are sampled and assigned to b. Figure 7.35 shows the waveforms of a and b with the above assignment and execution of the module in Figure 7.34. Changes in the values of a lasting less than 3 ns may be ignored. Specifically, in this case, a took the value of 1 during the interval 4th ns to the 5th ns which is not passed on to b.

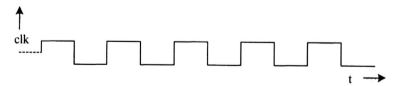

Figure 7.33 The clock waveform with an **always** block of one statement to generate a clock.

```
module del1;
reg a,b;
always #3 b=a;
Initial
begin
            a = 1'b1;
            b = 1'b0;
     #1     a = 1'b0;
     #3     a = 1'b1;
     #1     a = 1'b0;
     #2     a = 1'b1;
     #3     a = 1'b0;
end
initial $monitor($time, " a = %d,  b = %d", a, b);
initial #20 $finish;
endmodule
```

Figure 7.34 A module to illustrate delayed assignment.

The module of figure 7.36 is a modified version of that in Figure 7.34. The activities within the always block (of a single statement) are carried out whenever the value of a changes. The sole activity is that of assigning the value of a to b with a delay of 2 ns – that is, 2 ns after a changes sign. The waveform assigned to a as well as the resulting waveform of b is shown in Figure 7.37. If a were to remain invariant, b will have no assignment here. In contrast in the previous case (Figure 7.35), b is given an assignment (=a) at every 3rd ns.

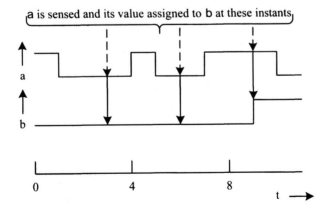

Figure 7.35 Waveforms of a and b with the simulation of the module in Figure 7.34.

```
module del2;
reg a,b;
always @(a) #2 b=a;

Initial
begin
              a = 1'b1;
              b = 1'b0;
       #1     a = 1'b0;
       #3     a = 1'b1;
       #1     a = 1'b0;
       #2     a = 1'b1;
       #3     a = 1'b0;
end
initial $monitor($time, " a = %d,  b = %d", a, b);
initial #20 $finish;
endmodule
```

Figure 7.36 A modified version of the module in Figure 7.34.

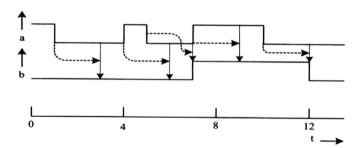

Figure 7.37 Waveforms of a and b obtained with the simulation of the module in Figure 7.36.

Consider a more detailed example – that of Figure 7.38. The **always** block has two assignments. These are carried out sequentially and repeatedly. At the 3rd ns the assignment b = a is executed. The assignment that follows is executed 1 ns later – that is, at the 4th ns. Again 3 ns later – that is, at the 7th ns – the first assignment is executed, and so on. The results obtained are shown in Table 7.1. Only the values of a, b, and c at the first few time step values are shown in the table.

```
module del3;
integer a,b, c;
always
begin
        # 3 b = a;
        # 1 c = a;
        end
initial
begin
                a = 0;
                b = 0;
                c = 0;
        #2   a = 1;
        #2   a = 2;
        #2   a = 3;
        #2   a = 4;
        #2   a = 5;
        #2   a = 6;
end
initial $monitor($time, " a = %d,  b = %d", a, b);
initial #20 $finish;
endmodule
```

Figure 7. 38 A module where b and c are versions of a with different delays.

7.7.1   Intra-assignment Delays

An assignment delay of the type discussed above, delays execution of the whole assignment by the specified time duration. In contrast, the "intra-assignment" delay carries out the assignment in two parts. An assignment with an intra-assignment has the form

A = # dl expression;

Here the expression is scheduled to be evaluated as soon as it is encountered. However, the result of the evaluation is assigned to the right-hand side quantity a

Table 7.1 Values of variables in the module of Figure 7.38

t	0	1	2	3	4	5	6	7	8	9	10	11	12
a	0	0	1	1	2	2	3	3	4	4	5	5	6
b	0	0	0	1	1	1	1	3	3	3	3	5	5
c	0	0	0	0	2	2	2	2	4	4	4	4	6

after a delay specified by dl. dl can be an integer or a constant expression [see Section 7.7.2]. Consider the example in Figure 7.39. b is assigned the value of a with an intra-assignment delay of 2 ns. The value of a is sensed at zero ns and assigned to b after 2 ns. Until that time, b retains its old value. Again at the 2nd ns, a is sensed and b is assigned the new value of a at the 4th ns, and so on. Partial results of simulation are shown in Table 7.2. The following points are to be noted here:

- The value of a is sensed at time instants 2, 4, 6, *etc.*
- Values at other instants of time are not sensed.
- All assignments are carried out with a delay of 2 ns.
- Changes in a which do not last for 2 ns may be ignored.

```
Module del4;
Integer a, b;
Always b = #2 a;
Initial
begin
          a = 0;     b = 0;   #2   a =1;  #2   a =2;  #2 a =3;
   #2     a =4;  #2 a = 5;  #2    a =6;  #2   a =7;  #2 a =8;
end
initial $monitor($time, "  a = %d,  b = %d", a, b);
initial #20 $finish;
endmodule
```

Figure 7.39 A module to illustrate delayed assignment.

Delays tied to different segments of an assignment have different effects. The subtle differences are brought out through two more examples crafted specifically for the purpose. Consider the module in Figure 7.40. The integer a is assigned the value 0 at 0th ns and the value 1 at 1 ns. Subsequently, it is incremented every 2 ns until the end of simulation. Values are assigned to b, c, and d – declared as integers. These assignments are done with specific delays. The results of the simulation are given in Table 7.3. Changes to b, c, and d and the reasons for the same in each case are explained in the remarks columns of the table. A few observations are in order here:

Table 7.2 Partial output with the simulation of the module in Figure 7.39

t	a	b	Remarks
0	0	0	There are two assignment statements to a at 2 ns intervals – namely
2	1	x	the one in the always block and the other one in the initial block;
4	2	1	both are concurrent. The simulator decides the precedence. The
6	3	2	output here shows that the assignment in the always block has the
8	4	2	precedence.

```
Module del_dem4;
Integer a,b,c,d,n;
Always
begin
        #2      b = a;
                c = #1 a;
                d = a;
end
initial
begin
        a = 0; b = 0; c = 0; d = 0;
        #1 a = 1;  #2 a = 2;  #2 a = 3;  #2 a = 4;
        #2 a = 5;  #2 a = 6;  #2 a = 7;  #2 a = 8; #2 a = 9; #2 a = 10;
end
initial $monitor ($time, "  a = %d,  b = %d,  c = %d,  d = %d", a, b, c, d);
endmodule
```

Figure 7.40 A module to illustrate combinations of delays.

- The always block extends for three time steps. Thereafter it is repeated cyclically.
- The assignment statements in the always block are sequential assignment statements.
- Precedence of assignments slotted for a specific time instant, when they are in one block, is clear. However, when they are in different blocks, the compiler decides the precedence. But this does not cause any discrepancy in the present case.

Table 7.3 Output obtained with the simulation of the module in Figure 7.40 (shown rearranged)

Time	a	b	c	d	Remarks
0	0	0	0	0	...
1	1	0	0	0	...
2	1	1	0	0	The value of a at 2nd ns is assigned to b; the same is stored for assignment to c, 1 ns later
3	2	1	1	2	c is assigned the value of a 1 ns earlier; the present value of a is assigned to d.
5	3	3	1	2	All assignments within the always block are done; the assignment sequential is repeated; no change at the 4th ns; at the 5th ns b is assigned the value of a and so on.
6	3	3	3	3	...
7	4	3	3	3	...

Consider the module of Figure 7.41, which is a slight variant of the above in Figure 7.40. The assignments to b and c in the module of Figure 7.40 have been interchanged to form the module here. The simulated results are shown in Table 7.4. The following additional observations are in order here:

- The always block is repeated after every 3 ns – the total assignment time for the sequential.
- At t = 0, a is sampled and the sampled value is stored for assignment to c at t = 1; the sampling precedes the assignment a = 0 at zero time. Hence the value of c at zero time is not decided.
- The increment to a and (the samples of a for subsequent assignment to c) at 0th, 3rd, 6th *etc.*, ns values are concurrent. The compiler decides their precedence. With the specific compiler used, the value of a is sampled and only then a is incremented. Hence the assignment to c at the 4th ns is the value of a sampled at the 3rd ns before its increment – that is, 1. Similar is the case with the subsequent assignment changes to c.
- At the 3rd, 6th, 9th, *etc.*, ns values, a is sampled and assigned to b as well as d. Hence changes in b and d are identical. Contrast this with the previous example where the assignment sequence

c = #1 a;
d = a;

results in different sampling instances and assignments to c and d.

```
module del_dem5;
integer a,b,c,d;
always
begin
                c = #1 a;
        #2      b = a;
                d = a;
end
initial
begin
        a = 0; b = 0; c = 0; d = 0;
        #1 a = 1; #2 a = 2; #2 a = 3; #2 a = 4; #2 a = 5; #2 a = 6; #2 a = 7; #2 a = 8;
        #2 a = 9; #2 a = 10;
end
initial $monitor ($time, "  a = %d,  b = %d,  c = %d,  d = %d", a, b, c, d);
endmodule
```

Figure 7.41 Another module to illustrate combinations of delays.

Table 7.4 Simulated results of the module of Figure 7.41

t	0	1	3	4	5	6	7	9	10	11
a	0	1	2	2	3	3	4	5	5	6
b	0	0	1	1	1	3	3	5	5	5
c	0	X	X	1	1	1	3	3	5	5
d	0	0	1	1	1	3	3	5	5	5

t	12	13	15	16	17	18	19	21	22
a	6	7	8	8	9	9	10	10	10
b	6	6	8	8	8	9	9	10	10
c	5	6	6	8	8	8	9	9	10
d	6	6	8	8	8	9	9	10	10

7.7.2   Delay Assignments

In all the illustrations above, delay was specified as a number. It may be a variable or a constant expression. In case it is an expression, it is evaluated and execution delayed by the number of time steps. If the number evaluates to a negative quantity, the same is interpreted as a 2's complement value. In the statement

always #b   a = a + 1;

a and b are variables. The execution incrementing a is scheduled at b ns. If b changes, the execution time also changes accordingly. As another example consider the procedural assignment

always #(b + c) a = a + 1;

Here a, b, and c are variables. The algebraic addition of variables b and c is to be done. The scheduler schedules the incrementing of a and reassigning the incremented values back to a with a time delay of (b + c) ns. As an additional example consider the assignment below with an intra-assignment delay.

always #(a + b) a = #(b + c) a +1;

Here the simulator evaluates (a + b) during simulation. After a lapse of (a + b) ns, execution of the statement is taken up; (a + 1) is evaluated and assigned as the new value of a – but the assignment is delayed by (b + c) ns.

7.7.3   Zero Delay

A delay of 0 ns does not really cause any delay. However, it ensures that the assignment following is executed last in the concerned time slot. Often it is used to avoid indecision in the precedence of execution of assignments.

7.8   wait CONSTRUCT

The **wait** construct makes the simulator wait for the specified expression to be true before proceeding with the following assignment or group of assignments. Its syntax has the form

wait (alpha) assignment1;

alpha can be a variable, the value on a net, or an expression involving them. If alpha is an expression, it is evaluated; if true, assignment1 is carried out. One can also have a group of assignments within a block in place of assignment1. The activity is level-sensitive in nature, in contrast to the edge-sensitive nature of event specified through @. Specifically the procedural assignment

@clk a = b;

assigns the value of b to a when clk changes; if the value of b changes when clk is steady, the value of a remains unaltered. In contrast, with

wait(clk) #2 a = b;

the simulator waits for the clock to be high and then assigns b to a with a delay of 2 ns. The assignment will be refreshed as long as the clk remains high. The use of wait construct is brought out here through two examples.

Example 7.6

Figure 7.42 shows one version of the up-down counter module along with a test bench. It is a modification of the up down counter of Figure 7.10 and uses a **wait** construct. It has an enable input En. The counter is active and counts only when En = 1, that is, from the 5th ns to the 25th ns. The simulation results reproduced in Figure 7.43 confirm this.

```
module ctr_wt(a,clk,N,En);
input clk,En;
input[3:0]N;
output[3:0]a;
reg[3:0]a;
initial a=4'b1111;
always
begin
        wait(En)
        @(negedge clk)
        a=(a==N)?4'b0000:a+1'b1;
end
endmodule
```

continued

continued

```
//TEST_BENCH
module tst_ctr_wt;
reg clk,En;
reg[3:0]N;
wire[3:0]a;
ctr_wt c1(a,clk,N,En);
initial
begin
clk=0;N=4'b1111;En=1'b0;#5 En=1'b1;#20 En=1'b0;
end
always
#2 clk=~clk;
initial #35 $stop;
initial $monitor($time,"clk=%h,En=%b,N=%b,a=%b",clk,En,N,a,);
endmodule
```

Figure 7.42 A counter module to illustrate the use of wait construct. The test bench is also shown in the figure.

```
//output
    #              0clk=0,En=0,N=1111,a=1111
    #              2clk=1,En=0,N=1111,a=1111
    #              4clk=0,En=0,N=1111,a=1111
    #              5clk=0,En=1,N=1111,a=1111
    #              6clk=1,En=1,N=1111,a=1111
    #              8clk=0,En=1,N=1111,a=0000
    #             10clk=1,En=1,N=1111,a=0000
    #             12clk=0,En=1,N=1111,a=0001
    #             14clk=1,En=1,N=1111,a=0001
    #             16clk=0,En=1,N=1111,a=0010
    #             18clk=1,En=1,N=1111,a=0010
    #             20clk=0,En=1,N=1111,a=0011
    #             22clk=1,En=1,N=1111,a=0011
    #             24clk=0,En=1,N=1111,a=0100
    #             25clk=0,En=0,N=1111,a=0100
    #             26clk=1,En=0,N=1111,a=0100
    #             28clk=0,En=0,N=1111,a=0101
    #             30clk=1,En=0,N=1111,a=0101
    #             32clk=0,En=0,N=1111,a=0101
    #             34clk=1,En=0,N=1111,a=0101
```

Figure 7.43 Simulation results of the module in Figure 7.42.p

Example 7.7

Figure 7.44 shows a rudimentary and crude version of a serial receiver module and its test bench. Simulation results are shown in Figure 7.45. The module receives serial data on the di line. The data are synchronized to the clock clk. The sequence of operations carried out by the module is as follows:

- Wait for recv input to go high.
- Once recv=1, latch the next 4 successive bits of incoming data into respective bit positions of the do register.

```
//Example for 'wait'
module sr_rec(do, ack, clk, di, recv);
output [3:0] do; output ack;
input clk, recv, di;
reg [3:0] do; reg ack;
initial ack = 1'b0;
always   begin
                wait(recv)
                @(negedge clk) do[0]=di;
                @(negedge clk) do[1]=di;
                @(negedge clk) do[2]=di;
                @(negedge clk) do[3]=di;
                @(negedge clk) ack = 1'b1;
        end
endmodule

module tst_sr_rec;
reg clk, di, recv;
wire [3:0]do; wire ack;
initial    begin
        clk=1'b0; recv=1'b0; di=1'b0;  #5 recv=1'b1;
        end
always #2clk = ~clk;
initial    begin
        #7di=1'b1; #4di=1'b0; #8di=1'b1; #8di=1'b0;
        end
initial $monitor($time, "clk=%d, recv=%b, di=%b, do=%b, ack=%b",
        clk, recv, di, do, ack);
sr_rec rrcc(do, ack, clk, di, recv);

initial #25 $stop;
endmodule
```

Figure 7.44 A rudimentary serial transmitter module.

```
//output
    #              0clk=0, recv=0, di=0, do=xxxx, ack=0
    #              2clk=1, recv=0, di=0, do=xxxx, ack=0
    #              4clk=0, recv=0, di=0, do=xxxx, ack=0
    #              5clk=0, recv=1, di=0, do=xxxx, ack=0
    #              6clk=1, recv=1, di=0, do=xxxx, ack=0
    #              7clk=1, recv=1, di=1, do=xxxx, ack=0
    #              8clk=0, recv=1, di=1, do=xxx1, ack=0
    #             10clk=1, recv=1, di=1, do=xxx1, ack=0
    #             11clk=1, recv=1, di=0, do=xxx1, ack=0
    #             12clk=0, recv=1, di=0, do=xx01, ack=0
    #             14clk=1, recv=1, di=0, do=xx01, ack=0
    #             16clk=0, recv=1, di=0, do=x001, ack=0
    #             18clk=1, recv=1, di=0, do=x001, ack=0
    #             19clk=1, recv=1, di=1, do=x001, ack=0
    #             20clk=0, recv=1, di=1, do=1001, ack=0
    #             22clk=1, recv=1, di=1, do=1001, ack=0
    #             24clk=0, recv=1, di=1, do=1001, ack=1
```

Figure 7.45 Simulation results of the module in Figure 7.44.

- Once the above nibble receipt is accomplished, set acknowledgment flag high.
- If recv continues to remain high, the subsequent serial bits will be loaded into the do nibble, again and again in groups of 4 bits.
- If at any time recv goes low, the receipt and the serial to parallel conversion will come to a stop.

7.9   MULTIPLE ALWAYS BLOCKS

All the activities within an always block are scheduled for sequential execution. The activities can be of a combinational nature, a clocked sequential nature, or a combination of these two. (A design description involving such combinations is conventionally called the 'Register Transfer Level' description.) Basically, any circuit block whose end-to-end operation can be described as a continuous sequence can be described within an **always** block. A typical circuit block conforming to the above description is shown in Figure 7.46. It has three activities termed A1, A2, and A3. These three are to be done in that order. Activity A1 accepts x as input, and it generates output B and p. p and y form inputs to activity A2. Similarly activity A2 generates outputs c and q after activity A1 is completed.

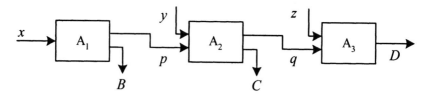

Figure 7.46 A module where execution proceeds through three blocks sequentially.

q and z form outputs of A2.   After activity A2 is completed, activity A3 is scheduled.   It accepts z and q as inputs and generates D as output.   Here if A1, A2, and A3 are logical activities, the whole block can be synthesized as a combinational logic unit.   If one or more of these are clocked events, execution may be sequential.   The design examples considered so far are broadly of this category.

In a comparatively bigger IC, the activity flow can be more complex.   One with an additional level of complexity is shown in Figure 7.47.   The activities are marked A1-A2-A3 and B1-B2-B3, These are the two streams in the circuit.   It is possible that the intermediate results of one may affect the flow of the other. Functioning of two timers – dependent on each other –is a typical example.   A processor servicing serial reception and serial transmission simultaneously is another example.   In all these cases, each sequential activity is described in a separate always block.

A design of the type in Figure 7.47 can be described with two always blocks. In some others, three or more always blocks may be called for.   Examples of such designs are considered later.

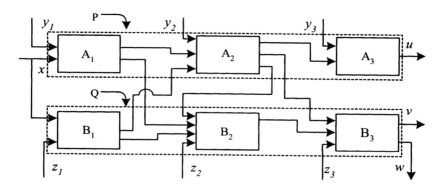

Figure 7.47 A module where execution proceeds concurrently through two groups of blocks.

Activities within one always block are normally sequential – as with the examples considered so far. If necessary, they can be made selectively concurrent. (see Section 7.11). But when designs are spread out in two or more always blocks (with design structures as in Figure 7.47), they are necessarily concurrent. Thus the blocks P and Q in Figure 7.47 are concurrent while the "sub-blocks" within each (namely A1, A2, and A3 within block P and B1, B2, and B3 within block Q) are sequential. In short, with behavioral level descriptions, one can organize the activities to be in concurrent form, in sequential form, or in combinations. In contrast, all design descriptions involving constructs at gate and data flow levels are necessarily concurrent.

7.10   DESIGNS AT BEHAVIORAL LEVEL

All simple algebraic as well as logical expressions can be described at the behavioral level. One can also mix them with blocks at the gate level as well as the data flow level to form composite as well as more involved modules. The simple A-O-I gate is taken as an example below to bring out the possibilities.

Example 7.8

Figure 7.48 shows a module of an AOI gate and its test bench; Figure 7.49 shows the simulation results, and the synthesized circuit is shown in Figure 7.50. The A-O-I gate module has two vector inputs – a and b – both being two bits wide. The bits of the two vectors are ANDed; the ANDed bits are subsequently used as the inputs to the following NOR gate to form the output. Note the following:

- All the input bits are to figure in the sensitivity list specified to trigger execution. If any one is left out, a change in that will not be reflected in the output immediately.
- The block becomes active, if any bit in the sensitivity list changes value.
- The assignments specified are executed out sequentially – but all at the same time step. Some elaboration is in order here. All the four assignments within the aoibeh module of Figure 7.48 are sequentially executed but at the same time step. The values of a and b displayed at the end of the respective time steps in Figure 7.49 confirm this. Concurrency of the assignments here also leads to a combinational circuit in synthesis.
- All quantities that appear to the left of the assignment statements have to be of the variable type; they have been declared as **reg** here.

```
module aoibeh(o,a,b);
output o;
input[1:0]a,b;
reg o,al,bl,ol;
always@(a[1] or a[0]or b[1]or b[0])
begin
      al=&a;
      bl=&b;
      ol=al||bl;
      o=~ol;
end
endmodule

module tst_aoibeh;
reg [1:0]a,b; /*  specicific values will be assigned to
al,a2,bl, and b2 and these connected
to input ports of the gate insatntiations;
hence these variables are declared as reg */
wire o;
initial
begin
    a[0]=1'b0;a[1] =1'b0;b[0]=1'b0;b[1] =1'b0;
    #3 a[0] =1'b1;
    #3 a[1] =1'b1;
    #3 b[0] =1'b1;
    #3 b[1] =1'b0;
    #3 a[0] =1'b1;
    #3 a[1] =1'b0;
    #3 b[0] =1'b0;
end
initial #100 $stop;//the simulation ends after running
for 100 tu's.
initial $monitor($time, "o =%b,a[0]=%b,a[1]=%b, b[0] =
%b ,b[1] = %b ",o,a[0],a[1],b[0],b[1]);
aoibeh gg(o,a,b);
endmodule
```

Figure 7.48 An A-O-l gate module at the behavioral level and its test bench.

```
# 0   o = 1,a[0]=0,a[1]=0,b[0]=0,b[1]=0
# 3   o = 1,a[0]=1,a[1]=0,b[0]=0,b[1]=0
# 6   o = 0,a[0]=1,a[1]=1,b[0]=0,b[1]=0
# 9   o = 0,a[0]=1,a[1]=1,b[0]=1,b[1]=0
#18   o = 1,a[0]=1,a[1]=0,b[0]=1,b[1]=0
#21   o = 1,a[0]=1,a[1]=0,b[0]=0,b[1]=0
```

Figure 7.49 Simulation results of the module in Figure 7.48.

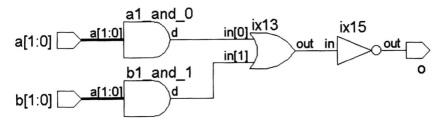

Figure 7.50 Synthesized circuit of the A-O-I module in Figure 7.48.

Example 7.9

Figure 7.51 shows an alternate but more compact description of the A-O-I gate again at the behavioral level.   Since the full assignment is realized in one line, no **begin-end** type construct is called for. Simulation results are identical to those of Figure 7.49 and are not repeated.

Example 7.10

The AOI gate in Figure 7.51 has again been described as a module in Figure 7.52. Here the AND functions are realized as and-gate primitives. The NOR function alone is realized in behavioral mode. The sensitivity list includes the two outputs of the AND gates. The gate primitives describe a set of two continuous AND functions. In contrast, the NOR function is activated only when **a**l or **b**l changes. Though conceptually different, the latter also results in outputs identical to the continuous assignments.   The test bench in Figure 7.51 can be used here by changing the instantiation statements suitably.

```
module aoibeh1(o,a,b);
output o;
input[1:0]a,b;
reg o;
always@(a[1]ora[0]or b[1]orb[0]) o=~((&a)||(&b));
endmodule
```

Figure 7.51 Another realization of the AOI gate at the behavioral level.

```
module aoibeh2(o,a,b);
output o;
input[1:0]a,b;
wire a1,b1;
reg o;
and g1(a1,a[1],a[0]),g2(b1,b[1],b[0]);
always@(a1 or b1)
o=~(a1||b1);
endmodule
```

Figure 7.52 AOI gate realization by the combined use of primitive instantiations and procedural assignments.

Example 7.11

Figure 7.53 shows another realization of the AOI gate. Here the AND functions are realized as continuous assignments. The NOR function is realized as an always block.

```
module aoibeh3(o,a,b);
output o;
input[1:0]a,b;
wire a1,b1;
reg o;
assign a1=&a,b1=&b;
always@(a1 or b1)o=~(a1||b1);
endmodule
```

Figure 7.53 The AOI gate realized by combining continuous assignments and procedural assignments.

Figure 7.54 shows another realization of the AOI gate where a gate primitive, a continuous assignment (at data flow level), and an always block are present.

The examples above bring out the variety of possibilities in design description. Designers' expertise as well as constraints and facilities in the simulation and synthesis tools often limit the choice. More often the same design may have to be described differently as one proceeds from a system level design and simulation to circuit synthesis [Navabi, Palnitkar].

```
module aoibeh4(o,a,b);
output o;
input[1:0]a,b;
wire al,bl;
reg o;
assign al=&a;
and g2(bl,b[1],b[0]);
always@(al or bl)
o=~(al||bl);
endmodule
```

Figure 7.54 The AOI gate realized by combining primitive instantiation, continuous assignment, and procedural assignment.

7.11   BLOCKING AND NONBLOCKING ASSIGNMENTS

All assignment within an initial or an always block considered so far are done through an equality ("=") operator. These are executed sequentially – that is, one statement is executed, and only then the following one is executed.   Such assignments block the execution of the following lot of assignments at any time step.   Hence they are called "blocking assignments".   Further, when such a blocking assignment has time delays associated with it, the delay is applicable to the following assignment or activity also.   Different examples of groups of blocking assignments have been considered in the preceding sections.

One comes across situations where assignments are to be effected concurrently (as with the continuous assignments considered in the preceding chapter).   A facility called the "nonblocking assignment" is available for such situations.   The symbol "<=" signifies a non-blocking assignment.   The same symbol signifies the "less than or equal to" operator in the context of an operation. The context decides the role of the symbol.   The main characteristic of a non-blocking assignment is that its execution is concurrent with that of the following assignment or activity.   A discussion of the features of nonblocking assignments and their comparison with blocking assignments are in order here.

Consider the set of nonblocking assignments in Figure 7.55.   All three assignments are executed concurrently – that is, A, B, and C are assigned the values 00 01 and 11concurrently and not sequentially. Figure 7.56 shows the same non-blocking assignments with time delays. All three assignments are taken up for execution concurrently. If the block is entered at time step t1,

- A is assigned the value 00 at time step t1.
- B is assigned the value 01 with a time delay of 2 ns – that is, at time t1 + 2 ns.
- C is assigned the value 11 with a delay of 1 ns – that is, at time t1 + 1 ns (and not at time 3 ns as happens with blocking assignments).

```
A <= 2'b00;
B <= 2'b01;
C <= 2'b11;
```

Figure 7.55 A group of nonblocking assignments.

```
A <= 2'b00;
#2 B <= 2'b01;
#1 C <= 2'b11;
```

Figure 7.56 A group of nonblocking assignments with time delays.

Nonblocking assignments are essentially two-step affairs. For all the non-blocking assignments in a block, the right-hand sides are evaluated first. Subsequently the specified assignments are scheduled. Consider the block of assignments in Figure 7.57. First A is assigned the binary value 00, and then B is assigned the value 01. These two assignments are sequential. The subsequent two assignments are concurrent. The assignment

A <= b

"reads" the value of B, stores it separately, and then assigns it to A. The new value of a is 01. The assignment

 B <= A ;

takes the value of A– i.e., 00 – stores it separately and assigns it to B. Thus the new value of B is 00. After the block is executed, A has the value 01 while B has the value 00. Contrast this with the set of blocking assignments in Figure 7.58. All four assignments here are sequential in nature. The third one, namely

A = B;

assigns the value 01 to a; subsequently the fourth and following assignment

B = A ;

assigns the present value of A (*i.e.*, 01) to b; The value of b remains at 01 itself. Consider the block of Figure 7.59. It has three nonblocking assignments. The sequence of execution of the three assignments is as follows:

1. At the positive edge of the clock, values of A, B, and C are read and stored and B &(~c) are computed.

```
A = 2'b00;
B = 2'b01;
A <= B;
B <= A;
```

Figure 7.57 Swapping variable values through nonblocking assignments.

```
A = 2'b00;
B = 2'b01;
A = B;
B = A;
```

Figure 7.58 Another group of blocking assignments.

```
initial
begin
    A= 1'b0;
    B= 1'b1;
    C = 1'b0;
end
always  @(posedge clk)
        begin
        A <= B;
        @(negedge clk) C <= B &(~c);
        #2 B< = C;
        end
```

Figure 7.59 Segment of a module involving blocking and nonblocking assignments.

2. A is assigned the stored value of B (=1); this and the activity in (1) above are carried out concurrently in the same time step.
3. At the next negative clk edge, C is assigned the value of B & (~C) evaluated and stored earlier (=1) – mentioned in (1) above.
4. Two nanoseconds after the positive edge of clk (*i.e.*, after the entry to the block), B is assigned the value of C stored earlier (=0).

In the segment in Figure 7.60, two always blocks do assignments concurrently; both of these are of the blocking variety. The values assigned to A and B are decided by the structure of the simulator. The block has the potential to create a race condition. In contrast, in the segment of Figure 7.61, the two assignments are of the nonblocking type; A is assigned the previous value of B, while B is assigned the previous value of A. The race condition is avoided here.

Observations :

- In a design whenever a number of concurrent data transfers take place after a common event, nonblocking assignments are preferred. The common event forms the sensitivity list followed by the nonblocking assignments.

```
always @(posedge clk)
A = B;
always @(posedge clk)
B = A;
```

```
always @(posedge clk)
A <= B;
always @(posedge clk)
B <= A;
```

Figure 7.60 A set of assignments with a potential race condition.

Figure 7.61 The assignments of Figure 7.60 modified to avoid race condition.

- All nonblocking assignments in a block are executed concurrently. However, the scheduling is done in the same order as the specified statements. If two assignments are done to a **reg** in a time step, the latter prevails. For example with the following sequence of statements in a block,

   A <= 1;
   A <= 0;

   A is assigned the value of zero.

- Although blocking and nonblocking assignment can be mixed in a block, many synthesis tools may not support such combinations.

7.11.1   Nonblocking Assignments and Delays

Delays – of the assignment type and the intra-assignment type – can be associated with nonblocking assignments also. The principle of their operation is similar to that with blocking assignments. As explained earlier, the delay values can be constant expressions. Blocking and nonblocking assignments, together with assignment and intra-assignment delays, open up a variety of possibilities. They can be used individually and in combinations to suit different situations. The subtle differences in their use are brought out here through a series of simple illustrations. Some further clarifications regarding assignments and time delays are in order here.

Example 7.12

Consider the module of Figure 7.62, which has a delay of 3 ns for the blocking assignment to c1. If a or b changes, the always block is activated. Three ns later, (a&b) is evaluated and assigned to c1. The event "(a or b)" will be checked for change or trigger again. If a or b changes, all the activities are frozen for 3 ns. If a or b changes in the interim period, the block is not activated. Hence the module does not depict the desired output.

```
module nil1 (c1, a, b);
output c1;
input a, b;
reg c1;
always @(a or b)
       #3      c1 = a&b;
endmodule
```

```
module nil2 (c2, a, b);
output c2;
input a, b;
reg c2;
always  @(a or b)
                c2 = #3  a&b;
endmodule
```

Figure 7.62 A time delay in an evaluation.   **Figure 7.63** An intra-assignment delay.

```
module nil3 (c3, a, b);
output c3;
input a, b;
reg c3;
always @(a or b)
        #3        c3 <= a&b;
endmodule
```

Figure 7.64 A time delay in a non-blocking assignment.

```
module nil4 (c4, a, b);
output c4;
input a, b;
reg c4;
always @(a or b)
        c4 <= #3 a&b;
endmodule
```

Figure 7.65 An intra-assignment delay in a nonblocking assignment.

Consider the module of Figure 7.63 with an intra-assignment delay of 3 ns to the assignment to c2. The always block is activated if a or b changes. (a & b) is evaluated immediately but assigned to c2 only after 3 ns. However, the behavior is not acceptable on two counts:

- The output assignment has to wait for 3 ns after the change.
- Only after the delayed assignment to c2, the event (a or b) checked for change. If a or b changes in the interim period, the block is not activated.

The module in Figure 7.64 has a blocking delay of 3 ns; but the assignment is of the nonblocking type. The block is entered if the value of a or b changes but the evaluation of a&b and the assignment to c3 take place with a time delay of 3 ns. If a or b changes in the interim period, the block is not activated. The module in Figure 7.65 possibly represents the best alternative with time delay. The always block is activated if a or b changes. (a&b) is evaluated immediately and scheduled for assignment to c4 with a delay of 3 ns. Without waiting for the assignment to take effect (*i.e.*, at the same time step as the entry to the block), control is returned to the event control operator. Further changes to a or b – if any – are again taken cognizance of. The assignment is essentially a delay operation. Figure 7.66 shows the waveforms for c1, c2, c3, and c4 in the modules of Figures 7.62 to 7.65 for representative waveforms of a and b. One can clearly see that c4 has a representation of a & b, which is the most acceptable of the lot.

7.12   THE case STATEMENT

The **case** statement is an elegant and simple construct for multiple branching in a module. The keywords **case**, **endcase**, and **default** are associated with the **case** construct. Format of the **case** construct is shown in Figure 7.67. First expression is evaluated. If the evaluated value matches ref1, statement1 is executed; and the simulator exits the block; else expression is compared with

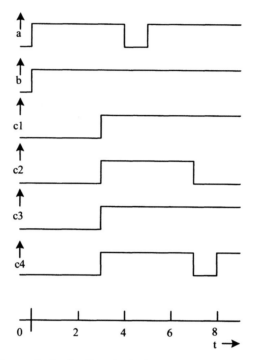

Figure 7.66 Waveforms of c1, c2, c3, and c4 of the modules in Figures 7.62 to 7.65 for representative values of a and b.

ref2 and in case of a match, statement2 is executed, and so on. If none of the ref1, ref2, *etc.*, matches the value of expression, the **default** statement is executed.

```
Case (expression)
Ref1 : statement1;
Ref2 : statement2;
Ref3 : statement3;
...
...
default: statementd;
endcase
```

Figure 7.67 Structure of the case statement.

Observations:

- A statement or a group of statements is executed if and only if there is an exact – bit by bit – match between the evaluated expression and the specified ref1, ref2, *etc.*
- For any of the matches, one can have a block of statements defined for execution. The block should appear within the **begin-end** construct.
- There can be only one **default** statement or **default** block. It can appear anywhere in the case statement.
- One can have multiple signal combination values specified for the same statement for execution. Commas separate all of them.

Example 7.13

Consider the module in Figure 7.68 for a 2-to-4 decoder. The test bench is also included in the figure. One of the 4 output bits goes high, depending on the binary value of {i1, i2}. If i1, i2, or both take x or z values, there is no match and the **default** block is executed. The simulation results are shown in Figure 7.69.

```
module dec2_4beh(o,i);
output[3:0]o;
input[1:0]i;
reg[3:0]o;
always@(i)
begin
case(i)
      2'b00:o=4'h0;
      2'b01:o=4'h1;
      2'b10:o=4'h2;
      2'b11:o=4'h4;
default:
begin
      $display ("error");
      o=4'h0;
end
endcase
end
endmodule
```

continued

continued

```
//test bench
module tst_dec2_4beh();
wire [3:0]o;
reg[1:0] i;
//reg en;
dec2_4beh dec(o,i);
initial
begin
        i  =2'b00;
     #2i =2'b01;
     #2i =2'b10;
     #2i =2'b11;
     #2i =2'b11;
     #2i =2'b0x;
end
initial $monitor ($time  ,   " output o =   %b   , input i
= %b " , o ,i);
endmodule
```

Figure 7.68 A 2-to-4 decoder using the **case** statement.

Example 7.14

Consider the module in Figure 7.70, which is a modified version of the decoder module in Figure 7.68. A test bench is also included in the figure. Here if either bit is at **x** state, all the output bits are in the **x** state. Default corresponds to one or both of the input bits being **z** or both the bits being at **x** state. In such a case an error message is also output by the simulator. The simulation results are shown in Figure 7.71.

```
output
# 0 output o =   0000   , input i   = 00
# 2 output o =   0001   , input i   = 01
# 4 output o =   0010   , input i   = 10
# 6 output o =   0100   , input i   = 11
# error
# 10 output o =   0000   , input i = 0x
```

Figure 7.69 Simulation results of the decoder module in Figure 7.69.

```
module dec2_4beh1(o,i);
output[3:0]o;
input[1:0]i;
reg[3:0]o;
always@(i)
begin
case(i)
      2'b00:o[0]=1'b1;
      2'b01:o[1]=1'b1;
      2'b10:o[2]=1'b1;
      2'b11:o[3]=1'b1;
      2'b0x,2'b1x,2'bx0,2'bx1:o=4'b0000;
default:    begin
                      $display ("error");
                      o=4'h0;
              end
endcase
end
endmodule

module tst_dec2_4beh1;//test bench
wire [3:0]o;
reg[1:0] i;
dec2_4beh1 dec(o,i);
initial
begin
       i =2'b00;
      #2i =2'b01;
      #2i =2'b10;
      #2i =2'b11;
      #2i =2'b11;
      #2i =2'b1x;
      #2i =2'b0x;
      #2i =2'bx0;
      #2i =2'bx1;
      #2i =2'bxx;
      #2i =2'b0z;
end
initial $monitor ($time  ,   " output o =   %b  , input i
= %b " , o ,i);
endmodule
```

Figure 7.70 A 2-to-4 decoder where all the outputs are forced to zero, if any of the inputs is at **x** state.

```
# 0 output o =   xxx1  , input i = 00
# 2 output o =   xx11  , input i = 01
# 4 output o =   x111  , input i = 10
# 6 output o =   1111  , input i = 11
# 10 output o = 0000  , input i = 1x
# 12 output o = 0000  , input i = 0x
# 14 output o = 0000  , input i = x0
# 16 output o = 0000  , input i = x1
# error
# 18 output o = 0000  , input i = xx
# error
# 20 output o = 0000  , input i = 0z
```

Figure 7.71 Results of the simulation run with the test bench in Figure 7.70.

Example 7.15 ALU

Figure 7.72 shows an ALU module along with a test bench. The ALU function has been realized through a block with a **case** construct. The ALU realization can be seen to be compact and elegant compared to the versions considered thus far. Additional functions can be added to the ALU by a direct expansion of the **case** block. The ALU size too can be altered to suit requirements. Results of the simulation run with the test bench in Figure 7.72 are shown in Figure 7.73 and Figure 7.74. The synthesized circuit is shown in Figure 7.75.


```
module  alubeh(c,s,a,b,f);
output[3:0]c;
output s;
input [3:0]a,b;
input[1:0]f;
reg s;
reg[3:0]c;
always@(a or b or f)
begin
      case(f)
      2'b00:      c=a+b;
      2'b01:      c=a-b;
      2'b10:      c=a&b;
      2'b11:      c=a|b;
      endcase
```

continued

continued

```
end
endmodule

module tst_alubeh;//test-bench
reg[3:0]a,b;
reg[1:0]f;
wire[3:0]c;
wire s;
alubeh aa(c,s,a,b,f);
initial
begin
f=2'b00;a=2'b00;b=2'b00;
end
always
begin
     #2 f=2'b00;a=4'b0011;b=4'b0000;
     #2 f=2'b01;a=4'b0001;b=4'b0011;
     #2 f=2'b10;a=4'b1100;b=4'b1101;
     #2 f=2'b11;a=4'b1100;b=4'b1101;
end
initial $monitor($time,"f=%b,a=%b,b=%b,c=%b",f,a,b,c);
initial #10 $stop;
endmodule
```

Figure 7.72 A simple ALU module along with its test bench.

```
0f=00,a=0000,b=0000,c=0000
#2f=00,a=0011,b=0000,c=0011
#4f=01,a=0001,b=0011,c=1110
#6f=10,a=1100,b=1101,c=1100
#8f=11,a=1100,b=1101,c=1101
#10f=00,a=0011,b=0000,c=0011
```

Figure 7.73 Results of the simulation run with the test bench in Figure 7.72.

/tst_alubeh/a	0000	0011	0001	1100	
/tst_alubeh/b	0000		0011	1101	
/tst_alubeh/f	00		01	10	11
/tst_alubeh/c	0000	0011	1110	1100	1101
/tst_alubeh/s					

| 10 ns | | | | | 4 | | | | 8 | |

Figure 7.74 Results of the simulation run with the test bench in Figure 7.73 – another view.

7.12.1   Casex and Casez

The **case** statement executes a multiway branching where every bit of the **case** expression contributes to the branching decision. The statement has two variants where some of the bits of the **case** expression can be selectively treated as don't cares – that is, ignored. **Casez** allows **z** to be treated as a don't care. "?" character also can be used in place of **z**. **casex** treats **x** or **z** as a don't care. An illustrative example using **casez** construct follows.

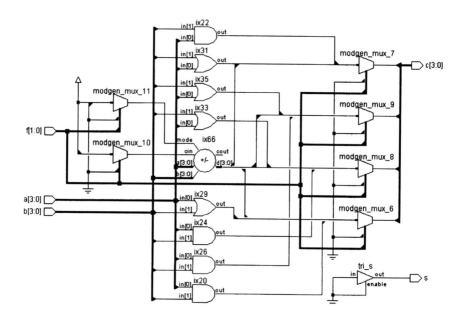

Figure 7.75 Synthesized circuit of the ALU module in Figure 7.73.

Example 7.16

A module for a priority encoder and a test bench for it are shown in Figure 7.76. The encoder gives a 2-bit output. The binary output represents the position of the first one bit in the 4-bit input combination. The simulation results are shown in Figure 7.77. The synthesized circuit is shown in Figure 7.78.

```
module pri_enc(a,b);
output[1:0]a;
input[3:0]b;
reg[1:0]a;
always@(b)
casez(b)
4'bzzz1:a=2'b00;
4'bzz10:a=2'b01;
4'bz100:a=2'b10;
4'b1000:a=2'b11;
endcase
endmodule

module pri_enc_tst;//test-bench
reg [3:0]b;
wire[1:0]a;
pri_enc pp(a,b);
initial b=4'bzzz0;
always
begin
     #2 b=4'bzzz1;
     #2 b=4'bzzz1;
     #2 b=4'bzz10;
     #2 b=4'bz100;
     #2 b=4'b1000;
end
initial $monitor($time, "input b =%b,a   =%b ",b,a);
initial #40 $stop;
endmodule
```

Figure 7.76 A design module for a 2-bit priority encoder using the **casez** statement; a test bench is also shown.

```
   0input   b   =   zzz0   ,a   =   01
   2input   b   =   zzz1   ,a   =   00
   6input   b   =   zz10   ,a   =   01
   8input   b   =   z100   ,a   =   10
  10input   b   =   1000   ,a   =   11
  12input   b   =   zzz1   ,a   =   00
  16input   b   =   zz10   ,a   =   01
  18input   b   =   z100   ,a   =   10
  20input   b   =   1000   ,a   =   11
  22input   b   =   zzz1   ,a   =   00
  26input   b   =   zz10   ,a   =   01
  28input   b   =   z100   ,a   =   10
  30input   b   =   1000   ,a   =   11
  32input   b   =   zzz1   ,a   =   00
  36input   b   =   zz10   ,a   =   01
  38input   b   =   z100   ,a   =   10
```

Figure 7.77 Results of simulating the test bench in Figure 7.76.

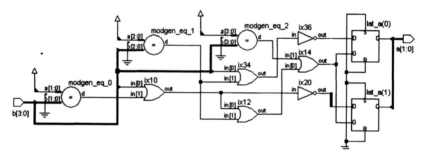

Figure 7.78 Synthesized circuit of the priority encoder in Figure 7.76.

7.13   SIMULATION FLOW

Different constructs for design description and simulation have been dealt with so far. These can be at different levels of abstraction – gate, data flow, or behavioral level. The constructs to be discussed in the following chapters add to the variety and flexibility. Such elements in different combinations make up the design and simulation modules in Verilog. Further, as an HDL, Verilog has to be an inherently parallel processing language. The fact that all the elements of a digital circuit (or any electronic circuit for that matter) function and interact continuously conforming to their interconnections demands parallel processing. In Verilog the parallel processing is structured through the following [IEEE]:

- Simulation time: Simulation is carried out in simulation time. The simulator functions with simulation time advancing in (equal) discrete steps.
- At every simulation step a number of active events are sequentially carried out.
- The simulator maintains an event queue – called the "Stratified Event Queue" – with an active segment at its top. The top most event in the active segment of the queue is taken up for execution next.
- The active event can be of an update type or evaluation type.
  The evaluation event can be for evaluation of variables, values on nets, expressions, *etc.*
  Refreshing the queue and rearranging it constitutes the update event.
- Any updating can call for a subsequent evaluation and *vice versa.*
- Only after all the active events in a time step are executed, the simulation advances to the next time step.

Completion of the sequence of operations above at any time step signifies the parallel nature of the HDL.

A number of active events can be present for execution at any simulation time step; all may vie for "attention." Amongst these, an event specified at #0 time is scheduled for execution at the end – that is, before simulation advances to the next time step. The order, in which the other events are executed, is essentially simulator-dependent.

7.13.1   Stratified Event Queue

The events being carried out at any instant give rise to other events – inherent in the execution process. All such events can be grouped into the following 5 types:

- Active events – explained above.
- Inactive events – The inactive events are the events lined up for execution immediately after the execution of the active events. Events specified with zero delay are all inactive events.
- Blocking Assignment Events – Operations and processes carried out at previous time steps with results to be updated at the current time step are of this category.
- Monitor Events – The Monitor events at the current time step – **$monitor** and **$strobe** – are to be processed after the processing of the active events, inactive events, and nonblocking assignment events.
- Future events – Events scheduled to occur at some future simulation time are the future events.

The simulation process conforming to the stratified event queue is shown in flowchart form in Figure 7.79.

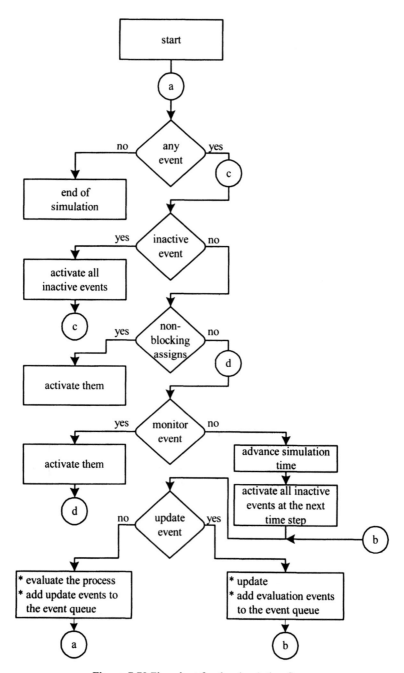

Figure 7.79 Flowchart for the simulation flow.

7.14   EXERCISES

Prepare design modules for the following operations [Sedra, Tocci, Wakerly]. In each case prepare a suitable test bench and test the design module.

1.   Add two BCD nibbles.

2.   Add two pairs of BCD nibbles – 2 decimal numbers each of two digits.

3.   Interrupt Service Routine (ISR): An ISR receives an Interrupt Request (IRQ). The PC content is saved on a stack – 4 bytes deep. Then a specific byte 5a5ah is loaded into the PC. The ISR sets an INTA flag high and returns. Use the 'wait' construct and design the module.

4.   Form an ALU for two input bytes. All the operations are to be carried out using the "case" construct. Use the algebraic and logic instructions available with 8085, 6805, 6502 z80, and the PIC series of processors as the basis [in all 5 ALUs]. Designate the two input vectors as ba and bb. Output is on ba. All the flags are to be on bb.

5.   Memory Block: Have a 1 kb size memory with a 10-bit Memory Address Register. Use clock beta for memory read and memory write. Use Wr and Rd as two separate control input lines. The operations to be realized are:

   - Wr=1: Write into the location specified by the MAR.
   - RD=1: Read from location specified by MAR.
   - Wr=0 & Rd=0: Condition to be satisfied to write into the MAR.

   Data input and data output are to be through an 8-bit-wide bus "ba."

6.   Change the always block in Example 7.6 (Figure 7.42) to the following:
```
     always
     begin
              @(negedge clk)
              a=(En)?((a==N)?4'b0000:a+1'b1):a);
     end
```
   How does the block here differ from that in the Example? Prepare a test bench, simulate, and explain.

7.   A priority encoder is used to prioritize service to interrupt requests in a microcontroller. The priority encoder in Example 7.15 can be expanded to suit the desired role here. It receives a byte (IRQ byte) and outputs a byte (vector address). The vector address has to be 32 times the serial number of the leading one bit in the IRQ byte. Prepare the necessary design module and synthesize it.

8

BEHAVIORAL MODELING II

8.1   INTRODUCTION

Comparatively simple and direct behavioral level constructs were discussed in the last chapter. They are essentially centered on the algebraic or logic operators. Different combinational and sequential circuits can be realized using them. The **case** construct and its variants enhance the possibilities of design description considerably. A few constructs are available for looping and branching. Their usage can make the design description compact and elegant. Further such constructs enhance the modeling capabilities substantially [Navabi]. These constructs mostly follow their counterparts in C language [Gottfried]. These constructs and certain other facilities that add to the flexibility of test benches are discussed here. Their use is illustrated through appropriate examples.

8.2   **if** AND **if-else** CONSTRUCTS

The **if** construct checks a specific condition and decides execution based on the result. Figure 8.1 shows the structure of a segment of a module with an **if** statement. After execution of assignment1, the condition specified is checked. If it is satisfied, assignment2 is executed; if not, it is skipped. In either case the execution continues through assignment3, assignment4, *etc*. Execution of assignment2 alone is dependent on the condition. The rest of the sequence remains. The flowchart equivalent of the execution is shown in Figure 8.2. If the

```
. . .
assignment1;
if (condition) assignment2;
assignment3;
assignment4;
. . .
```

Figure 8.1 Use of **if** construct.

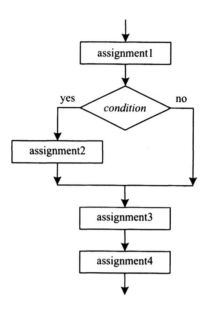

Figure 8.2 Flowchart of the **if** loop.

number of assignments associated with the **if** condition is more than 1, the whole set of them can be grouped within a **begin-end** block. Figure 8.3 shows a segment of a design using the **if** construct. It is a ring counter, which shifts one bit right at every clock pulse. The shift operation shifts the **a** byte right by one bit and fills the vacated bit – a[7] – with a zero. It is set to 1 if the bit shifted out last – a[0] – was a 1. The same is carried out through the **if** statement. The **if-else** construct is more common and turns out to be more useful than the **if** construct taken alone. Figure 8.4 shows the use in a typical design description. Figure 8.5 shows the same in flowchart form. The design description has two branches; the alternative taken is decided by the *condition*:

```
Reg[7:0] a;
Reg c;
always@(posedge clk)
begin
        c = a[0];
        a = a>>1'b1; // Since the vacated bit of a is filled with a zero, it need be
        if( c ) a[7] = c;// set only if a[0] =1
end
```

Figure 8.3 A Ring counter description using the **if** construct.

```
. . .
assignment1;
if(condition)
        begin           // Alternative 1
                assignment2;
                assignment3;
        end
else
        begin           //alternative 2
                assignment4;
                assignment5;
        end
assignment6;
. . .
. . .
```

Figure 8.4 Use of the **if-else** construct.

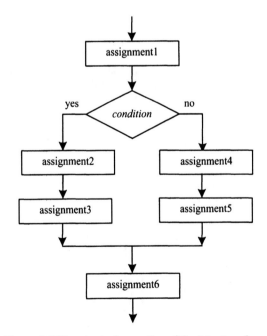

Figure 8.5 Flowchart of execution of the **if-else** loop.

- After the execution of assignment1, if the *condition* is satisfied, alternative1 is followed and assignment2 and assignment3 are executed. Assignment4 and assignment 5 are skipped and execution proceeds with assignment6.
- If the *condition* is not satisfied, assignment2 and assignment3 are skipped and assignment4 and assignment5 are executed. Then execution continues with assignment6.

Example 8.1

Figure 8.6 shows a 2 to 4 demux module. The whole demux module is realized through the **if-else-if** sequence of constructs. The selected channel is connected to the output, and all other channels are tri-stated. A test bench for the demux module is also shown in the figure. Partial results of simulation are shown in Figure 8.7; the synthesized circuit is shown in Figure 8.8.

In fact the use of **case** statement to realize mux, demux, direct encoders, and decoders makes the design description simple and direct – in contrast to the use of **if-else-if** construct. But the **if-else-if** construct is more general. It can accommodate different types of conditions at each branching. In contrast the **case** construct does a direct multiway branching.

```
module demux(a,b,s);
output [3:0]a;
input b;
input[1:0]s;
reg[3:0]a;
always@(b or s)
begin
        if(s==2'b00)
        begin
                a[2'b0]=b;
                a[3:1]=3'bZZZ;
        end
        else if(s==2'b01)
        begin
                a[2'd1]=b;
                {a[3],a[2],a[0]}=3'bZZZ;
        end
else if(s==2'b10)
        begin
                a[2'd2]=b;
                {a[3],a[1],a[0]}=3'bZZZ;
```

continued

continued

```
        end
else
        begin
                a[2'd3]=b;
                a[2:0]=3'bZZZ;
        end
end
endmodule

//tst_bench
module tst_demux();
reg b;
reg[1:0]s;
wire[3:0]a;
demux d1(a,b,s);
initial
b=1'b0;
always
begin
        #2 s=2'b00;b=1'b1;
        #2 s=2'b00;b=1'b0;
        #2 s=2'b01;b=1'b0;
        #2 s=2'b10;b=1'b1;
        #2 s=2'b11;b=1'b0;
end
initial
$monitor("t=%0d, s=%b,b=%b,output =%b",$time,s,b,a);
initial #30 $stop;
endmodule
```

Figure 8.6 A 2-to-4 demux module using the if-else-if construct: A testbench is also shown in the figure.

```
# t=0,   s=xx,b=0,output a=0zzz
# t=2,   s=00,b=1,output a=zzz1
# t=4,   s=00,b=0,output a=zzz0
# t=6,   s=01,b=0,output a=zz0z
# t=8,   s=10,b=1,output a=z1zz
# t=10,  s=11,b=0,output a=0zzz
```

Figure 8.7 Partial results of the simulation of the testbench in Figure 8.6.

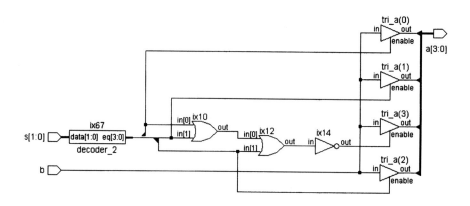

Figure 8.8 The synthesized circuit of the 2-to-4 demux module in Figure 8.6.

Example 8.2

Figure 8.9 shows the design description of a mod-*n* up-counter module along with a test-bench for it. At every clock pulse the counter advances by one bit. As soon as the count reaches the binary value n, the counter is reset to zero. The initial value of n is specified within the module itself. It is changed at a later stage. The simulation results are shown in Figure 8.10 (only partial results are shown).

Observations:

- The **$write** is a system task; it is similar to the **$display** task except in one respect: When $write is executed, the simulator does not advance to the new line after the specified display [see Chapter 11 for details].
- The value of n can be changed only from within the module. If necessary, the constraint can be removed by making n as an input to the module.
- The character set '%0d' within the $write statement ensures that the concerned quantity is displayed in decimal form with the minimum number of digits necessary for it. It makes the display elegant.
- For convenience the value of time is displayed in decimal form. Other quantities are in hex form.
- The counter can be easily modified to function as a down counter, a clock divider, or an up / down counter. It can be made more versatile with additional control inputs for Preset, Reset, and Enable.

```
//counter using if else if;
module countif(a,clk);
output[7:0]a;
input clk;
reg[7:0]a,n;
initial
begin
      n=8'h0a;
      a=8'b00000000;
  #45 n=8'h23;
end
always@(posedge clk)
begin
      $write ("time=%0d ",$time);
      if(a==n)
      a=8'h00;
      else a=a+1'b1;
end
endmodule

module tst_countif();//test-bench
reg clk;
wire[7:0]a;
countif c1(a,clk);
initial clk =1'b0;
always
#2clk=~clk;
initial
$monitor(" n=%h, a=%h",c1.n,a);
initial #200 $stop;
endmodule
```

Figure 8.9 A counter realized using the **if-else** construct.

8.3   assign–deassign CONSTRUCT

A behavior block is activated by the event at the beginning. A proper operation demands that all variables with assignments within the block are to be included in the sensitivity list. The **assign – deassign** constructs allow continuous assignments within a behavioral block. By way of illustration, consider the following simple block:

n=0a, a=00	
time=2	n=0a, a=01
time=6	n=0a, a=02
time=10	n=0a, a=03
time=14	n=0a, a=04
time=18	n=0a, a=05
time=22	n=0a, a=06
time=26	n=0a, a=07
time=30	n=0a, a=08
time=34	n=0a, a=09
time=38	n=0a, a=0a
time=42	n=0a, a=00
n=23, a=00	
time=46	n=23, a=01

time=50	n=23, a=02
time=54	n=23, a=03
time=58	n=23, a=04
time=62	n=23, a=05
time=66	n=23, a=06
time=70	n=23, a=07
time=74	n=23, a=08
time=78	n=23, a=09
time=82	n=23, a=0a
time=86	n=23, a=0b
time=90	n=23, a=0c
time=94	n=23, a=0d
time=98	n=23, a=0e

Figure 8.10 Partial results of running the test bench in Figure 8.9.

```
always@(posedge clk) a = b;
```

By way of execution, at the positive edge of clk the value of b is assigned to variable a, and a remains frozen at that value until the next positive edge of clk. Changes in b in the interval are ignored.

As an alternative, consider the block

```
always@(posedge clk) assign c = d;
```

Here at the positive edge of clk, c is assigned the value of d in a continuous manner; subsequent changes in d are directly reflected as changes in variable c: The assignment here is akin to a direct (one way) electrical connection to c from d established at the positive edge of clk.

Again consider an enhanced version of the above block as

```
Always
Begin
        @(posedge clk) assign c = d;
        @(negedge clk) deassign c;
end
```

The above block signifies two activities:

1. At the positive edge of clk, c is assigned the value of d in a continuous manner (as mentioned above).
2. At the following negative edge of clk, the continuous assignment to c is removed; subsequent changes to d are not passed on to c; it is as though c is electrically disconnected from d.

The above sequence of twin activities is repeated cyclically.

In short, assign allows a variable or a net change in the sensitivity list to mandate a subsequent continuous assignment within. **deassign** terminates the assignment done through the **assign-**based statement. The assignment to c in the above two cases is referred to as a "Procedural Continuous Assignment."

Example 8.3 A 2 to 4 Demux through Procedural Continuous Assignment

Consider the mux module in Figure 8.11. It is activated whenever s changes. But the assignment is continuous to **reg b**. It is achieved through the use of the

```
//an alternate realization of the demux using the assign construct
module demux1(a0,a1,a2,a3,b,s);
output a0,a1,a2,a3;
input b;
input [1:0]s;
reg a0,a1,a2,a3;
always@(s)
        if(s==2'b00)
        assign {a0,a1,a2,a3}={b,3'oz};
        else if(s==2'b01)
        assign {a0,a1,a2,a3}={1'bz,b,2'bz};
        else if(s==2'b10)
        assign {a0,a1,a2,a3}={2'bz,b,1'bz};
        else if(s==2'b11)
        assign {a0,a1,a2,a3}={3'oz,b};
endmodule

module tst_demux1();
reg b;
reg[1:0]s;
demux1 d2(a0,a1,a2,a3,b,s);
initial begin b=1'b0;s=2'b0; end
always
begin
#1 s=s+1'b1;
$display("t=%0d, s=%b, b=%b, {a0,a1,a2,a3} =%b",$time,s,b,{a0,a1,a2,a3});
#1b=~b;
$display("t=%0d, s=%b, b=%b, {a0,a1,a2,a3} =%b",$time,s,b,{a0,a1,a2,a3});
end
initial #14 $stop;
endmodule
```

Figure 8.11 An alternate realization of the demux using the **assign** construct.

```
# t=1, s=01, b=0,   {a0,a1,a2,a3} =0zzz
# t=2, s=01, b=1,   {a0,a1,a2,a3} =z0zz
# t=3, s=10, b=1,   {a0,a1,a2,a3} =z1zz
# t=4, s=10, b=0,   {a0,a1,a2,a3} =zz1z
# t=5, s=11, b=0,   {a0,a1,a2,a3} =zz0z
# t=6, s=11, b=1,   {a0,a1,a2,a3} =zzz0
# t=7, s=00, b=1,   {a0,a1,a2,a3} =zzz1
# t=8, s=00, b=0,   {a0,a1,a2,a3} =1zzz
# t=9, s=01, b=0,   {a0,a1,a2,a3} =0zzz
# t=10, s=01, b=1,  {a0,a1,a2,a3} =z0zz
# t=11, s=10, b=1,  {a0,a1,a2,a3} =z1zz
# t=12, s=10, b=0,  {a0,a1,a2,a3} =zz1z
# t=13, s=11, b=0,  {a0,a1,a2,a3} =zz0z
```

Figure 8.12 Results of simulating the test bench of Figure 8.11.

"**assign**" construct. Specifically, if s = 2'b01, a[1] is connected to b and remains so connected so long as s remains unchanged. If b changes value, a[1] follows it even though b is not included in the sensitivity list. A test bench is also included in Figure 8.11. Simulation results are shown in Figure 8.12.

Example 8.4 A D Flip-Flop through assign – deassign *Constructs*

Consider the module Figure 8.13, which represents a D flip-flop with Preset and Clear. If Clear or Preset becomes true, the output is forced to the Preset or Set condition, respectively. It is ensured by the first always block with the quasi-continuous assignments. If both Preset and Clear are false, the quasi-continuous assignment is removed. The second always block provides the assignment to q at every positive edge of the clk. It can take effect only if the asynchronous set–reset block is not active. Thus the asynchronous set/reset through Preset/Clear override the synchronous set/reset decided by the value of di at the clock edge. A test bench for the D flip-flop module is also included in the figure.

Observations:

- Some (many) synthesizers may not support the quasi-continuous **assign-deassign** constructs.
- The quasi-continuous assignment is made only to a variable (**reg** type); it can be a scalar or a full vector but not a part vector.
- The quasi-continuous assignment overrides all other assignments to the variable.

```
module dffassign(q,qb,di,clk,clr,pr);
output q,qb;
input di,clk,clr,pr;
reg q;
assign qb=~q;
always@(clr or pr)
begin
      if(clr)assign q = 1'b0;//asynchronous clear and
      if(pr) assign q = 1'b1;// preset of FF overrides
      else deassign q;// the synchronous behaviour
end
always@(posedge clk)
      q = di;//synchronous (clocked)value assigned to q
endmodule

//test-bench
module dffassign_tst();
reg di,clk,clr,pr;
wire q,qb;
dffassign dd(q,qb,di,clk,clr,pr);
initial
begin
      clr=1'b1;pr=1'b0;clk=1'b0;di=1'b0;
end
always
begin
      #2 clk=~clk;clr=1'b0;
end
always
# 4 di =~di;
always
#16 pr=1'b1;
always
#20 pr =1'b0;
initial  $monitor("t=%0d, clk=%b, clr=%b, pr=%b,
di=%b, q=%b ",     $time,clk,clr,pr,di,q);
initial #46 $stop;
endmodule
```

Figure 8.13 Design description of a D_flip-flop with Preset and Clear facilities: The module illustrates the use of the **assign-deassign** construct.

```
# t=0,  clk=0, clr=1, pr=0,  di=0, q=0
# t=2,  clk=1, clr=0, pr=0,  di=0, q=0
# t=4,  clk=0, clr=0, pr=0,  di=1, q=0
# t=6,  clk=1, clr=0, pr=0,  di=1, q=1
# t=8,  clk=0, clr=0, pr=0,  di=0, q=1
# t=10, clk=1, clr=0, pr=0,  di=0, q=0
# t=12, clk=0, clr=0, pr=0,  di=1, q=0
# t=14, clk=1, clr=0, pr=0,  di=1, q=1
# t=16, clk=0, clr=0, pr=1,  di=0, q=1
# t=18, clk=1, clr=0, pr=1,  di=0, q=1
# t=20, clk=0, clr=0, pr=0,  di=1, q=1
# t=22, clk=1, clr=0, pr=0,  di=1, q=1
# t=24, clk=0, clr=0, pr=0,  di=0, q=1
# t=26, clk=1, clr=0, pr=0,  di=0, q=0
# t=28, clk=0, clr=0, pr=0,  di=1, q=0
# t=30, clk=1, clr=0, pr=0,  di=1, q=1
# t=32, clk=0, clr=0, pr=1,  di=0, q=1
# t=34, clk=1, clr=0, pr=1,  di=0, q=1
# t=36, clk=0, clr=0, pr=1,  di=1, q=1
# t=38, clk=1, clr=0, pr=1,  di=1, q=1
# t=40, clk=0, clr=0, pr=0,  di=0, q=1
# t=42, clk=1, clr=0, pr=0,  di=0, q=0
# t=44, clk=0, clr=0, pr=0,  di=1, q=0
```

Figure 8.14 Simulation results of the test bench in Figure 8.13.

Example 8.5 Another D Flip-Flop with `if` and `if-else`

Figure 8.15 shows a module of a flip-flop again using the **`if-else-if`**
construct. clr, pr, and clk are all included in the sensitivity list itself. A test bench
is also included in the figure. The synthesized circuit of the module is shown in
Figure 8.16. Simulation results are in Figure 8.17.

```
module dffalter(q,qb,di,clk,clr,pr);
output q,qb;
input di,clk,clr,pr;
reg q;
assign qb =~q;//continous assignment
always@(posedge clr or posedge pr or posedge clk)
      begin
            if(clr) q=1'b0;
```

continued

continued

```
            else if(pr) q=1'b1;
            else q=di;
        end
endmodule

//test-bench
module dffalter_tst();
reg di,clk,clr,pr;
wire q;
dffalter dff(q,qb,di,clk,clr,pr);
initial
begin
        clr=1'b1;pr=1'b0;clk=1'b0;di=1'b0;
end
always
begin
        #2 clk=~clk;clr=1'b0;
end
always  # 4 di =~di;
always  #16 pr=1'b1;
always  #20 pr =1'b0;
initial  $monitor("t=%0d, clk=%b, clr=%b, pr=%b,
di=%b, q=%b ",   $time,clk,clr,pr,di,q);
initial #46 $stop;
endmodule
```

Figure 8.15 An alternate description of the D_FF module and its test bench.

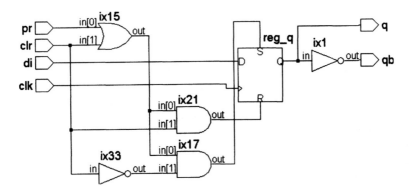

Figure 8.16 Synthesized circuit of the flip-flop of Example 8.5.

```
# 0clk = 0,  clr = 1, pr = 0, di = 0, q = 0
# 2clk = 1,  clr = 0, pr = 0, di = 0, q = 0
# 4clk = 0,  clr = 0, pr = 0, di = 1, q = 0
# 6clk = 1,  clr = 0, pr = 0, di = 1, q = 1
# 8clk = 0,  clr = 0, pr = 0, di = 0, q = 1
# 10clk = 1, clr = 0, pr = 0, di = 0, q = 0
# 12clk = 0, clr = 0, pr = 0, di = 1, q = 0
# 14clk = 1, clr = 0, pr = 0, di = 1, q = 1
# 16clk = 0, clr = 0, pr = 1, di = 0, q = 1
# 18clk = 1, clr = 0, pr = 1, di = 0, q = 1
# 20clk = 0, clr = 0, pr = 0, di = 1, q = 1
# 22clk = 1, clr = 0, pr = 0, di = 1, q = 1
# 24clk = 0, clr = 0, pr = 0, di = 0, q = 1
# 26clk = 1, clr = 0, pr = 0, di = 0, q = 0
# 28clk = 0, clr = 0, pr = 0, di = 1, q = 0
# 30clk = 1, clr = 0, pr = 0, di = 1, q = 1
# 32clk = 0, clr = 0, pr = 1, di = 0, q = 1
# 34clk = 1, clr = 0, pr = 1, di = 0, q = 1
# 36clk = 0, clr = 0, pr = 1, di = 1, q = 1
# 38clk = 1, clr = 0, pr = 1, di = 1, q = 1
# 40clk = 0, clr = 0, pr = 0, di = 0, q = 1
# 42clk = 1, clr = 0, pr = 0, di = 0, q = 0
# 44clk = 0, clr = 0, pr = 0, di = 1, q = 0
```

Figure 8.17 Simulation results for the test bench of Figure 8.15.

Examle 8.6 A Counter with a Continuous Procedural Assignment

Figure 8.18 shows the module of an up counter with Preset and Clear facilities. Preset and Clear are carried out through Procedural Continuous Assignments. If clr goes high, a is reset to zero. If pr goes high, a is set to the number specified as n. Either of these assignments will remain as long as either Clear or Preset is active as the case may be. If both these asynchronous control signals go low, the module increments the value of a at every positive edge of the clock. The module can easily be modified to function as a down counter, an up–down counter, or a counter to any other modulus.

```
module ctr_a(a,n,clr,pr,clk);
output [7:0]a;
input [7:0]n;
input clr,pr,clk;
reg[7:0]a;
initial a =8'h00;
always@ (posedge clk)
a=a+1'b1;
always@(clr or pr)
        if (clr)assign a =7'h00;
                else if(pr)assign a =n;
                        else deassign a;
endmodule


module counprclrasgn_tst();//test-bench
reg [7:0]n;
reg clr,pr,clk;
wire[7:0] a;
ctr_a cc(a,n,clr,pr,clk);
initial
begin
        n=8'h55; clr=1'b1;
        pr=1'b0;clk=1'b0;
end
always
begin
        #2 clk=~clk;clr=1'b0;
end
always #16 pr=1'b1;
always #20 pr =1'b0;
initial $monitor(   $time  ,  "clk  = %b , clr =   %b
, pr  =   %b  ,   a  =  %b ", clk,clr,pr,a);
initial #44 $stop;
endmodule
```

Figure 8.18 Design description of an up counter with Preset and Clear facilities.

```
0clk  = 0 , clr = 1 , pr =  0 ,  a = 00000000
2clk  = 1 , clr = 0 , pr =  0 ,  a = 00000001
4clk  = 0 , clr = 0 , pr =  0 ,  a = 00000001
6clk  = 1 , clr = 0 , pr =  0 ,  a = 00000010
8clk  = 0 , clr = 0 , pr =  0 ,  a = 00000010
10clk = 1 , clr = 0 , pr =  0 ,  a = 00000011
12clk = 0 , clr = 0 , pr =  0 ,  a = 00000011
14clk = 1 , clr = 0 , pr =  0 ,  a = 00000100
16clk = 0 , clr = 0 , pr =  1 ,  a = 01010101
18clk = 1 , clr = 0 , pr =  1 ,  a = 01010101
20clk = 0 , clr = 0 , pr =  0 ,  a = 01010101
22clk = 1 , clr = 0 , pr =  0 ,  a = 01010110
24clk = 0 , clr = 0 , pr =  0 ,  a = 01010110
26clk = 1 , clr = 0 , pr =  0 ,  a = 01010111
28clk = 0 , clr = 0 , pr =  0 ,  a = 01010111
30clk = 1 , clr = 0 , pr =  0 ,  a = 01011000
32clk = 0 , clr = 0 , pr =  1 ,  a = 01010101
34clk = 1 , clr = 0 , pr =  1 ,  a = 01010101
36clk = 0 , clr = 0 , pr =  1 ,  a = 01010101
38clk = 1 , clr = 0 , pr =  1 ,  a = 01010101
40clk = 0 , clr = 0 , pr =  0 ,  a = 01010101
42clk = 1 , clr = 0 , pr =  0 ,  a = 01010110
```

Figure 8.19 Simulation results of the test bench in Figure 8.18.

Example 8.7

Consider the module in Figure 8.20 which is a variant of the flip-flop in Figure 8.15. A test bench for the flip-flop is also included in the figure. Here the **always** block is activated at every positive edge of the clock. At that instant if clr = 1, the flip-flop is cleared. If pr = 1, the flip-flop is set. If clr = 0 and pr = 0, the flip-flop output takes on the value of d. Here all the assignments to q take effect at the positive edge of the clock. Hence the behavior is fully synchronous. This is not necessarily the case with the flip-flop of Figure 8.15. The synthesized circuit of the flip-flop is shown in Figure 8.21.

```
module dff_1beh(q,qb,di,clk,clr,pr);
output q,qb;
input di,clk,clr,pr;
reg q;
assign qb=~q;
always@(posedge clk)
begin
        if(clr)q = 1'b0;
```

continued

continued

```
        else if(pr) q = 1'b1;
        else    q=di;
end
endmodule

//test-bench
module dff_1beh_tst();
reg di,clk,clr,pr;
wire q,qb;
dff_1beh dd(q,qb,di,clk,clr,pr);
initial
begin
clr=1'b1;pr=1'b0;clk=1'b0;di=1'b0;
end
always
begin
        #2 clk=~clk;clr=1'b0;
end
always # 4 di =~di;
always #16 pr=1'b1;
always #20 pr =1'b0;
always #24 clr=1'b1;
always #28 clr =1'b0;
initial  $monitor( "t=%0d, clk=%b, clr=%b, pr=%b,
di=%b,   q=%b ", $time, clk,clr,pr,di,q);
initial #46 $stop;
endmodule
```

Figure 8.20 Another design description of a flip-flop and its test bench.

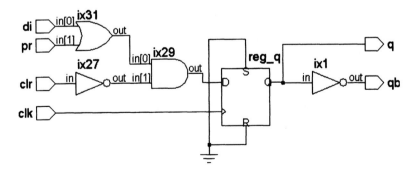

Figure 8.21 Synthesized circuit of the flip-flop in Figure 8.20.

8.4   repeat CONSTRUCT

The repeat construct is used to repeat a specified block a specified number of times. Typical format is shown in Figure 8.22. The quantity a can be a number or an expression evaluated to a number.   As soon as the repeat statement is encountered, a is evaluated. The following block is executed "a" times. If "a" evaluates to 0 or x or z, the block is not executed.

Example 8.8

The **repeat** construct is well-suited to repeat a block of assignments a fixed number of times. Figure 8.23 shows a block in a module using it. The block has a set of 16 registers each 8 bits wide. A **repeat** block is used to load a set of numbers into them.   Subsequently, the content of each register is displayed sequentially again through a **repeat** block. The simulation results are shown in Figure 8.24.

```
...
repeat (a)
        begin
                assignment1;
                assignment2;
                ...
        end
        ...
```

Figure 8.22 Structure of a **repeat** block.

```
module trial_8b;
reg[7:0] m[15:0];
integer i;
reg clk;
always
begin
        repeat(8)
        begin
                @(negedge clk)
                m[i]=i*8;
                i=i+1;
        end
        repeat(8)
        begin
                @(negedge clk)
                i=i-1;
                $display("t=%0d, i=%0d, m[i]=%0d", $time,i,m[i]);
```

continued

continued

```
        end
end
initial
begin
        clk = 1'b0;
        i=0;
        #70 $stop;
end
always #2 clk=~clk;
endmodule
```

Figure 8.23 A module to illustrate the use of the **repeat** construct.

```
# t=32, i=7, m[i]=56
# t=36, i=6, m[i]=48
# t=40, i=5, m[i]=40
# t=44, i=4, m[i]=32
# t=48, i=3, m[i]=24
# t=52, i=2, m[i]=16
# t=56, i=1, m[i]=8
# t=60, i=0, m[i]=0
```

Figure 8.24 Results of simulating the test bench in Figure 8.23.

Example 8.9

The module in Figure 8.25 outputs n successive words. The data to be output are available in n successive locations of memory. out is the output port. The output activity takes place at the positive edge of clk and is completed in n cycles of clk.

```
. . .
always
        begin
                repeat(n-1'b1)
                begin
                        @(posedge clk)
                        begin
                                out = m(mar);
                                mar = mar + 1'b1;
                        end
                end
        end
```

Figure 8.25 A block in a module to output *n* successive bytes using the repeat construct.

8.5   for LOOP

The **for** loop in Verilog is quite similar to the **for** loop in C; the format of the **for** loop is shown on Figure 8.26. It has four parts; the sequence of execution is as follows:

1. Execute assignment1.
2. Evaluate *expression*.
3. If the *expression* evaluates to the true state (1), carry out statement. Go to step 5.
4. If *expression* evaluates to the false state (0), exit the loop.
5. Execute assignment2. Go to step 2.

Operation of the loop is shown in Figure 8.27 in flowchart form. It may be compared with Figure 8.5 for the **if-else-if** construct. In general, whenever one has to accommodate alternatives for execution, the **if** and **if-else** constructs are preferred. Whenever a sequence of assignments is to be done repeatedly with an index for termination, the **for** construct is preferred.

```
. . . .
for(assignment1; expression; assignment 2)
statement;
    . . .
```

Figure 8.26 Structure of the **for** loop.

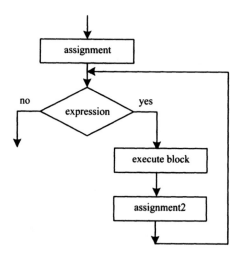

Figure 8.27 Flowchart of execution of the **for** loop.

Examle 8.10

The earlier memory-load example – Example 8.8 – has been redone here with the **for** loop. The changed module and the simulation results are shown in Figure 8.28. The simulation results can be compared with those in Figure 8.24.

```
module trial_8a;
reg[7:0] m[15:0];
integer i;
reg clk;
always
begin
        for(i=0;i<8;i=i+1)
        @(negedge clk)
        m[i]=i*8;
        for(i=0;i<8;i=i+1)
        @(negedge clk)
        $display("t=%0d, i=%0d, m[i]=%0d", $time,i,m[i]);
end
initial clk = 1'b0;
always #2 clk=~clk;
initial #70 $stop;
endmodule

//Simulation results
# t=32, i=0, m[i]=0
# t=36, i=1, m[i]=8
# t=40, i=2, m[i]=16
# t=44, i=3, m[i]=24
# t=48, i=4, m[i]=32
# t=52, i=5, m[i]=40
# t=56, i=6, m[i]=48
# t=60, i=7, m[i]=56
```

Figure 8.28 A module to illustrate the use of the **for** construct to load a memory block.

Example 8.11

Figure 8.29 shows the design description of an 8-bit adder module using the **for** loop. The module waits for En to go high; then the adder block is executed. Addition is carried out sequentially on a bit-by-bit basis starting with the 0th bit. Carry bit $c[1]$ is generated when adding the bits in the 0th position. It is the carry input to the addition in the first bit position, and so on. Since all the assignments are of the blocking type, execution is sequential; but all are carried out at the same time step. A test bench is also included in the figure. Simulation results are in Figure 8.30.

```verilog
module addfor(s,co,a,b,cin,en);
output[7:0]s;
output co;
input[7:0]a,b;
input en,cin;
reg[8:0]c;
reg co;
reg[7:0]s;
integer i;
always@( posedge en )
begin
        c[0] =cin;
        for(i=0;i<=7;i=i+1)
        begin
                {c[i+1],s[i]}=(a[i]+b[i]+c[i]);
        end
        co=c[8];
end
endmodule

//testbench
module tst_addfor();
wire [7:0]s;
wire co;
reg [7:0]a,b;
reg en,cin;
addfor add(s,co,a,b,cin,en);
always #2 en=~en;
initial
begin
        #0 en=1'b0;
        #1 cin=1'b0;a=8'h01;b=8'h00;
        #2 cin=1'b0;a=8'h01;b=8'h00;
        #2 cin=1'b0;a=8'h01;b=8'h01;
        #2 cin=1'b0;a=8'h01;b=8'h01;
        #2 cin=1'b1;a=8'h01;b=8'h02;
        #2 en=1'b1;cin=1'b1;a=8'h01;b=8'h03;
        #2 cin=1'b0;a=8'h01;b=8'h09;
        #2 cin=1'b1;a=8'h01;b=8'h09;
        #2 cin=1'b0;a=8'hff;b=8'hff;
        #2 cin=1'b1;a=8'hff;b=8'hff;
        #2 cin=1'b1;a=8'hff;b=8'hff;
end
initial $monitor( "t=%0d, en = %b, cin = %b, a = %0h, b
= %0h, s = %0h, co = %b ",$time,en,cin,a,b,s,co);
initial #30 $stop;
endmodule
```

Figure 8.29 An adder module using the **for** loop.

```
# t=0,   en = 0,  cin = x,  a = x,  b = x,  s = x,  co = x
# t=1,   en = 0,  cin = 0,  a = 1,  b = 0,  s = x,  co = x
# t=2,   en = 1,  cin = 0,  a = 1,  b = 0,  s = 1,  co = 0
# t=4,   en = 0,  cin = 0,  a = 1,  b = 0,  s = 1,  co = 0
# t=5,   en = 0,  cin = 0,  a = 1,  b = 1,  s = 1,  co = 0
# t=6,   en = 1,  cin = 0,  a = 1,  b = 1,  s = 2,  co = 0
# t=8,   en = 0,  cin = 0,  a = 1,  b = 1,  s = 2,  co = 0
# t=9,   en = 0,  cin = 1,  a = 1,  b = 2,  s = 2,  co = 0
# t=10,  en = 1,  cin = 1,  a = 1,  b = 2,  s = 4,  co = 0
# t=11,  en = 1,  cin = 1,  a = 1,  b = 3,  s = 4,  co = 0
# t=12,  en = 0,  cin = 1,  a = 1,  b = 3,  s = 4,  co = 0
# t=13,  en = 0,  cin = 0,  a = 1,  b = 9,  s = 4,  co = 0
# t=14,  en = 1,  cin = 0,  a = 1,  b = 9,  s = a,  co = 0
# t=15,  en = 1,  cin = 1,  a = 1,  b = 9,  s = a,  co = 0
# t=16,  en = 0,  cin = 1,  a = 1,  b = 9,  s = a,  co = 0
# t=17,  en = 0,  cin = 0,  a = ff, b = ff, s = a,  co = 0
# t=18,  en = 1,  cin = 0,  a = ff, b = ff, s = fe, co = 1
# t=19,  en = 1,  cin = 1,  a = ff, b = ff, s = fe, co = 1
# t=20,  en = 0,  cin = 1,  a = ff, b = ff, s = fe, co = 1
# t=22,  en = 1,  cin = 1,  a = ff, b = ff, s = ff, co = 1
# t=24,  en = 0,  cin = 1,  a = ff, b = ff, s = ff, co = 1
# t=26,  en = 1,  cin = 1,  a = ff, b = ff, s = ff, co = 1
# t=28,  en = 0,  cin = 1,  a = ff, b = ff, s = ff, co = 1
```

Figure 8.30 Results of simulating the test bench in Figure 8.29.

Example 8.12

Figure 8.31 shows an alternate realization of the adder along with a test bench. Here again the addition proceeds sequentially starting with the 0th bit. The 0th bits are added at the first positive edge of clk. The next set of bits is added at the subsequent positive edge of clk, and so on. The adder is realized as a one-bit adder doing the 8-bit addition. The synthesis tool will minimize hardware but will demand maximum time for execution as the price. The simulation results are shown in Figure 8.32.

```
module addfor1(s,co,a,b,cin,en,clk);
output[7:0]s;
output co;
input[7:0]a,b;
input en,cin,clk;
reg[8:0]c;
reg co;
reg[7:0]s;
integer i;

//assign c[0]=cin;
always@(posedge en)
begin
        for(i=0;i<=7;i=i+1)
        @(posedge clk)
        begin
                if(i==0)c[0]=cin;
                {c[i+1],s[i]}=(a[i]+b[i]+c[i]);
        end
        co=c[8];
end
endmodule

//testbench
module tst_addfor1();
wire [7:0]s;
wire co;
reg [7:0]a,b;
reg en,cin,clk;
addfor1 add1(s,co,a,b,cin,en,clk);
initial
begin
        clk=1'b0;en=1'b0;cin=1'b0;a=8'h00;b=8'h00;
end
always #2 clk =~clk;
initial
begin
        #1   en=1'b1; #34 en=1'b0;
        #1   cin=1'b0;a=8'h01;b=8'h00;
        #1   en=1'b1; #34 en=1'b0;
        #1   cin=1'b1;a=8'h05;b=8'h02;
        #1   en=1'b1; #34 en=1'b0;
        #1   cin=1'b1;a=8'h06;b=8'h03;
```

continued

continued

```
      #1   en=1'b1; #34 en=1'b0;
      #1   cin=1'b0;a=8'h07;b=8'h09;
      #1   en=1'b1; #34 en=1'b0;
      #1   cin=1'b1;a=8'h01;b=8'h09;
      #1   en=1'b1; #34 en=1'b0;
      #1   cin=1'b0;a=8'hff;b=8'hff;
      #1   en=1'b1; #34 en=1'b0;
      #1   cin=1'b1;a=8'hff;b=8'hff;
end
always@(negedge en)
$display("t=%0d, clk=%0b, en=%0b, cin=%0b, a=%0h,
b=%0h, s=%0h, co=%0b",$time,clk,en,cin,a,b,s,co);
initial #300 $stop;
endmodule
```

Figue 8.31 Another module for byte addition using the **for** construct.

```
# t=0, clk=0, en=0, cin=0, a=0, b=0, s=x, co=x
# t=35, clk=1, en=0, cin=0, a=0, b=0, s=0, co=0
# t=71, clk=1, en=0, cin=0, a=1, b=0, s=1, co=0
# t=107, clk=1, en=0, cin=1, a=5, b=2, s=8, co=0
# t=143, clk=1, en=0, cin=1, a=6, b=3, s=a, co=0
# t=179, clk=1, en=0, cin=0, a=7, b=9, s=10, co=0
# t=215, clk=1, en=0, cin=1, a=1, b=9, s=b, co=0
# t=251, clk=1, en=0, cin=0, a=ff, b=ff, s=fe, co=1
```

Figure 8.32 Simulation output with the test-bench in Figure 8.31.

Example 8.13

Figure 8.33 shows a segment of a test bench to test the adder module for all input combinations. At every time step, one out of a total of 8 bits (4 of **a**, 4 of **b**, and one **cin**) changes. The test is carried out for a total of 2^9 possibilities. The test bench uses nested **for** loops as well as the **if** construct along with the **for** loop. The test bench description can be seen to be compact.

```
. . .
initial
begin
        a = 4'h0;
        b = 4'h0;
        cin = 1'b0;
end
initial
begin
        for (k = 0; k <=1; k = k + 1'b0)
        begin
        #1      if (k)  cin = 1'b1;
                    else  cin = 1'b0;
                    for (l = 0; l <= 3'o7; l = l+1'b1)
                    begin
                            #1 a[l] = a[l] + 1'b1;
                            for (j = 0; j <= 3'o7; j = j + 1'b0)
                            begin
                                    #1  b[j] = b[j] + 1'b0;
                            end
                    end
            end
end
. . .
. . .
```

Figure 8.33 A segment of a test bench for the 8-bit adder of Example 8.6.

8.6   THE disable CONSTRUCT

There can be situations where one has to break out of a block or loop. The **disable** statement terminates a named block or task. Control is transferred to the statement immediately following the block. Conditional termination of a loop, interrupt servicing, *etc.*, are typical contexts for its use. Often the disabling is carried out from within the block itself. The **disable** construct is functionally similar to the *break* in C [Gotttfried].

Example 8.14

Figure 8.34 shows a module that uses a **disable** statement. The module realizes an OR gate in an elegant manner. The OR gate output b is assigned the value 0 initially. All bits of the input a are examined sequentially within a **for** loop. If any bit is 1, the OR gate output is set to 1 and execution is terminated (since examining the other input bits is superfluous). A master enable signal (en) is also included in the module. The simulation results are in Figure 8.35. NOR, AND, and NAND gates too can be realized in a similar manner.

```
module or_gate(b,a,en);
input [3:0]a;
input en;
output b;
reg b;
integer i;
always@(posedge en)
      begin:OR_gate
      b=1'b0;
              for(i=0;i<=3;i=i+1)
                    if(a[i]==1'b1)
                    begin
                          b=1'b1;
                    disable OR_gate;
                    end
      end
endmodule

//test-bench
module tst_or_gate();
reg[3:0]a;
reg en;
wire b;
or_gate gg(b,a,en);
initial
begin
      a   = 4'h0;
      en = 1'b0;
end
initial begin
      #2 en=1'b1; #2 a =4'h1; #2 en=1'b0;
      #2 en=1'b1; #2 a =4'h2; #2 en=1'b0;
      #2 en=1'b1; #2 a =4'h0; #2 en=1'b0;
      #2 en=1'b1; #2 a =4'h3; #2 en=1'b0;
      #2 en=1'b1; #2 a= 4'h4; #2 en=1'b0;
      #2 en=1'b1; #2 a=4'hf;
          end
initial $monitor("t=%0d, en = %b, a = %b, b =
%b",$time,en,a,b);
initial #60 $stop;
endmodule
```

Figure 8.34 An OR gate module to demonstrate the use of the **disable** construct. A test bench is also included in the figure.

```
# t=0,  en = 0, a = 0000, b = x
# t=2,  en = 1, a = 0000, b = 0
# t=4,  en = 1, a = 0001, b = 0
# t=6,  en = 0, a = 0001, b = 0
# t=8,  en = 1, a = 0001, b = 1
# t=10, en = 1, a = 0010, b = 1
# t=12, en = 0, a = 0010, b = 1
# t=14, en = 1, a = 0010, b = 1
# t=16, en = 1, a = 0000, b = 1
# t=18, en = 0, a = 0000, b = 1
# t=20, en = 1, a = 0000, b = 0
# t=22, en = 1, a = 0011, b = 0
# t=24, en = 0, a = 0011, b = 0
# t=26, en = 1, a = 0011, b = 1
# t=28, en = 1, a = 0100, b = 1
# t=30, en = 0, a = 0100, b = 1
# t=32, en = 1, a = 0100, b = 1
# t=34, en = 1, a = 1111, b = 1
```

Figure 8.35 Simulation results of the test bench in Figure 8.34.

The synthesized circuit of the module is in Figure 8.36. Since the OR activity is triggered at the edge of en, the output is made available through a latch; the latching is done at the positive edge of en as specified. The circuit does not respond to the subsequent changes in the input quantities until en is made to go through 0 and 1 once again and latches the OR gate output.

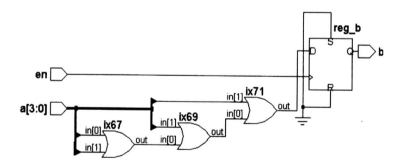

Figure 8.36 Synthesized circuit of the OR gate module in Figure 8.34.

Example 8.15

Figure 8.37 is a module to illustrate the conditional termination of a **for** loop. a is a byte of pending interrupt vectors. The b0 position represents the highest-priority interrupt, and b7 represents the lowest-priority one. The module is activated by en going high. Each of the bits of a is examined in succession. The module returns n as the serial number of the first interrupt flag that is active. If no interrupt flag is active, n takes the value 8. Simulation results are shown in Figure 8.38. Whenever en changes from 0 to 1 (positive edge) the value of a is updated – specifically at the 5th, 20th, and 35th ns as can be seen from the test bench included in the figure. The synthesized circuit is shown in Figure 8.39.

```
module int(n,a,en);
output [3:0]n;
input en;
input[7:0]a;
reg [3:0]n;
integer i;
always@(posedge en)
begin:source
        n=4'b0001;
        for(i=0;i<=7;i=i+1'b1)
                if (a[i]==1'b0)
                begin
                        n=n+1'b1;
                        if(n==4'b1001)
                        n=1'b0;
                end
                else disable source;
end
endmodule

//test-bench
module tst_int();
reg en;
reg [7:0]a;
wire [3:0]n;
int ii(n,a,en);
initial
begin
        en=1'b0;
        a=8'h00;
end
```

continued

continued

```
initial
begin
        #5 en=1'b1;  #5 a=8'h02;  #5 en=1'b0;
        #5 en=1'b1;  #5 a=8'hb0;  #5 en=1'b0;
        #5 en=1'b1;
end
initial  $monitor(  "t=%0d,  n=  %b,  a  =  %b,  en=%b
",$time,n,a,en);
initial #50 $stop;
endmodule
```

Figure 8.37 A module identify the highest-priority pending Interrupt; a test bench is also included.

```
# t=0,  n= xxxx,  a = 00000000,  en=0
# t=5,  n= 0000,  a = 00000000,  en=1
# t=10, n= 0000,  a = 00000010,  en=1
# t=15, n= 0000,  a = 00000010,  en=0
# t=20, n= 0010,  a = 00000010,  en=1
# t=25, n= 0010,  a = 10110000,  en=1
# t=30, n= 0010,  a = 10110000,  en=0
# t=35, n= 0101,  a = 10110000,  en=1
```

Figure 8.38 Simulation results of the test bench in Figure 8.37.

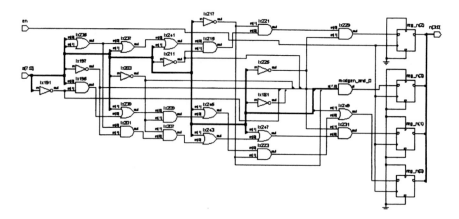

Figure 8.39 Synthesized circuit of the design module in Figure 8.37.

Observations:

- The **disable** statement has to have a block (or task) identifier tagged to it – in this respect it differs from "*break*" in C.
- Once encountered, it terminates execution of the block; the following statements within the block are not executed.
- Typically it can be used to handle exceptions to regularly assigned activities for example, Interrupt, Hold, Reset, *etc*.

8.7   while LOOP

The format for the while loop is shown in Figure 8.40. The Boolean *expression* is evaluated. If it is **true**, the statement (or block of statements) is executed and expression evaluated and checked. If the *expression* evaluates to **false**, the loop is terminated and the following statement is taken for execution. If the *expression* evaluates to **true**, execution of statement (block of statements) is repeated. Thus the loop is terminated and broken only if the *expression* evaluates to false. The flowchart for the while loop is shown in Figure 8.41.

while (*expression*) statement ;

Figure 8.40 Structure of the **while** loop.

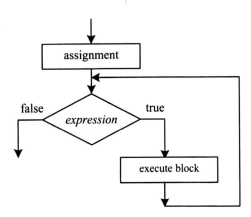

Figure 8.41 Flowchart for the execution of the **while** loop.

Observations :

- Whenever the **while** construct is used, event or time-based activity flow within the block has to be ensured.
- With the **while** construct the expression associated with the keyword **while** must become **false** through the execution of assignments inside the block. Otherwise we end up with an endless looping within the block, causing a deadlock.
- There may be situations where we have to **wait** in a loop while an **event** external to it changes to trigger an activity. The **wait** construct is to be used for such situations and not **while**. With the **wait** construct the activity is scheduled and execution continued with the other activities. With the **while** construct until the associated loop is not complete, other activities are not taken up.

Example 8.16

Figure 8.42 shows a module which illustrates the use of the **while** construct for the generation of a pulse of definite width. It accepts clk and an 8-bit number n as inputs and gives out a single-bit output – b. b is normally low. n represents the desired pulse width. It is loaded into a register a maintained within the module. As soon as en goes high, b becomes 1 and countdown of a starts within a **while** loop. As a becomes 0, the loop is terminated and b brought back to 0. The pulse width represented by the high state of b can be changed by changing the value of n. Simulation results are shown in Figure 8.43(a) in tabular form and in Figure 8.43(b) as waveforms.

```
module while2(b,n,en,clk);
input[7:0]n;
input clk,en;
output b;
reg[7:0]a;
reg b;
always@(posedge en)
begin
      a=n;
      while(|a)
      begin
            b=1'b1;
            @(posedge clk)
            a=a-1'b1;
```

continued

continued

```
      end
      b=1'b0;
end
initial b=1'b0;
endmodule

module tst_while2();
reg[7:0]n;
reg en,clk;
wire b;
while2 ww(b,n,en,clk);
initial
begin
      n  = 8'h10;clk = 1'b1;en = 1'b0;
  #3  en = 1'b1;
  #60 en = 1'b0;
end
initial $monitor( " t=  %0d, output b = %b ,ww.a = %0d
,en = %b ,clk = %b ",$time,b,ww.a,en,clk);
always
#2 clk =~clk;
initial #80 $stop;
endmodule
```

Figure 8.42 A module to illustrate the use of while construct: It generates a pulse of definite width.

```
t= 0, output b = 0 ,ww.a = x ,en = 0 ,clk = 1
t= 2, output b = 0 ,ww.a = x ,en = 0 ,clk = 0
t= 3, output b = 1 ,ww.a = 16 ,en = 1 ,clk = 0
t= 4, output b = 1 ,ww.a = 15 ,en = 1 ,clk = 1
t= 6, output b = 1 ,ww.a = 15 ,en = 1 ,clk = 0
t= 8, output b = 1 ,ww.a = 14 ,en = 1 ,clk = 1
t= 10, output b = 1 ,ww.a = 14 ,en = 1 ,clk = 0
t= 12, output b = 1 ,ww.a = 13 ,en = 1 ,clk = 1
t= 14, output b = 1 ,ww.a = 13 ,en = 1 ,clk = 0
t= 16, output b = 1 ,ww.a = 12 ,en = 1 ,clk = 1
t= 18, output b = 1 ,ww.a = 12 ,en = 1 ,clk = 0
t= 20, output b = 1 ,ww.a = 11 ,en = 1 ,clk = 1
t= 22, output b = 1 ,ww.a = 11 ,en = 1 ,clk = 0
```

continued

continued

```
t= 24, output b = 1 ,ww.a = 10 ,en = 1 ,clk = 1
t= 26, output b = 1 ,ww.a = 10 ,en = 1 ,clk = 0
t= 28, output b = 1 ,ww.a = 9 ,en = 1 ,clk = 1
t= 30, output b = 1 ,ww.a = 9 ,en = 1 ,clk = 0
t= 32, output b = 1 ,ww.a = 8 ,en = 1 ,clk = 1
t= 34, output b = 1 ,ww.a = 8 ,en = 1 ,clk = 0
t= 36, output b = 1 ,ww.a = 7 ,en = 1 ,clk = 1
t= 38, output b = 1 ,ww.a = 7 ,en = 1 ,clk = 0
t= 40, output b = 1 ,ww.a = 6 ,en = 1 ,clk = 1
t= 42, output b = 1 ,ww.a = 6 ,en = 1 ,clk = 0
t= 44, output b = 1 ,ww.a = 5 ,en = 1 ,clk = 1
t= 46, output b = 1 ,ww.a = 5 ,en = 1 ,clk = 0
t= 48, output b = 1 ,ww.a = 4 ,en = 1 ,clk = 1
t= 50, output b = 1 ,ww.a = 4 ,en = 1 ,clk = 0
t= 52, output b = 1 ,ww.a = 3 ,en = 1 ,clk = 1
t= 54, output b = 1 ,ww.a = 3 ,en = 1 ,clk = 0
t= 56, output b = 1 ,ww.a = 2 ,en = 1 ,clk = 1
t= 58, output b = 1 ,ww.a = 2 ,en = 1 ,clk = 0
t= 60, output b = 1 ,ww.a = 1 ,en = 1 ,clk = 1
t= 62, output b = 1 ,ww.a = 1 ,en = 1 ,clk = 0
t= 63, output b = 1 ,ww.a = 1 ,en = 0 ,clk = 0
t= 64, output b = 0 ,ww.a = 0 ,en = 0 ,clk = 1
t= 66, output b = 0 ,ww.a = 0 ,en = 0 ,clk = 0
t= 68, output b = 0 ,ww.a = 0 ,en = 0 ,clk = 1
t= 70, output b = 0 ,ww.a = 0 ,en = 0 ,clk = 0
t= 72, output b = 0 ,ww.a = 0 ,en = 0 ,clk = 1
t= 74, output b = 0 ,ww.a = 0 ,en = 0 ,clk = 0
t= 76, output b = 0 ,ww.a = 0 ,en = 0 ,clk = 1
t= 78, output b = 0 ,ww.a = 0 ,en = 0 ,clk = 0
```

Figure 8.43(a) Simulation results of the test bench in Figure 8.42.

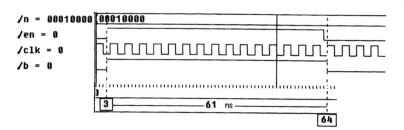

Figure 8.43(b) Simulation results of the test bench in Figure 8.42 showing the signal waveforms.

Example 8.17

Figure 8.44 shows a module that uses the **while** loop for the "memory load" function considered in Example 8.8. The test bench is also included in the figure. The simulation results are given in Figure 8.45. The loading is done on successive negative edges of clk; the loaded values are displayed with an "initializing" tag preceding. Subsequently the memory is read, and the read value is displayed with a "reading" tag preceding. The reading is done through a **for** loop again at successive negative edges of clk.

```
module trial_8c;
reg[7:0] m[15:0];
integer i;
reg clk;
always
begin
#0      while(i<8)
        @(negedge clk)
        begin
                m[i]=i*8;
                $display("initializing: tt=%0d, mm[%0d]=%0d", $time,i,m[i]);
                i=i+1;
        end
#3      begin
                for(i=7;i>=0;i=i-1)
                @(negedge clk)
                $display("reading:t=%0d, m[%0d]=%0d", $time,i,m[i]);
        end
end
initial
begin
        clk = 1'b0; i=0; #65 $stop;
end
always #2 clk=~clk;
endmodule
```

Figure 8.44 A module to illustrate the use of **while** and **for** loops to load a memory and read the same.

```
# initializing: tt=4,  mm[0]=0
# initializing: tt=8,  mm[1]=8
# initializing: tt=12, mm[2]=16
# initializing: tt=16, mm[3]=24
# initializing: tt=20, mm[4]=32
# initializing: tt=24, mm[5]=40
# initializing: tt=28, mm[6]=48
# initializing: tt=32, mm[7]=56
# reading:t=36, m[7]=56
# reading:t=40, m[6]=48
# reading:t=44, m[5]=40
# reading:t=48, m[4]=32
# reading:t=52, m[3]=24
# reading:t=56, m[2]=16
# reading:t=60, m[1]=8
# reading:t=64, m[0]=0
```

Figure 8.45 Results of simulating the module of Figure 8.44.

8.7.1   Selection for Conditional Execution

Conditional execution can be directly described in a module using a conditional operator, the **case** construct, or the **if-else-if** construct. Looping can be effected with **for** or **while**. The conditional operator too can be employed here, though it makes the description a bit cumbersome. Depending upon the context or application, design description with one may be simpler compared to that with others. Practice makes the choice easier. Often, personal preferences too dictate choice.

8.8   **forever** LOOP

Repeated execution of a block in an endless manner is best done with the **forever** loop (compare with repeat where the repetition is for a fixed number of times). Typical illustrative examples follow.

Example 8.18

Consider the module in Figure 8.46. It uses a **forever** block to generates a clock waveform (Compare with the clock using the **always** construct in Example 7.5). The clock toggles every 4 time steps as decided by the **forever** block. A code segment of this type appears typically in a test bench. A code segment of the type in Example 7.5 which generates the clock with the **always** construct appears typically in a design description.

```
module clk;
reg clk, en;
always @(posedge en)
forever#2 clk=~clk;
initial
begin
        clk=1'b0; en=1'b0;#1 clk=1'b1; #4 en=1'b1;#30 $stop;
end
initial $monitor("clk=%b, t=%0d, en=%b ", clk,$time,en);
endmodule
```

Figure 8.46 A module to generate a clock waveform using the **forever** construct.

Example 8.19

Figure 8.47 shows a module wherein the memory load and read operations done in earlier examples are carried out in **forever** loops. In either case the loop is terminated through **disable** statements. The test bench is also included in the figure. Simulation results are as in Figure 8.45 and not shown again.

```
module trial_8d;
reg[7:0] m[15:0];
integer i;
reg clk;
always
begin:load
        forever@(negedge clk)
        begin
                if(i>=8)disable load;
                m[i]=i*8;
                $display("initializing :tt=%0d, mm[%0d]=%0d", $time,i,m[i]);
                i=i+1;
        end
end
always#36
begin:mem_dsply
        forever
        @(negedge clk)
        begin
```

continued

continued

```
                        if(i>15)disable mem_dsply;
                        $display("reading: t=%0d, m[%0d]=%0d", $time,i-8,m[i-8]);
                        i=i+1;
            end
    end
    initial
    begin
            clk = 1'b0;
            i=0;
            #70 $stop;
    end
    always #2 clk=~clk;
    endmodule
```

Figure 8.47 A module that uses **disable** with **forever** to load and read a memory file.

Example 8.20

During normal operation a microprocessor fetches an instruction from a program memory pointed by the PC, increments the PC, fetches the next instruction, and so on. The cycle is repeated eternally [Hill & Peterson, Heuring & Jordan]. An interrupt input breaks the sequence and shifts execution to a different program segment. Figure 8.48 shows a module using the **forever** type of loop; it links the PC, the IR, and the program memory in the normal cyclic operation. The cycle is interrupted only by the external Interrupt input. The module uses a look-up-table (LUT) type of decoder. The instruction is decoded as part of the loop execution. The program memory and the LUT are initialized before program execution commences. The module has three inputs clk, en, and int. The loop operation commences with en going high; it continues until int goes high and then stops. The interrupt service has to be organized separately. A test bench is also included in the figure. The simulation results are in Figure 8.49.

```
module mup_opr(clk,int,en);//mup operation
input clk, int,en;
reg[7:0] pgm_mem[15:0], irdc[255:0],ir,pc,dcop; //pgm_mem : program memory
integer i; // irdc: IR decoder output
//ir : Instruction register; pc : Program counter;  dcop : decoded output
```

continued

continued

```
always@(posedge en )
begin
        forever
        begin:mup_work
                if(int) disable mup_work;
                wait(clk)ir=pgm_mem[pc];//fetch instruction
                wait(!clk)
                begin
                        dcop=irdc[ir];//execute instruction
                        pc=pc+1;//increment program counter
                end
        end
end
initial
begin
        pc=0;
        for(i=0;i<16;i=i+1)pgm_mem[i]=i*8;
        for(i=0;i<255;i=i+1)irdc[255-i]=i;
end
endmodule


module tst_mup;
reg clk,en,int;
initial
begin
        int=1'b0;clk=1'b0;en=1'b0;
  #5    en=1;
  #34   int=1'b1;
end
always #2 clk=~clk;
initial $monitor("clk=%0d, t=%0d, en=%b, int=%b,  pgm_mem[%0d] =%0d,
dcop=%0d", clk,$time,en,int,rr.pc,rr.ir,rr.dcop);
mup_opr rr(clk,int,en);
initial #40 $stop;
endmodule
```

Figure 8.48 A module to control basic operation of a microprocessor.

```
# clk=0,  t=0,   en=0,  int=0,  pgm_mem[0] =x, dcop=x
# clk=1,  t=2,   en=0,  int=0,  pgm_mem[0] =x, dcop=x
# clk=0,  t=4,   en=0,  int=0,  pgm_mem[0] =x, dcop=x
# clk=0,  t=5,   en=1,  int=0,  pgm_mem[0] =x, dcop=x
# clk=1,  t=6,   en=1,  int=0,  pgm_mem[0] =0, dcop=x
# clk=0,  t=8,   en=1,  int=0,  pgm_mem[1] =0, dcop=x
# clk=1,  t=10,  en=1,  int=0,  pgm_mem[1] =8, dcop=x
# clk=0,  t=12,  en=1,  int=0,  pgm_mem[2] =8, dcop=247
# clk=1,  t=14,  en=1,  int=0,  pgm_mem[2] =16, dcop=247
# clk=0,  t=16,  en=1,  int=0,  pgm_mem[3] =16, dcop=239
# clk=1,  t=18,  en=1,  int=0,  pgm_mem[3] =24, dcop=239
# clk=0,  t=20,  en=1,  int=0,  pgm_mem[4] =24, dcop=231
# clk=1,  t=22,  en=1,  int=0,  pgm_mem[4] =32, dcop=231
# clk=0,  t=24,  en=1,  int=0,  pgm_mem[5] =32, dcop=223
# clk=1,  t=26,  en=1,  int=0,  pgm_mem[5] =40, dcop=223
# clk=0,  t=28,  en=1,  int=0,  pgm_mem[6] =40, dcop=215
# clk=1,  t=30,  en=1,  int=0,  pgm_mem[6] =48, dcop=215
# clk=0,  t=32,  en=1,  int=0,  pgm_mem[7] =48, dcop=207
# clk=1,  t=34,  en=1,  int=0,  pgm_mem[7] =56, dcop=207
# clk=0,  t=36,  en=1,  int=0,  pgm_mem[8] =56, dcop=199
# clk=1,  t=38,  en=1,  int=0,  pgm_mem[8] =64, dcop=199
# clk=1,  t=39,  en=1,  int=1,  pgm_mem[8] =64, dcop=199
```

Figure 8.49 Results of simulating the test bench in Figure 8.48.

8.9   PARALLEL BLOCKS

All the procedural assignments within a **begin-end** block are executed
sequentially.   The **fork-join** block is an alternate one where all the
assignments are carried out concurrently (The nonblocking assignments too can be
used for the purpose.).   One can use a fork-join block within a **begin-end** block
or vice versa.   The examples below bring out some possible combinations and
their subtle differences.   In each case the module and the simulation results are
shown within the same figure.

Example 8.21

Figure 8.50(a) shows a module with assignments to the integer **a** within a **begin-
end** block. All the assignments are carried out sequentially.   The time values
specified within the block are intervals for the following assignments.   Figure
8.50(b) shows the same block of assignments within a **fork-join** block.   The

```
module fk_jn_a;                          module fk_jn_b;
integer a;                               integer a;
initial                                  initial
begin                                    fork
          a=0;                                     a=0;
#1        a=1;                            #1        a=1;
#2        a=2;                            #2        a=2;
#3        a=3;                            #3        a=3;
#4        $stop;                          #4        $stop;
end                                      join
initial $monitor ("a=%0d,                initial $monitor ("a=%0d,
t=%0d",a,$time);                         t=%0d",a,$time);
endmodule                                endmodule

//Simulatiom results                     //Simulation results
# a=0, t=0                               # a=0, t=0
# a=1, t=1                               # a=1, t=1
# a=2, t=3                               # a=2, t=2
# a=3, t=6                               # a=3, t=3
```

                    (a)   (b)

Figure 8.50 A simple illustrative example to bring out the difference between
begin–end and **fork–join** blocks: (a) A module with a **begin–end**
block and the simulation results (b) A module with a **fork–join** block and
the simulation results.

assignments take effect at 0, 1, 2, and 3 time steps after entry to the block. The
time values specified are interpreted as being delays with respect to the time of
entry to the loop, in contrast to the previous case where they are treated as
successive time intervals. The last assignment in Figure 8.50(b) is at the third
time step; in Figure 8.50(a) it is at the sixth time step.

Example 8.22

Figure 8.51 shows an enhanced version of the modules in Figure 8.50. It has a
fork-join block within a **begin–end** block. The integer a is assigned the
value 5 at entry time to the **begin–end** block; it is followed by a set of
assignments to it (within the **fork–join** block) all carried out concurrently. The
last assignment is at the 9th time step. Execution stops at the 10th time step. The
begin–end and **fork–join** blocks in Figure 8.51 have been interchanged and
shown in Figure 8.52. The entry to the **begin–end** block is concurrent with the
first assignment at the fifth time step. All the assignments within the begin-end
block are sequential. The last of the assignments is at the 10th time step.
Execution stops at the 15th time step.

```
module fk_jn_c;
integer a;
initial
begin
#5      a=5;
        fork
        #1      a=0;
        #2      a=1;
        #3      a=2;
        #4      a=3;
        #5      $stop;
        join
end
initial $monitor ("a=%0d, t=%0d",a,$time);
endmodule

//Simulation results
# a=x, t=0
# a=5, t=5
# a=0, t=6
# a=1, t=7
# a=2, t=8
# a=3, t=9
```

Figure 8.51 An example of a **fork-join** block within a **begin-end** block.

```
module fk_jn_d;
integer a;
initial
fork
#5      a=5;
        begin
        #1      a=0;
        #2      a=1;
        #3      a=2;
        #4      a=3;
        #5      $stop;
        end
join
initial $monitor ("a=%0d, t=%0d",a,$time);
endmodule
```

continued

continued

//Simulation results
a=x, t=0
a=0, t=1
a=1, t=3
a=5, t=5
a=2, t=6
a=3, t=10

Figure 8.52 An example of a **begin–end** block within a **fork–join** block.

8.10   Force–release CONSTRUCT

When debugging a design with a number of instantiations, one may be stuck with an unexpected behavior in a localized area. Tracing the paths of individual signals and debugging the design may prove to be too tedious or difficult. In such cases suspect blocks may be isolated, tested, and debugged and *status quo ante* established. The **force–release** construct is for such a localized isolation for a limited period. Figure 8.53 shows the use of a **force–release** construct in a test bench. The assignment

force a = 1'b0;

forces the variable a to take the value 0.

force b = c&d;

forces the variable b to the value obtained by evaluating the expression c&d. Subsequently a few assignments are made in the test bench. At a later part of the test bench, a and b are released that is, their original assignments are restored. The assignments here have specific characteristics:

```
. . .
force a = 1'b0;
force b = c&d;
assignment1;
assignment2;
. . .
release a;
release b;
. . .
```

Figure 8.53 Use of the **force–release** construct in a test bench.

- They are temporary, for a limited time and for test purposes only.
- Both nets and **reg**s can be forced in this manner; that is, their regular values can be overridden.
- When a net is forced to a value, it takes the new value assigned. On release, its previous assignment comes back into effect.
- When a **reg** is forced to a value, it takes the newly assigned value. Even after the release, the newly assigned value continues to hold good until another procedural assignment changes its value.

- Figure 8.54 illustrates a test case for different uses of the **force–release** construct. CUT is a circuit block under test. The design has the following input connections:

- Input x connected to combinational circuit g1
- Input y connected to combinational circuit g2
- Input u connected to combinational circuit g3
- Input v connected to **reg1**

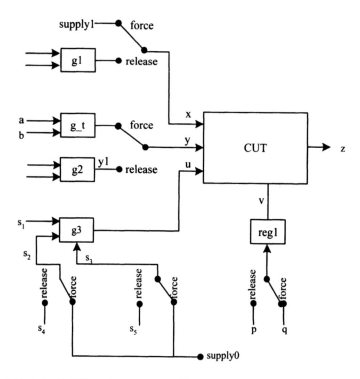

Figure 8.54 A circuit CUT under test with different possibilities of forcing test signals on it.

The circuit is identified to have a fault (unexpected behavior). To debug the circuit, specific signals are to be forced at its inputs; after debugging the connections are to be restored. The **force-release** construct can be used here effectively. The testing and debugging activity is carried out in a test bench routine. Typically, CUT may be the instantiation of a module defined elsewhere.

Consider a hypothetical situation where x, y, u, and v are to be forced to specific signals for testing purposes as shown in Figure 8.54 itself. Different possible test situations are brought out through the figure. The relevant program segments in the test bench are shown in Figure 8.55; pertinent explanations follow:

- Figure 8.55(a) shows one segment of the test bench. Until execution of the assignment 1 is complete, x (cut.x) is connected to the output of g1 as in the design description. At this stage, x is forced to supply1 10 ns after assignment1. The testing is continued with assignment2, assignment3, etc. Subsequently – 20 ns later – x is released, that is, its connection to the output of g1 is restored. The test-bench execution continues with the next assignment. x being a net, the restoration takes effect immediately after the **release** x command. Note that the **force** and **release** are to be done through appropriate dereferencing.

	(a)
. . .	
assignment1;	
#10	**force** cut.x = **supply1**;
	assignment2;
	assignment3;
#20	**release** cut.x;
. . .	

	(b)
. . .	
assignment4;	
#10	force cut.y = a & b;
	assignment5;
	assignment6;
#20	release cut.y;
. . .	

	(c)
. . .	
assignment7;	
#10	force cut.s2 = supply0;
	force cut.s3 = supply0;
	assignment8;
	assignment9;
#20	release cut.s2;
	release cut.s3;
	. . .

	(d)
. . .	
assignment10;	
#10 force cut.v = q;	
assignment11;	
assignment12;	
#20 release cut.v;	
#5 assignment13;	
cut.v = p;	
. . .	

Figure 8.55 Different segments of the test bench to force test signals at the input points of CUT in Figure 8.54.

- Figure 8.55(b) shows another segment of the test bench. 10 ns after the execution of assignment4 in the test-bench, y (cut.y) is disconnected from y1, the output of g2. It is assigned a new (temporary value) through g_t. here the signals a and b are ANDed to form the input to y. Assignment5, assignment6, *etc.*, are executed. 20 ns later, y is released. Immediately, y – being a net – takes its normally assigned value y1 and execution of the test bench continues.

- In a typical simple case, g3 may be an OR gate with the continuous assignment

  assign u = s1 | s2 | s3;

  Forcing s2 and s3 to **supply0** amounts to (bypassing signals s2 and s3 and) connecting u directly to signal s1. The corresponding test segment in the test bench is shown in Figure 8.55(c). Ten ns after assignment7, s2 (cut.s2) and s3 (cut.s3) are forced to supply0. assignment8 and assignment9 are executed at this stage; s2 and s3 are released. Subsequently, assignment10 is executed.

- Figure 8.55(d) shows one more segment of the test bench. Ten time steps after execution of assignment10, v (cut.v) is given a new assignment (=q) through the **force** construct. Testing continues through assignment11, assignment12 *etc.* Twenty time steps later, cut.v is released. Cut.v being a **reg** type of variable, the value assigned to it continues as q itself. With this assignment being still valid, 5 time steps later, assignment 12 is executed. Subsequently, cut.v is assigned the value p. The new value of cut.v = p is valid only for the test segment that follows assignment 12.

Observations:

- The **force–release** construct is similar to the **assign–deassign** construct. The latter construct is for conditional assignment in a design description. The **force–release** construct is for "short time" assignments in a test-bench. Synthesis tools will not support the **force–release** constructs.

- The **force–release** construct is equally valid for net-type variables and **reg**-type variables. The net–type variables revert to their normal values on release. With **reg**-type variables the value forced remains until another assignment to the reg.

- The variable, on which the values are forced during testing, must be properly dereferenced.

- In the illustration above, each variable was forced one at a time. It was done only to simplify the illustration sequence and focus attention on the possible use of the construct. In practice, different variables can be forced together before the special debug sequence. Their release too can be together.

Example 8.23

Use of the **force-release** pair is brought out here through a simple example. Figure 8.56 shows a module of an OR gate with two inputs along with a test bench for the same. Simulation results are shown in Figure 8.57. Input c toggles every 3 ns between 0 and 1; but input b is kept at 0 value throughout the test period. Hence in the normal course, output a will follow the input c and toggle along with it. Input b is forced to 1 at 7 ns and released at 14 ns; correspondingly, the gate output a too goes to 1 state in the interval 7 ns to 14 ns; these can be seen from the values of a, b, and c displayed in Figure 8.57 at the 1st, 8th, and 15th ns of simulation.

```
module or_fr_rl(a,b,c);
input b,c; output a; wire a,b,c;
assign  a= b|c;
initial begin
#1 $display("display:time=%0d, b=%b, c=%b, a=%b", $time,b,c,a);
#6 force  b=1'b1;
#1 $display("display:time=%0d, b=%b, c=%b, a=%b", $time,b,c,a);
#6 release b;
#1 $display("display:time=%0d, b=%b, c=%b, a=%b", $time,b,c,a);
end
endmodule

module orfr_tst;
reg b,c;wire a;
initial begin b=1'b0; c=1'b0; #20 $stop; end
always #3 c = ~c;
or_fr_rl dd(a,b,c);
endmodule
```

Figure 8.56 An OR gate module and its test bench to illustrate the use of **force-release** construct..

```
# display:time=1, b=0, c=0, a=0
# display:time=8, b=1, c=0, a=1
# display:time=15, b=0, c=0, a=0
```

Figure 8.57 Waveforms of the inputs and output of the OR gate module in Figure 8.56 during its test.

8.11   EVENT

The keyword **event** allows an abstract event to be declared. The event is not a data type with any specific values; it is not a variable (**reg**) or a net. It signifies a change that can be used as a trigger to communicate between modules or to synchronize events in different modules. Figure 8.58 shows a segment of a module to bring out its use. change has been declared as an **event**. In the course of execution of an **always** block, the event is triggered. The operator "→" signifies the triggering. Subsequently, another activity can be started in the module by the event change. The **always**@(change) block activates this. The event change can be used in other modules also by proper dereferencing; with such usage an activity in a module can be synchronized to an event in another module.

. . .
event change;
. . .
always
. . .
. . . → change;
. . .
.**always**@change
. . .

Figure 8.58 Use of the event construct in a module.

The **event** construct is quite useful, especially in the early stages of a design. It can be used to establish the functionality of a design at the behavioral level; it allows communication amongst different instantiated modules without associated inputs or outputs.

Example 8.24

Figure 8.59 illustrates an application of an event construct for a skeletal serial receiver. Module rec is the serial receiver and the module rec_tst is its test bench. The test bench – rec_tst –has an 8-bit register aa into which a sequence of bytes (their values decided at random) is loaded. The bytes are converted into a serial data stream di synchronized to the positive edge of the clock. The test bench – rec_tst – instantiates the module rec with the name rrcc, gives di and clk as input to rrcc, and receives the buffer output from it. The receiver converts the serial data into parallel form by loading successive bits into a register designated "a" at the negative edges of the clock. Once the a register is full, the "buf-ful" event is activated. The test bench uses the event to read the buffer a and display its content along with that of aa.

```
module rec_tst;
reg clk,di; integer n,i;
reg[8:1] aa;wire [8:1] a;
always #2 clk = ~clk;
rec rrcc(a,di,clk);
always @(rrcc.buf_ful) $display("t=%0d, aa=%h, a=%h",$time,aa,a);
initial
        for (n=1;n<3000;n=n+113) begin
                        aa=n;i=0;
                        repeat(8)@(posedge clk)
                                begin
                                        i=i+1;
                                        di=aa[i];
                                        //$write("bb=%b",aa[i]);
                                end
                #3      i=0;
                end //Why '#3'?
initial clk=1'b0; initial #400  $stop;
endmodule


module rec(a,ddi,clk);
output[8:1]a; input ddi,clk;reg[8:1] a;integer j,jj;
event buf_ful;
always for (j=0;j<20;j=j+1)              begin
                        #0 jj=0;
                        repeat(8)@(negedge clk) begin
                                                jj=jj+1;
                                                a[jj]=ddi;
                                                //$display("b=%b",a[jj]);
                                                end
                #0      ->buf_ful;
                        end
endmodule
```

Figure 8.59 A module to illustrate the **event** construct: A serial data receiver and a test bench for the same.

```
# t=32, aa=01, a=01
# t=64, aa=72, a=72
# t=96, aa=e3, a=e3
# t=128, aa=54, a=54
# t=160, aa=c5, a=c5
# t=192, aa=36, a=36
# t=224, aa=a7, a=a7
# t=256, aa=18, a=18
# t=288, aa=89, a=89
# t=320, aa=fa, a=fa
# t=352, aa=6b, a=6b
# t=384, aa=dc, a=dc
```

Figure 8.60 Simulation results of the test bench in Figure 8.59.

8.12   EXERCISES

Prepare design modules for the Exercises 1 to 10 below. In each case prepare a suitable test bench and test the design module [Arnold, Tocci]].

1. An adder to add two eight-digit numbers in BCD form.
2. Add two BCD digits using a look-up table.
3. Multiply two BCD digits using a look-up table.
4. An 8-digit multiplier all the digits being in BCD form.
5. A multiplier to multiply two 32-bit numbers.
6. A module to convert angle in radians to one in degrees.
7. A module to convert a 48-bit number into a decimal one in BCD form.
8. Combine the above two: Form a module to convert an angle in radians into one in degrees in decimal form.
9. A table to give the sines of angles. The given angle is a four-digit decimal number – in degrees in the range 0 to 90 degrees. The given table has two parts – a main table of four digits and a table of mean differences of one digit.
10. The outputs of a set of shift registers are designated as q1, q2, q3, *etc.* A selected set of these is exor'ed and the exor output fed as data input to q1. As the set of registers is clocked, the state vector representing the shift register outputs goes on changing state. With a properly selected set as the input to the exor gates, one can ensure that the state vector sequences through all the possible states in a "pseudo-random" manner. Thus an n stage shift register sequences through $2n - 1$ states.
11. Consider the code block in Figure 8.61(a). Complete the module and test it with the inputs a and b in Figure 8.61(b). Explain the difference in the waveforms of c and d.

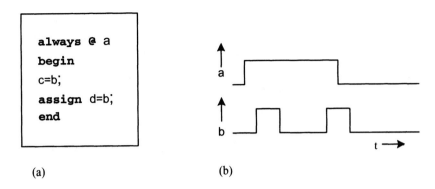

Figure 8.61 The behavioral block and the input waveforms for it for Exercise 11.

12. *A Serial Receiver with* **event**: A serial data stream is coming on an input data line. It is synchronized with a clock signal. Do the following in a module
    - Receive 8 bits and fill a byte-wide receive register.
    - Set **event** REC.
    - Use REC to transfer the received byte to the top of a FIFO.

13. *A Serial Transmitter with* **event**: Tr_buf is a byte-wide buffer. Serially output its content on a serial line. When Tr_buf is empty, set event TR. On TR event, load Tr_buf from bottom of FIFO.

14. Prepare modules to realize the priority encoder using the "**if**" and "**for**" constructs. Simulate and synthesize each.

15. In Example 8.8 the **event** @(**negedge** clk) succeeds **repeat**. Interchange the two and suitably modify the block with additional **begin** and **end** lines. Simulate, compare the results with those in the example, and explain the difference.

16. Complete the "block memory output" module in Figure 8.25. Test it with a suitable test bench.

17. Prepare modules for the following and simulate each with a test-bench:
    - Clear a block of memory.
    - Input a block of bytes to a register file.
    - Move a set of bytes from one to another page of memory with specified starting and ending addresses.

18. Use the **disable** construct and prepare modules for AND, NAND, and NOR functions. Follow the approach in Figure 8.34. Test each with corresponding test benches.

19. Use the **repeat** construct along with the **disable** construct to realize an AND gate. Synthesize the module and compare the synthesized circuits.

20. Repeat the above Exercise with **casez** and **if-else-if** constructs.

21. Repeat the above two Exercises for OR, NOR, and NAND functions.

22. What is the functional difference between the two blocks in Figure 8.62? Illustrate through suitable test benches.

    If the combination

    @(**posedge** en1)
    @(**posedge** en2)

    is replaced by

    @ (**posedge** en1 or **posedge** en2)

    how will the performance differ? Explain through test benches.

```
Initial                          Initial    begin
 begin                                       #1 a=0;
        #1 a=0;                              fork
        @(posedge en1)                          @(posedge en1)
        @(posedge en2)                          @)(posedge en2)
        a=1;                                 join
 end                                         a=1;
                                             end
```

Figure 8.62 Two functional blocks to illustrate the difference between **begin–end** and **fork–join** pairs of constructs in Exercise 22.

23. Compare the behavior of the blocks in the above Exercise with one using the **if** construct.

24. A serial link has a clock rate of 1 MHz and a bit rate 1/32 times the clock rate. Set up a receiver to receive 8 successive bytes of data and to load them into a register file. The expected functioning of the unit is to be on the following lines:

    ▪ A clock to function at 1 MHz. A bit rate clock derived from the main clock.
    ▪ A flag **En** to enable serial reception.
    ▪ A serial data input stream.
    ▪ At the first positive edge of the main clock following **En**, transmission starts.
    ▪ At every 4th pulse of the main clock, the input data line is to be sensed. A polling of 4 consecutive data bits decides the received output bit value and the status of an error bit.
    ▪ Whenever the error flag goes high, the corresponding byte is made ff.

    Set up a test bench and test the functioning of the link.

25.   Figure 8.63 shows a module. Get the waveforms of a and b by simulation.

```
module pulses;
reg [8:0] I;
reg a,b;
initial
while (I<100)
begin
    #1        a= I(0);
              b= I(1);
              I = I + 1;
end
initial I=0;
initial #100 $stop;
endmodule
```

Figure 8.63 A module to generate simple waveforms.

26. Generate three waveforms with the following characteristics (see Figure 8.64):

   - All have a time period of 21 time steps.
   - All are identical.
   - All have a continuous ON period of 5 ns.
   - All are equally phase-shifted.

Generate the waveforms using **case**, **if-else-if** and **for** constructs.

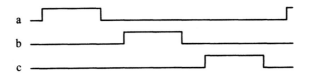

Figure 8.64 Three phase clock waveforms.

27. Generate the waveforms in Figure 8.65 using the **case**, **if**, and **for** loops. Use **repeat** and **forever** constructs for cyclic repetition.

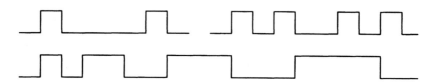

Figure 8.65 Different waveforms for Exercise 27.

28. A module and other modules instantiated within it can have a number of events scheduled for execution at the same time step. The sequence of execution is simulator-dependent. If any particular statement is assigned for execution with a zero time delay, it is executed as the last one in the concerned time step. Consider Example 8.23: The event buf_ful is assigned a zero time delay; delete the delay, simulate the module, and explain the difference in results, if any. The commented $write and $display statements may be activated for this.

29. Again consider example 8.24: The last statement in the block used to generate the serial data stream is assigned a 3 ns time delay. Delete the delay, simulate the module, and explain the difference in results, if any. The commented $write and $display statements may be activated for this.

9

FUNCTIONS, TASKS, AND USER-DEFINED PRIMITIVES

9.1   INTRODUCTIUON

Bigger designs are better arranged in small functional blocks; it facilitates debugging and any reorganization. Thus a module can have well-defined sub-modules inside, treated as separate entities. Functions and Tasks are such entities inside modules. They play three broad roles:

- A well-defined structure with a separate identity.
- They can hide some variables.
- They can be repeatedly invoked within the module.

User-defined primitive (UDP) provides an alternative form of a submodule; it can realize specific outputs. The UDP has a specific format. It can be defined by the user and used wherever necessary. The fact that the UDP has a specific format allows a straightforward definition – often at the expense of flexibility.

9.2   FUNCTION

A function is like a subroutine or a procedure in a program. It is defined separately within a module and can be called whenever necessary. When a function is declared with a function name, the system allocates a register for it. The name of the register is that of the function; and its type (as well as size) is also that of the function. When a function is called, the system executes the functional activity and generates the output. Eventually the output is assigned to the register identified for the function. The quantity returned by the function can be used as an operand in an assignment or in an expression. The structure of a function definition is shown in Figure 9.1. The significance of each of the quantities as well as the rules of using them is also explained in the figure. The use of functions is brought out through a set of examples.

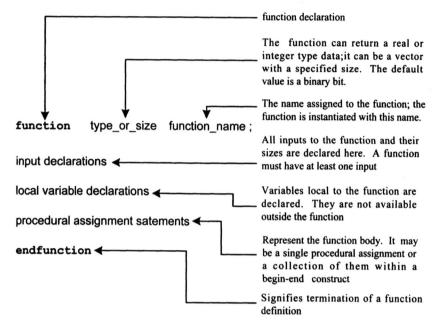

function declaration

The function can return a real or integer type data;it can be a vector with a specified size. The default value is a binary bit.

The name assigned to the function; the function is instantiated with this name.

```
function    type_or_size   function_name ;
```

All inputs to the function and their sizes are declared here. A function must have at least one input

input declarations

local variable declarations

Variables local to the function are declared. They are not available outside the function

procedural assignment satements

endfunction

Represent the function body. It may be a single procedural assignment or a collection of them within a begin-end construct

Signifies termination of a function definition

Figure 9.1 Structure for function definition.

Example 9.1

The function odd-parity is defined within the module **parity-check** in Figure 9.2. It generates a parity bit. The parity bit is 1 if the number of one-bits in the byte is odd. Otherwise it is zero. The module has an 8-bit vector input and a flag input – en. It has an output **chk**. Whenever the flag goes high, the function **odd-parity** is called. It returns the parity bit value and assigns it to **chk** in the module. **parity-check** is an example with a single-bit output-type function in it. The function has no local variables in it.

```
module parity_chk(a,en,chk);
input[7:0]a;
input en;
output chk;
wire[7:0] a;
reg chk;
always @(posedge en)
begin
        chk=pb(a);
        $display("t=%0d, a = %b, en = %0b, pb = %0b ",$time,a,en,chk);
end
```

continued

continued

```
function pb;
input[7:0]a;
pb=^a;
endfunction

endmodule
module tst_pchk;
reg [7:0]a;
reg en;
wire chk;
integer i;
parity_chk pchk(a,en,chk);
initial #0 en=1'b0;
always #2 en = ~en;
initial
begin
        #1 a=8'h00;
        for(i=0;i<8;i=i+1)
        begin
                #4 a=a+3'o6;
        end
end
initial #40 $stop;
endmodule
```

Figure 9.2 A module for parity generation through a function.

```
# t=2, a = 00000000, en = 1, pb = 0
# t=6, a = 00000110, en = 1, pb = 0
# t=10, a = 00001100, en = 1, pb = 0
# t=14, a = 00010010, en = 1, pb = 0
# t=18, a = 00011000, en = 1, pb = 0
# t=22, a = 00011110, en = 1, pb = 0
# t=26, a = 00100100, en = 1, pb = 0
# t=30, a = 00101010, en = 1, pb = 1
# t=34, a = 00110000, en = 1, pb = 0
# t=38, a = 00110000, en = 1, pb = 0
```

Figure 9.3 Simulation results of the test bench in Figure 9.2.

Example 9.2

Figure 9.4 shows another module for parity generation. The module has a function to count the number of one-bits in the input byte. In the module the parity bit is decided by mod-2 division of the number returned by the function. The function has an integer declared and used within it. (In contrast, in the last example the parity bit was generated directly within the function defined.)

```
module parity(p,a,En);
input[7:0]a;
input En;
output p;
reg p;
always @(posedge En)
begin
        p=n1(a)%2;  //Use n1 & generate the parity bit.
        $display("t=%0d, a = %b, en = %b, p = %b ",$time,a,en,p);
end

function integer n1;  //A function to count the number of 1 bits in a byte
input[7:0]a;
integer i;
        for(i=0;i!=8;i=i+1)
        begin
                if(i==0) n1=0;
                if(a[i]) n1=n1+1;
        end
endfunction
endmodule
```

Figure 9.4 A module to generate a parity bit: The parity bit is generated by counting the number of one-bits in a function and doing a mod-2 division.

Example 9.3

In the module of Figure 9.5 the number of one-bits is decided by shifting out the bits of the input vector and counting the ones in them. Otherwise the module is similar to the one in Figure 9.4. The module (as well as the previous ones) can be easily extended to generate the parity bit for wider binary streams.

```
module parity_a(p,a,En);
input[7:0]a;
input En;
output p;
reg p;
always @(posedge En)
begin
        p=nn(a)%2;
        $display("t=%0d, a = %b, En = %b, p = %b ",$time,a,En,p);
end

function integer nn;
input[7:0]a;
integer i;
begin
        for(i=0;i!=8;i=i+1)
        begin
        if(i==0) nn=0;
        if(a[i]) nn=nn+1;
        a=a>>1;
        end
end
endfunction
endmodule
```

Figure 9.5 Another module to generate a parity bit similar to that in Figure 9.4.

Example 9.4

Figure 9.6 shows an adder module to add two 2-bit numbers. The module has two functions defined in it – a half-adder and a full-adder. Further, one can see that the full-adder function itself calls the half-adder function within it. The module calls the full-adder function repeatedly within itself. A test bench for the adder is also included in the figure. The simulation results are shown in Figure 9.7.

```
module adderfun(r,p,q,En);
input[1:0] p,q; input En; output [2:0] r; reg[2:0]r,c; integer i;
always@(posedge En)
```

continued

continued

```verilog
begin
        for(i=0;i<2;i=i+1)
        begin
                if(i==0) c[i]=1'b0;
                {c[i+1'b1],r[i]}=fa(p[i],q[i],c[i]);
        end

                r[2]=c[2];
                $display("t=%0d, En = %b, p = %b, q = %b, r = %b ",$time
                ,En,p,q,r);
end

function[1:0] ha;
input a,b;
ha={a&b,a^b};
endfunction

function [1:0]fa;
input a,b,c; reg[1:0]a1,a2,aa2;
begin
        a1=ha(a,b);
        aa2=ha(a1[0],c);
        a2[1] =  (aa2[1]|a1[1]);
        a2[0] = aa2[0];
        fa=a2;
end
endfunction
endmodule

module tst_adder_fun; //testbench;
reg [1:0] p,q; reg En; wire [2:0] r;
adderfun aa(r,p,q,En);
always #2 En=~En;
initial    begin
                        En=1'b0; p=2'b01;q=2'b00;
                        #5 p=2'b10;q=2'b10;
                        #4 p=2'b10;q=2'b11;
                        #4 p=2'b11;q=2'b11;
                        #4 p=2'b01;q=2'b01;
           end
initial #30 $stop;
endmodule
```

Figure 9.6 A module to illustrate a function calling another one; a test bench is also included in the figure.

```
# t=2, En = 1, p = 01, q = 00, r = 001
# t=6, En = 1, p = 10, q = 10, r = 100
# t=10, En = 1, p = 10, q = 11, r = 101
# t=14, En = 1, p = 11, q = 11, r = 110
# t=18, En = 1, p = 01, q = 01, r = 010
# t=22, En = 1, p = 01, q = 01, r = 010
# t=26, En = 1, p = 01, q = 01, r = 010
```

Figure 9.7 Results of running the test bench in Figure 9.6.

Example 9.5

A module to add two 32-bit numbers is shown in Figure 9.8. It is essentially a scaled-up version of the one in Figure 9.6. The addition is initiated by the En input going high; it is carried out in one time step. A test bench is also included in the figure. The simulation results for a specific set of input number combinations are shown in Figure 9.9.

```
module add32(r,p,q,En);
input[31:0] p,q; input En; output [32:0] r; reg[32:0]r,c; integer i;
always@(posedge En) begin
                    for(i=0;i<32;i=i+1)
                    begin
                    if(i==0) c[i]=1'b0;
                    {c[i+1'b1],r[i]}=fa(p[i],q[i],c[i]);
                    end
                    r[32]=c[32];
                    $display( "t=%0d, En = %b, p = %0h, q = %0h, r =
                    %0h ",$time, En,p,q,r);
               end

function[1:0] ha;
input a,b;
ha={a&b,a^b};
endfunction

function [1:0]fa;
input a,b,c; reg[1:0]a1,a2,aa2;
begin
          a1=ha(a,b);
```

continued

continued

```
        aa2=ha(a1[0],c);
        a2[1] =  (aa2[1]|a1[1]);
        a2[0] =  aa2[0];
        fa=a2;
end
endfunction
endmodule

module tst_add32; //testbench;
reg [31:0] p,q; reg En; wire [32:0] r;
add32 aa(r,p,q,En);
always #2 En=~En;
initial   begin
            #0 En  = 1'b0;
            #3 p   = 32'h1234;        q = 32'h4321;
            #4 p   = 32'h12345678; q = 32'h98765432;
            #4 p   = 32'habcdef12;  q = 32'hbbccddee;
            #4 p   = 32'hfedcba39;  q = 32'h13579bdf;
            #4 p   = 32'h9876abcd; q = 32'hfedc8765;
            #4 p   = 32'hf0e0d0c0;  q = 32'h11020304;
          end
initial #30 $stop;
endmodule
```

Figure 9.8 A scaled-up version of the 2-bit adder in Figure 9.6 to add 32-bit numbers.

```
# t=2, En = 1, p = x, q = x, r = x
# t=6, En = 1, p = 1234, q = 4321, r = 5555
# t=10, En = 1, p = 12345678, q = 98765432, r = aaaaaaaa
# t=14, En = 1, p = abcdef12, q = bbccddee, r = 1679acd00
# t=18, En = 1, p = fedcba39, q = 13579bdf, r = 112345618
# t=22, En = 1, p = 9876abcd, q = fedc8765, r = 197533332
# t=26, En = 1, p = f0e0d0c0, q = 11020304, r = 101e2d3c4
```

Figure 9.9 Results of running the test bench in Figure 9.8.

Example 9.6

A variant of the adder in Example 9.4 is shown in Figure 9.10: After the enable input **en** goes high, the full-adder function is called repeatedly in successive clock pulses and bit-wise addition is carried out. The figure also includes a test bench. As can be seen from the simulation results in Figure 9.11, each addition is spread over two clock periods.

```
module adderfunb(clk,r,p,q,En);
input[1:0] p,q; input En,clk; output [2:0] r; reg[2:0]r,c; integer i;
always@(posedge En) begin
                        for(i=0;i<2;i=i+1) begin
                                        @(posedge clk)
                                        if(i==0) c[i]=1'b0;
                                        {c[i+1'b1],r[i]}=fa(p[i],q[i],c[i]);
                                        end
                        r[2]=c[2];
                        $display(" t=%0d, clk = %b, En = %b, p = %b, q = %b,
                        r = %b ",$time,clk,En,p,q,r);
                end
function[1:0] ha;
input a,b;
ha={a&b,a^b};
endfunction

function [1:0]fa;
input a,b,c; reg[1:0]a1,a2,aa2;
begin
        a1=ha(a,b);
        aa2=ha(a1[0],c);
        a2[1] =  (aa2[1]|a1[1]);
        a2[0]=aa2[0];
        fa=a2;
end
endfunction
endmodule

module tst_adder_funb();
reg [1:0] p,q; reg En,clk; wire [2:0] r;
adderfunb bb(clk,r,p,q,En);
always #2 clk=~clk;
initial    begin
                        clk=1'b0;   En=1'b0; p=2'b01; q=2'b00;
                        #1 En=1'b1; #6 En=1'b0; p=2'b01; q=2'b10;
                        #1 En=1'b1; #7 En=1'b0; p=2'b01; q=2'b01;
                        #1 En=1'b1; #7 En=1'b0; p=2'b10; q=2'b01;
                        #1 En=1'b1; #7 En=1'b0; p=2'b10; q=2'b10;
                        #1 En=1'b1; #7 En=1'b0; p=2'b10; q=2'b11;
                        #1 En=1'b1; #7 En=1'b0; p=2'b11; q=2'b11;
                        #1 En=1'b1; #7 En=1'b0;
        end
initial #60 $stop;
endmodule
```

Figure 9.10 A variant of the 2-bit adder in Figure 9.6; bit-wise addition is carried out in successive clock pulses.

```
# t=6, clk = 1, En = 1, p = 01, q = 00, r = 001
# t=14, clk = 1, En = 1, p = 01, q = 10, r = 011
# t=22, clk = 1, En = 1, p = 01, q = 01, r = 010
# t=30, clk = 1, En = 1, p = 10, q = 01, r = 011
# t=38, clk = 1, En = 1, p = 10, q = 10, r = 100
# t=46, clk = 1, En = 1, p = 10, q = 11, r = 101
# t=54, clk = 1, En = 1, p = 11, q = 11, r = 110
```

Figure 9.11 Simulation results of the test bench for the adder module in Figure 9.10.

Example 9.7

A module to add 32-bit numbers is shown in Figure 9.12. It is a scaled-up version of that in the last example. The addition commences after the enable bit En goes high. Starting with the LSB, one bit is added at every succeeding clock pulse. Addition is completed in 32 clock pulses. The simulation results with a set of 32-bit numbers is shown in Figure 9.13.

```
module add32_a(clk,r,p,q,En);
input[31:0] p,q;input En,clk; output [32:0] r; reg[32:0]r,c; integer i;
always@(posedge En)        begin
        for(i=0;i<32;i=i+1)
        begin
                @(posedge clk)   begin
                                if(i==0) c[i]=1'b0;
                                {c[i+1'b1],r[i]}=fa(p[i],q[i],c[i]);
                                end
        end
        r[32]=c[32];
        $display( "t=%0d,  En = %b,  p = %0h, q = %0h, r = %0h
",$time,En,p,q,r);            end
function[1:0] ha;
input a,b; ha={a&b,a^b};
endfunction

function [1:0]fa;
input a,b,c; reg[1:0]a1,a2,aa2;
begin
        a1    = ha(a,b);
        aa2   = ha(a1[0],c);
        a2[1] = (aa2[1]|a1[1]);
```

continued

continued

```
        a2[0] =  aa2[0];
        fa    = a2;
end
endfunction
endmodule

module tst_add32a();
reg [31:0] p,q; reg En,clk; wire [32:0] r;
add32_a bb(clk,r,p,q,En);
always #1 clk=~clk;
initial   begin
          clk=1'b0;En=1'b0;p=32'h1234;q=32'h4321;
#1        En=1'b1;#100 En=1'b0;p=32'h12345678;q=32'h98765432;
#1        En=1'b1;#99 En=1'b0;p=32'habcdef12;q=32'hbbccddee;
#1        En=1'b1;#99 En=1'b0;p=32'hfedcba39;q=32'h13579bdf;
#1        En=1'b1;#99 En=1'b0;p=32'h9876abcd;q=32'hfedc8765;
#1        En=1'b1;#99 En=1'b0;p=32'hf0e0d0c0;q=32'h11020304;
#1        En=1'b1;#99 En=1'b0;
          end
initial #900 $stop;
endmodule
```

Figure 9.12 A 32-bit adder with the addition done in successive clock pulses.

```
# t=65,  En = 1,  p = 1234, q = 4321, r = 5555
# t=165, En = 1,  p = 12345678, q = 98765432, r = aaaaaaaa
# t=265, En = 1,  p = abcdef12, q = bbccddee, r = 1679acd00
# t=365, En = 1,  p = fedcba39, q = 13579bdf, r = 112345618
# t=465, En = 1,  p = 9876abcd, q = fedc8765, r = 197533332
# t=565, En = 1,  p = f0e0d0c0, q = 11020304, r = 101e2d3c4
```

Figure 9.13 Simulation results of the test bench for the adder in Figure 9.12.

9.2.1   Trade-off Between Hardware and Speed

Examples 9.5 and 9.7 represent two extreme cases of a trade-off between speed and hardware. Minimal hardware is used in Example 9.7 to carry out the addition, but the execution time is a maximum here due to the repeated and sequential use of the same hardware block. In contrast, in Example 9.5 the same hardware is replicated to the maximum extent and the addition is carried out "at one go", that is, in minimum time. Circuit-wise, it is a trade-off between silicon area and speed. One can have nibble or byte adders and do nibble-wise or byte-wise addition; these represent intermediate levels of trade-offs. Algebraic or logic operations, register-based operations, *etc.*, are other examples calling for similar trade-off decisions. Buswidth, memory organization, and ALU sizing all call for such trade-off decisions. In all such cases a decision may have to be based on considerations of speed of operation, power consumption, development time, cost, *etc.*

9.2.2   Scope of Functions

A few observations on functions and their use are in order here [IEEE].

- A function has only input arguments. It is to have at least one input. When a function with multiple input ports is called, the order of arguments in the calling statement should match that of the input declarations within the function definition.
- A function returns an output. It has no separate output ports.
- A function can have variables declared and used within it – these are variables local to the function.
- A function can be defined anywhere within the module.
- Event or timing based controls are not possible within a function. This restricts the function to be of a combinational logic type.
- A function can be called from within another function. Both the functions are to be defined within the module.
- A function in a module can be called from another module through proper hierarchical referencing.
- A function can be called repeatedly within the module of definition.
- Expressions can be used as arguments while calling a function.
- Definition of a function should not be within any initial or always block. or within another function.
- A function uses a register of the declared type and size to return the value of the output. Such a returned value can be **real**, **integer**, **time**, or **realtime** type. It can also be a vector with a range.
- Every variable declared inside a function has a corresponding location inside. These locations are physical entities. Each time a function is called, the same set of locations is reused. This is in contrast to the instantiation of a module where with every instantiation, a fresh set of locations is assigned.

9.2.3   Recursive Functions

Consider a function to compute the sum of the squares of the first n natural numbers: The sum designated as S_n can be expressed as

$S_n = .n^2 + (n - 1)^2 + \cdots\cdots + 3^2 + 2^2 + 1^2$
S_n can be expressed as
$S_n = n^2 + S_{n-1}$

where S_{n-1} represents the sum of the squares of the first $(n - 1)$ natural numbers. Thus if S_{n-1} were known, S_n can be obtained by adding n^2 to it. Continuing the same argument one can recursively arrive at the following:

$S_{n-1} = (n - 1)^2 + S_{n-2}$
$S_{n-2} = (n - 2)^2 + S_{n-3}$
...

...
$S_2 = 2^2 + S_1$

We know that

$S_1 = 1.$

The actual computation is carried out in the reverse order; that is, one computes S_1 directly and the subsequent sums S2, S3, *etc.*, are computed from it recursively – every sum by adding an increment to the previous sum.

A similar procedure can be adopted to compute factorials, infinite series and so on. Latest version of the LRM (2001) has expanded the scope of Functions to accommodate recursive functions. The keyword **automatic** following the keyword **function** implies it to be recursive. A recursive function can be called in the same manner as a nonrecursive function. Recursive function call is explained here through an example.

Example 9.8

The module sum_sq in Figure 9.14 computes the sum of the squares the first n natural numbers.

```
function automatic integer sum_sq;
input n;
begin
        if(n==1) sum_sq =1;
        else sum_sq = sum_sq + n*n;
end
endfunction
```

Figure 9.14 A module to compute the sum of squares of the first n natural numbers.

The term "**automatic**" in the function declaration statement ensures recursive computation. Thus if n is assigned the value 4, during compilation sum_sq (4) will be successively replaced by

sum_sq $(3) + 4^2$,
sum_sq $(4) + 3^2 + 4^2$,
sum_sq $(4) + 2^2 + 3^2 + 4^2$ and finally by
$1^2 + 2^2 + 3^2 + 4^2$.

9.3   TASKS

The role of a task in a module is similar to that of a subroutine in a program. It is defined within a module and can be called as many times as desired within a procedural block. Its scope and role are wider than those of a function.

9.3.1   Task Definition

The task definition is brought out in Figure 9.15. The first statement starts with the keyword task; it is followed by an identifier name and the customary semicolon. The input, inout, and the output declarations follow. Their order is not rigid. The body of the task comprises of a number of behavioral level statements. They may be executed in zero time or at specified time intervals or events. Thus the time of exit from a task can differ from that of entry to it.

9.3.2   Task Enabling

A task is enabled through a statement akin to the instantiation of a gate. It is enabled like a procedural assignment by specifying the task name followed by the list of arguments within brackets followed by the semicolon. A typical enabling statement has the form

Do_it (Expression1, Expression2, . .);

where
Do_it is the name of the task being enabled,
Expression1 is the first argument,
Expression2 is the second argument,
and so on.

The type and order of the arguments should match those of the respective declarations within the definition of the task. In a general case, an argument can be an expression. The following are characteristic of a task:

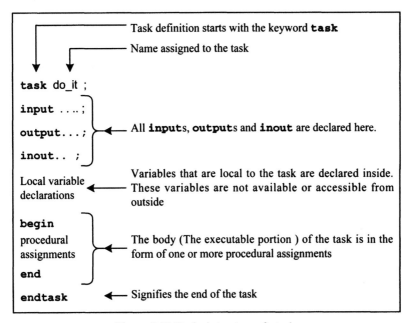

Figure 9.15 Typical structure of a task.

- A task can be activated by an event, sensitivity list, *etc.*
- A task can have activities assigned within it which are event-controlled or time-controlled.
- A task can have input, output and inout; however it need not necessarily have any of these; it can be complete in itself.
- A task can enable other tasks and functions.
- A task can call itself. The latest version of the LRM supports recursion. The keyword **automatic** is added to the keyword task to make it recursive.
- All assignments to a task are passed to it by value and not through a pointer to the argument.
- A task in a module can be invoked from another module through a hierarchical reference.

The arguments passed to a task retain their type within their environment of use. Thus a **wire**-type argument passed to a task as input cannot have its value altered within the task through an assignment.

There are no apparent restrictions on the input arguments of a task. They can be nets, regs, or expressions involving them. But any argument of inout or output type has to be a variable or of a similar type; the restrictions are similar to those on the quantities on the left side of procedural assignments. The use of tasks is illustrated through a set of four examples here.

Example 9.9

Figure 9.16 shows a module to count the total number of 1 bits in a nibble. A task has been defined to do the counting; the task has vector-type input and output; it has an integer defined within. The task has been invoked in the main module. A test-bench is also included in the figure. The simulated results for a set of inputs are given in Figure 9.17.

```
module oness_counter;
reg [3:0]x;reg [2:0]y;
always@(x)onescounter(x,y);

task onescounter;
input [3:0]x; output[2:0]y; integer i;
begin
        y=0;
        for(i=0;i<=3;i=i+1) if (x[i])y= y+1;
end
endtask

initial x=3'b000;
always #3 x=x+2'b11;
initial $monitor(" t=%0d, y= %b, x = %b ",$time,y,x);
initial #30 $stop;
endmodule
```

Figure 9.16 A module to count the number of 1 bits in a nibble.

```
#  t=0, y= 000, x = 0000
#  t=3, y= 010, x = 0011
#  t=6, y= 010, x = 0110
#  t=9, y= 010, x = 1001
#  t=12, y= 010, x = 1100
#  t=15, y= 100, x = 1111
#  t=18, y= 001, x = 0010
#  t=21, y= 010, x = 0101
#  t=24, y= 001, x = 1000
#  t=27, y= 011, x = 1011
```

Figure 9.17 Simulated results with the test bench in Figure 9.16.

Example 9.10

Figure 9.18 shows a module to divide a given clock with a given number. The scaling number can be changed if necessary. The task uses input, output and inout type of quantities. The waveforms of the input clock and the slower output clock obtained by simulating the test bench are shown in Figure 9.19.

```
module clk_tst;
reg clk,sclk;reg [3:0] n,nn;
always #2 clk=~clk;

task sl_clk;
input clk; input[3:0]nn; inout[3:0] n;
output sclk;
begin
        if(n!=4'h0)         begin
                                n    = n-1'b1;
                                sclk = 1'b0;
                            end
        else      begin
                                n    = nn;
                                sclk = 1'b1;
                      end
end
endtask

always @(negedge clk) sl_clk(clk,n,nn,sclk);
initial
begin
        clk=1'b0;nn=4'h2;n=nn; #45$stop;
end
initial $monitor($time, "n=%0d, clk=%0b, sclk=%0b",n,clk,sclk);
endmodule
```

Figure 9.18 A module to generate a slower clock from a given clock input.

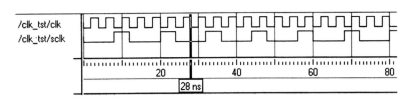

Figure 9.19 Simulation results of the module in Figure 9.18.

Example 9.11

The adder in Example 9.4 has been modified and shown in Figure 9.20. The half-
and full-adders have been defined as tasks and invoked to carry out the vector
addition. The half-adder has been invoked twice within the full-adder task. The
test-bench and simulation results are not repeated here. The module can be
directly expanded to add wider numbers.

```
module addertsk(r,p,q,En);
input[1:0] p,q; input En; output [2:0] r;
reg[2:0]r,c; integer i;
always@(posedge En)
        begin
                for(i=0;i<2;i=i+1)
                begin
                        if(i==0) c[i]=1'b0;
                        fa(p[i],q[i],c[i],{c[i+1'b1],r[i]});
                end
                r[2]=c[2];
                $display("t=%0d, En = %b, p = %b, q = %b, r = %b ",$time
                ,En,p,q,r);
        end

task ha;
input a,b; output[1:0] hfsum;
hfsum={a&b,a^b};
endtask

task fa;
input a,b,c; output[1:0]a2; reg[1:0]a1,aa2;
begin
        ha(a,b,a1);
        ha(a1[0],c,aa2);
        a2[1] = (aa2[1]|a1[1]);
        a2[0] =  aa2[0];
end
endtask
endmodule
```

Figure 9.20 A 2-bit adder using half-adder and full-adder tasks.

Example 9.12

The half-adder and full-adder tasks in Example 9.11 have been used to carry out addition of 2-bit numbers in the module of Figure 9.21. The addition has been carried out in successive clock pulses as with Example 9.6. The test bench and simulation results have been omitted. Once again the module can be redone easily to add wider numbers.

```
module addertskb(clk,r,p,q,En);
input[1:0] p,q;input En,clk;output [2:0] r;
reg[2:0]r,c;integer i;
always@(posedge En)
        begin
                for(i=0;i<2;i=i+1)
                begin
                        @(posedge clk)
                        if(i==0) c[i]=1'b0;
                        fatsk(p[i],q[i],c[i],{c[i+1'b1],r[i]});
                end
                r[2]=c[2];
                $display(" t=%0d, clk = %b, En = %b, p = %b, q = %b, r = %b
",$time,clk,En,p,q,r);
        end

task hatsk;
input a,b;output[1:0]ha;
ha={a&b,a^b};
endtask

task fatsk;
input a,b,c;output[1:0]a2;reg[1:0]a1,aa2;
begin
        hatsk(a,b,a1);
        hatsk(a1[0],c,aa2);
        a2[1] = (aa2[1]|a1[1]);
        a2[0] = aa2[0];
end
endtask
endmodule
```

Figure 9.21 Another 2-bit adder using half-adder and full-adder tasks.

9.4   USER-DEFINED PRIMITIVES (UDP)

The primitives available in Verilog are all of the gate or switch types. Verilog has the provision for the user to define primitives – called "user defined primitive (UDP)" and use them. A UDP can be defined anywhere in a source text and instantiated in any of the modules. Their definition is in the form of a table in a specific format. It makes the UDP types of functions simple, elegant, and attractive. UDPs are basically of two types – combinational and sequential. A combinational UDP is used to define a combinational scalar function and a sequential UDP for a sequential function.

9.4.1   Combinational UDPs

A combinational UDP accepts a set of scalar inputs and gives a scalar output. An **inout** declaration is not supported by a UDP. The UDP definition is on par with that of a module; that is, it is defined independently like a module and can be used in any other module. The definition cannot be within any other module.

Definition of a combinational type of UDP is illustrated through an example in Figure 9.22; it shows a simple UDP for an AND operation. The following are noteworthy:

- The first statement starts with the keyword "primitive", it is followed by the name assigned to the primitive and the port declarations.
- A UDP can have only one output port. It has to be the first in the port list.
- All the ports following the first are input ports and are all scalars.
- **inout** ports are not permitted in a UDP definition.
- Output and input are declared in the body of the UDP.

```
primitive udp_and (out, in1, in2);
output out;
input in1, in2;
table
//              In1             In2             Out
                0               0:              0;
                0               1:              0;
                1               0:              0;
                1               1:              1;
endtable
endprimitive
```

Figure 9.22 A two-input AND gate defined as a UDP.

- The behavior block of the primitive is given in the form of a table. It is specified between keywords **table** and **endtable**.
- The combinational function is defined as a set of rows (akin to the truth table).
- All the input values are specified first – each in a separate field in the same order as they appear in the port declaration.
- A colon and then the output value follow the set of input values. The statement ends with a semicolon – as with every statement in Verilog.
- A comment line is inserted in the example following the "**table**" entry. It facilitates understanding the tabular entries.
- All the inputs are nets – **wire**-type. Hence there is no need for a separate type definition.
- Output can be of the net or **reg** type depending upon the type of primitive – explained later.
- The last keyword statement – "**endprimitive**" – signifies the end of the definition.

9.4.2   More General Combinational UDPs

The UDP for the AND gate in Figure 9.22 specifies output values only for definite values of the inputs but not for their **x** states. A full and general definition of a UDP is characterized by the following additional factors:

- The output can take on only three values – **0, 1,** or **x**. It cannot take the value **z**.
- Outputs can be defined for **0, 1,** or **x** values of the inputs but not for the **z** state. However if an input takes the value **z**, it is taken as **x**.
- All the undefined input combinations lead to **x** state in the output. Hence it is desirable to specify outputs for all the possible input combinations.

Figure 9.23 shows the UDP definition of an AND gate with all the input combinations included. A test-bench for the UDP and the simulation results are shown in Figure 9.24.

A two-input UDP has nine rows of tabular entries; their number increases rapidly as the number of input logic variables increases. LRM has the provision to make the UDP definition more compact. The symbol "?" can be used to signify all the possible values – that is, 0, 1, or **x**. Figure 9.25 shows the elaborate AND gate UDP of Figure 9.23 made compact in this manner. Wherever possible, one can use the symbol "**b**" to signify "0" or "1" values and reduce the table size further.

Primitive udp_and (out, in1, in2);
Output out; //UDP of an AND gate defined fully
Input in1, in2;
Table

//	In1	In2	Out
	0	0:	0;
	0	1:	0;
	1	0:	0;
	1	1:	1;
	X	0:	0;
	X	1:	X;
	X	X:	X;
	0	X:	0;
	1	X:	X;

Endtable
Endprimitive

Figure 9.23 A more exhaustive definition of the two2-input AND gate UDP of Figure 9.21.

```
module tst_udp_and();
reg in1,in2; wire out;
udp_and uand(out,in1,in2);
initial begin in1=1'b0;in2=1'b0; end
always   begin
        #2 in1=1'b0;in2=1'b1;
        #2 in1=1'b1;in2=1'b0;
        #2 in1=1'b1;in2=1'b1;
        end
initial $monitor($time ,"in1 = %b ,in2 = %b ,out = %b ",in1,in2,out);
initial #18 $stop;
endmodule
```

Simulation results
```
//#          0in1 = 0 , in2 = 0 , out = 0
//#          2in1 = 0 , in2 = 1 , out = 0
//#          4in1 = 1 , in2 = 0 , out = 0
//#          6in1 = 1 , in2 = 1 , out = 1
//#          8in1 = 0 , in2 = 1 , out = 0
//#          10in1 = 1 , in2 = 0 , out = 0
//#          12in1 = 1 , in2 = 1 , out = 1
//#          14in1 = 0 , in2 = 1 , out = 0
//#          16in1 = 1 , in2 = 0 , out = 0
```

Figure 9.24 A test bench for the UDP module of Figure 9.23 and the simulation results.

```
Primitive udp_and_b (out, in1, in2);
Output out; // UDP of an AND gate defined compactly
Input in1, in2;
Table
```

//	In1	In2	Out
	?	0:	0;
	0	?:	0;
	x	X	x
	1	1:	1;

```
Endtable
Endprimitive
```

Figure 9.25 The UDP of Figure 9.22 made compact using the symbol "?".

9.4.3   Instantiation of an UDP

UDPs are instantiated in the same manner as gate primitives (see the test bench in Figure 9.24). It is further illustrated here through an example.

Example 9.13

The full adder accepts three input bits and outputs two bits – a sum bit and a carry bit. Figure 9.26 shows UDPs for the sum and the carry bits as well as a full adder module using them. Figure 9.27 shows a test-bench for the Full Adder as well as the simulation results.

```
primitive udpsum(sum, in1,in2,carryi);
output sum;
input in1, in2, carryi;
table
```

//	in1	in2	carryi:	sum
	0	0	0:	0;
	1	1	0:	0;
	0	1	1:	0;
	1	0	1:	0;
	1	0	0:	1;
	0	1	0:	1;
	0	0	1:	1;
	1	1	1:	1;

```
endtable
endprimitive
```

continued

continued

```
primitive udpcar(caro,in1,in2,cari); // This udp is  for carryout
output caro; input in1, in2, cari;

table
//      in1     in2     cari    caro
        0       0       ? :     0 ;
        0       ?       0 :     0 ;
        ?       0       0 :     0 ;
        b       1       1 :     1 ;
        1       b       1 :     1 ;
        1       1       b :     1 ;
endtable
endprimitive


module fa (car_o, sum_o, in1, in2, car_i);
input in1, in2, car_i; output car_o, sum_o;
udpcar aa(car_o,in1,in2,car_i);
udpsum bb(sum_o, in1,in2,car_i);
endmodule
```

Figure 9.26 A full adder module with the sum and carry bits generated through UDPs.

```
module fa_tst;
reg [2:0] a;wire c,s;integer i;
fa cc(c,s,a[0],a[1],a[2]);
initial for(i=1;i<8;i=i+1)
begin
        a=i;
#1      $display($time, "a=%b, cs=%b%b",a, c, s);
end
initial #10 $stop;
endmodule
```

Simulation results
```
#               1a=001, cs=01
#               2a=010, cs=01
#               3a=011, cs=10
#               4a=100, cs=01
#               5a=101, cs=10
#               6a=110, cs=10
#               7a=111, cs=11
```

Figure 9.27 A test bench for the full adder module of Figure 9.26 and the simulation results for the same.

Observations:

- With three inputs and three states for each input (0, 1, and **x**), the full table of definition has 27 entries. Such definitions become cumbersome as the number of inputs increase to even moderate values – say 4 or 5.
- Only the entries essential to the definition of the primitive are included here. Others which lead to **x** output are left out intentionally. Thus with the carry primitive if any two inputs have **x** values, the output car_o too has **x** value. Hence such a row has not been specified.
- "?" and "**b**" have been used in the primitive definition to make the tables more compact

9.4.4   Combinational UDP and Function

Definition-wise, UDP and function are similar, though their formats differ (*i.e.*, a UDP definition is in the form of a table while the function definition is as a sequence of procedural assignments). UDPs are stand-alone-type primitives and can be instantiated in any module. In contrast, a function is defined within a module; it cannot be accessed anywhere outside the module of definition.

9.4.5   Sequential UDPs

Any sequential circuit has a set of possible states. When it is in one of the specified states, the next state to be taken is described as a function of the input logic variables and the present state [Wakerly]. A positive or a negative going edge or a simple change in a logic variable can trigger the transition from the present state of the circuit to the next state. A sequential UDP can accommodate all these. The definition still remains tabular as with the combinational UDP. The next state can be specified in terms of the present state, the values of input logic variables and their transitions. The definition differs from that of a combinational UDP in two respects:

- The output has to be defined as a **reg**. If a change in any of the inputs so demands, the output can change.
- Values of all the input variables as well as the present state of the output can affect the next state of the output. In each row the input values are entered in different fields in the same sequence as they are specified in the input port list. It is followed by a colon (:). The value of the present state is entered in the next field which is again followed by a colon (:). The next state value of the output occupies the last field. A semicolon (;) signifies the end of a row definition (see the examples below).

As can be seen from the UDPs considered so far, its definition apparently calls for the use of a large number of tabular statements; it is all the more true of the sequential UDPs. Some shorthand notations are possible to make the UDP table more compact. All the notations that can be used are given in Table 9.1. Judicious selection and use of the symbols can make the tables compact.

Two examples of sequential UDPs are considered here – one being level-sensitive and the other edge-sensitive.

Example 9.14 A UDP for a D Latch

Figure 9.28 shows a UDP for a D latch (and a test bench for the same). It is an example of a level sensitive sequential UDP. The tabular description for the latch has been made succinct with the use of symbols – and ?. Any undefined input combination results in **x** value for the output; hence the output has not been separately defined for the **x** value of input in the table. Repeated use of the symbol ? has made the UDP table compact. The three rows of the table signify the following:

1. When clk = 1, if din = 0, the next state (qn) is also at 0 whatever be the value of present state (qp).
2. When clk = 1, if din = 1, the next state (qn) is also at 1 whatever be the value of present state (qp).
3. When clk = 0, the output (next state) does not change even if din changes.

Simulation results are shown in Figure 9.29.

Table 9.1 Symbols for UDP tabular rows

Symbol	Significance	Restrictions of use
B or **b**	0 and 1 values	Only in the input or current state fields
?	0, 1 or ,**x** value	Only in the input or current state fields
–	No change	Only in the output field of sequential UDP
(*mn*)	Change of value from *m* to *n*	Only in the input field. *m* & *n* can be 0, 1, **x**, **b**, or ?
*	Same as (??)	
r	Same as (01)	
f	Same as (10)	Only in the input field
p	Rise from 0 or **x** to **x** or 1	
n	Fall from 1 or **x** to **x** or 0	

```
primitive dlatch(q,din,clk);
output q; input din,clk; reg q;

table
//      din     clk     qp      qn
        0       1 :     ? :     0;        // If clk is at 1 state, the output
        1       1 :     ? :     1;        //follows the input.  If clk is at 0
        ?       0 :     ? :     -;        // state, the output remains frozen
endtable
endprimitive

module dlatch_tst;
wire q; reg din,clk;
dlatch ll(q,din,clk);
initial
begin
        clk=1'b1;din=1'b0;
        repeat (2)begin #4 din=1'b1; #4 din=1'b0; end
        clk=1'b0;repeat (2)begin #4 din=1'b1; #4 din=1'b0; end
        $stop;
end
initial $monitor($time ,"clk = %b, din = %b, q = %b ",clk,din,q);
endmodule
```

Figure 9.28 A D-latch module described as a level-sensitive UDP and a test bench for it.

```
#               0clk = 1, din = 0, q = 0
#               4clk = 1, din = 1, q = 1
#               8clk = 1, din = 0, q = 0
#              12clk = 1, din = 1, q = 1
#              16clk = 0, din = 0, q = 0
#              20clk = 0, din = 1, q = 0
#              24clk = 0, din = 0, q = 0
#              28clk = 0, din = 1, q = 0
```

Figure 9.29 Simulation results of running the test bench for the UDP of Figure 9.28.

Example 9.15 A UDP for an Edge-Triggered Flip-Flop

Figure 9.30 shows the UDP definition of a positive edge-triggered flip-flop with a
clear facility. In the table, (01) signifies the 0-to-1 transition edge of the clk – that
is, its positive edge. Other edge transitions too can be interpreted in a similar
manner. The simulation results are shown in Figure 9.31. From the simulation
results, one can see that as long as the Clear input is low, data input is latched in at
the positive going edge of the clock. But if the Clear input is high, its effect
prevails and the flip-flop output remains low and does not respond to changes in
the input data line.

```
primitive dff_pos(q,din,clk,clr);
output q;
input din,clk,clr;
reg q;
//initial q = 1'b0;

table
//     din      clk      clr      qp       qn        Whatever be the present
        0       (01)     0:       ?:       0;        // state of the output, at the
        1       (01)     0:       ?:       1;        // positive edge of clk input
        ?       (10)     0:       ?:       -;        // value is latched and
       (??)      ?       0:       ?:       -;        // output made equal to
        ?        ?       1:       ?:       0;        // that if clr = 0.  IF clr=1,
        ?        ?       *:       ?:       0;        // q .is made 0.
endtable
endprimitive
module dff_pos_tst;
wire q;
reg din,clk,clr;
dff_pos ll(q,din,clk,clr);
initial
begin
        clr=1'b0;din=1'b0;clk=1'b0;#3din=1'b1;
        repeat (2)begin #4 din=1'b1; #4 din=1'b0; end
        clr=1'b1;repeat (2) begin#4 din=1'b1; #4 din=1'b0; end
        $stop;
end
always  #2 clk=~clk;
initial $monitor($time ,"clr=%b, clk = %b, din = %b, q = %b ",clr,clk,din,q);
endmodule
```

Figure 9.30 An UDP for an edge-triggered flip-flop with clear facility: A test bench is also
included in the figure.

#	0clr=0, clk = 0, din = 0, q = 0
#	2clr=0, clk = 1, din = 0, q = 0
#	3clr=0, clk = 1, din = 1, q = 0
#	4clr=0, clk = 0, din = 1, q = 0
#	6clr=0, clk = 1, din = 1, q = 1
#	8clr=0, clk = 0, din = 1, q = 1
#	10clr=0, clk = 1, din = 1, q = 1
#	11clr=0, clk = 1, din = 0, q = 1
#	12clr=0, clk = 0, din = 0, q = 1
#	14clr=0, clk = 1, din = 0, q = 0
#	15clr=0, clk = 1, din = 1, q = 0
#	16clr=0, clk = 0, din = 1, q = 0
#	18clr=0, clk = 1, din = 1, q = 1
#	19clr=1, clk = 1, din = 0, q = 0
#	20clr=1, clk = 0, din = 0, q = 0
#	22clr=1, clk = 1, din = 0, q = 0
#	23clr=1, clk = 1, din = 1, q = 0
#	24clr=1, clk = 0, din = 1, q = 0
#	26clr=1, clk = 1, din = 1, q = 0
#	27clr=1, clk = 1, din = 0, q = 0
#	28clr=1, clk = 0, din = 0, q = 0
#	30clr=1, clk = 1, din = 0, q = 0
#	31clr=1, clk = 1, din = 1, q = 0
#	32clr=1, clk = 0, din = 1, q = 0
#	34clr=1, clk = 1, din = 1, q = 0

Figure 9.31 Simulation results of running the test bench for the UDP of Figure 9.30.

There can be situations where an edge sensitive entry in a UDP table clashes with a level-sensitive entry. In such situations of conflict, the level-sensitive entry dominates and decides the next state. The UDP in Figure 9.29 is sensitive to the level changes in one input (clr) and the edge in the other (clk). One can also have UDPs sensitive only to the edges in the inputs.

Observations:

- Only one edge transition can be specified in one line of the UDP definition. All other inputs are to be defined as state levels.
- If one edge of an input is used to specify a transition in the output, the output transition has to be defined for all possible edges of all the inputs.

A sequential UDP specifies the next state in terms of the present state and inputs. If necessary, one can specify an initial state and avoid ambiguity in operation at start. The initial declaration can be used here. Such an initial statement has to be a single procedural assignment. It can assign a 1(1'b1), a 0(1'b0), or an x value to the output reg of the UDP.

9.4.6   Sequential UDPs and Tasks

Sequential UDPs and tasks are functionally similar. Tasks are defined inside modules and used inside the module of definition. They are not accessible to other modules. In contrast, sequential UDPs are like other primitives and modules. They can be instantiated in any other module of a design.

9.4.7   UDP Instantiation with Delays

Outputs of UDPs also can take on values with time delays. The delays can be specified separately for the rising and falling transitions on the output. For example, an instantiation as

udp_and_b # (1, 2) g1(out, in1, in2);

can be used to instantiate the UDF of Figure 9.25 for carry output generation. Here the output transition to 1 (rising edge) takes effect with a time delay of 1 ns. The output transition to 0 (falling edge) takes effect with a time delay of 2 ns. If only one time delay were specified, the same holds good for the rising as well as the falling edges of the output transition.

9.4.8   Vector-Type Instantiation of UDP

UDP definitions are scalar in nature. They can be used with vectors with proper declarations. For example, the full-adder module fa in Figure 9.26 can be instantiated as an 8-bit vector to form an 8-bit adder. The instantiation statement can be

fa [7:0] aa(co, s, a, b, {co[6:0],1'b0});

s (sum), co (carry output), a (first input), and b(second input) are all 8-bit vectors here. The vector type of instantiation makes the design description compact; however, it may not be supported by some simulators.

9.5   EXERCISES

1. Define half-adder and full-adder as tasks and prepare a 32-bit adder using them. Test it through a suitable test bench.
2. Form a UDP for an A-O-I gate and test it through a test bench.

3. Form a UDP for a 3-to-1 mux and test it through a test bench.

4. b0, b1, and b2 represent the three bits of a mod-8 counter. The counter is to count at the positive edge of a clock input. Form UDPs for b0, b1, and b2; instantiate them in a module to form a counter. Test the counter using a test bench.

5. A 3-bit number is to advance through the following cyclic sequence:

   $0 - 3 - 5 - 4 - 1 - 0 - 3 \ldots$

   Form UDPs for the 3 bits; form the sequencer module by instantiating the UDPs. Test the module through a test bench.

6. Form a microcontroller core as follows:
   - Have a set of 4 registers designated r1, r2, r3, and r4.
   - Define a set of 6 algebraic / logic operations – Add, 1's complement, NAND, EXOR, left shift, and right shift
   - Have an 8-bit instruction opcode as *ssddpaaa*. Here *ss*, *dd* and *aaa* specify the source address, the destination address and the 3-bit code for the algebraic/logic opreration, respectively. P is a single-bit mode selector – if $p = 0$, data are to be transferred from source to the destination; if $p = 1$, the algebraic/logic operation is to be done.
   - Define each of the operations above as a function or as a task.

   Realize the ALU functions as UDPs. Realize the whole module using the **case** statement. For example, 01111101 stands for taking data from r1 and r3, adding them and putting the result in r1. Use r1 to store the result. Have a separate status register with carry bit and zero bit: set them whenever necessary. Write a test bench for the microcontroller, and test each of the instructions and instruction sequences.

7. Consider Figure 9.12: Shift the statement r[32] = c[32]; ahead by one line. Include a **$display** statement in both cases: Simulate the test bench. Explain any difference.

10

SWITCH LEVEL MODELING

10.1   INTRODUCTION

In today's environment the MOS transistor is the basic element around which a VLSI is built. Designers familiar with logic gates and their configurations at the circuit level may choose to do their designs using MOS transistors. Verilog has the provision to do the design description at the switch level using such MOS transistors, which is the theme of the present chapter. Switch level modeling forms the basic level of modeling digital circuits. The switches are available as primitives in Verilog; they are central to design description at this level. Basic gates can be defined in terms of such switches. By repeated and successive instantiation of such switches, more involved circuits can be modeled – on the same lines as was done with the gate level primitives in Chapters 4 and 5.

Designers familiar with logic gates, digital functional blocks, and their interplay can successfully carry out a complete VLSI design without any involvement at the switch level. Hence the switch level design was deferred to the present chapter.

10.2   BASIC TRANSISTOR SWITCHES

Consider an NMOS transistor of the depletion type. When used in a digital circuit, it can be in one of three modes:

- $V_G < V_S$ where V_G and V_S are the gate and source voltages with respect to the drain: The transistor is OFF and offers very high impedance across the source and the drain. It is in the z state.
- $V_G \cong V_S$: The transistor is in the active region. It presents a resistance between the source and the drain. The value depends on the technology. Such a resistive state of the transistor can be modeled in Verilog. A transistor in this mode can be represented as a resistance in Verilog – as **pull1** or **pull0** depending on whether the drain is connected to **supply1** or source is connected to **supply0**.

- $V_G > V_S$: The transistor is fully turned on. It presents very low resistance (~0 Ω) between the source and drain.
- An enhanced-mode NMOS transistor also has the above three modes of operation. It is OFF when $V_G \cong V_S$. It is moderately ON or in the active region when V_G is slightly greater than V_S, representing a resistive (**pull1** or **pull0**) mode of operation. When V_G is sufficiently greater than V_S, the transistor is in the on state representing very low (~0 Ω) resistance. Similar modes are possible for the PMOS transistor also. The modes and the voltage levels for each are summarized in Table 10.1.

The table is more for information and not of direct relevance to design description in Verilog. Whenever a switch primitive is present in a design, necessary biasing will be done automatically. The designer need not worry about it – at least at this stage.

10.2.1   Basic Switch Primitives

Different switch primitives are available in Verilog. Consider an **nmos** switch. A typical instantiation has the form

nmos (out, in, control);

nmos – a keyword – represents an NMOS transistor functioning as a switch. The switch has three terminals (see Figure 10.1) – in, out, and control. When the control input is at 1 (high) state, the switch is on. It connects the input lead to the output side and offers zero impedance. When the control input is low, the switch is OFF and output is left floating (**z** state). If the control is in the **z** or the **x** state, output may take corresponding values. Table 10.2 summarizes the input / output combinations. In the table the symbol "**L**" stands for 0 or z state. The symbol **H** stands for the 1 or z state.

The keyword **pmos** represents a PMOS transistor functioning as a switch. The PMOS switch has three terminals (see Figure 10.2). A typical instantiation of the switch has the form

Table 10.1 Operating voltages for different modes of operation of MOS switches

Mode		NMOS		PMOS	
		Depletion	Enhancement	Depletion	Enhancement
$V_D - V_S$ for normal operation (Range: 1.5V to 5V)		Positive	Positive	Negative	Negative
Range of $V_G - V_S$ for	OFF (**z**) state	Negative	$\cong 0$	Positive	Positive
	Resistive state (**pull1**, **pull0**)	$\cong 0$	Mildly positive	$\cong 0$	Mildly negative
	ON state (0Ω)	Positive	Fully positive	Negative	Fully negative

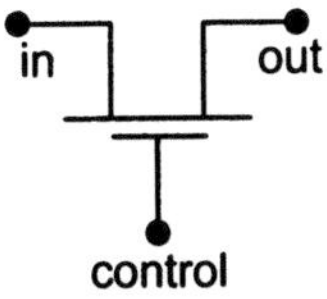

Figure 10.1 An NMOS switch with terminals.

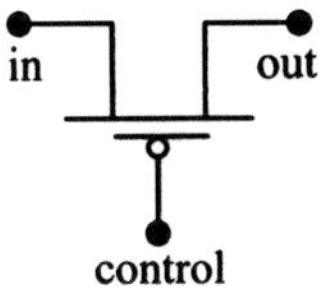

Figure 10.2 A PMOS switch with terminals.

pmos (out, in, control);

When the control is at 1 (high) state, the switch is off. Output is left floating. When control is at 0 (low) state, the switch is on, input is connected to output, and output is at the same state as input. For other input values, output is at other values. The output values for all possible input and control values are shown in Table 10.3. The symbols **L** and **H** have the same significance as in Table 10.2.

Observations:

- When in the on state, the switch makes its output available at the same strength as the input. There is only one exception to it: When the input is of strength **supply**, the output is of strength **strong**. It is true of **supply1** as well as **supply0**.
- When instantiating an **nmos** or a **pmos** switch, a name can be assigned to the switch. But the name is not essential. (The same is true of the other primitives discussed in the following sections as well.)
- The **nmos** and **pmos** switches function as unidirectional switches.

Table 10.2 Output values of an nmos switch for different values of signal and control inputs

		Control (input)			
		0	1	X	z
(Data) input	0	z	0	L	L
	1	z	1	H	H
	X	z	X	X	X
	z	z	z	z	z

Table 10.3 Output values of a pmos switch for different values of signal and control inputs

		Control (input)			
		0	1	X	z
(Data) input	0	z	0	L	L
	1	z	1	H	H
	X	z	X	X	X
	z	z	z	z	z

10.2.2   Resistive Switches

nmos and pmos represent switches of low impedance in the on-state. rnmos and rpmos represent the resistive counterparts of these respectively. Typical instantiations have the form

rnmos (output1, input1, control1);
rpmos (output2, input2, control2);

    With rnmos if the control1 input is at 1 (high) state, the switch is ON and functions as a definite resistance. It connects input1 to output1 through a resistance. When control1 is at the 0 (low) state, the switch is OFF and leaves output1 floating. The set of output values for all combinations of input1 and control1 values remain identical to those of the nmos switch given in Table 10.2.

    The rpmos switch is ON when control2 is at 0 (low) state. It inserts a definite resistance between the input and the output signals but retains the signal value. The output values for different input values remain identical to those in Table 10.3 for the pmos switch.

Observations:

- Because rpmos and rnmos are resistive switches, they reduce the signal strength when in the on state. The reduced strength is mostly one level below the original strength. The only exceptions are small and hi-z. For these the strength and the state remain unaltered (see Table 10.4).
- The rpmos and rnmos switches function as unidirectional switches; the signal flow is from the input to the output side.

Table 10.4 Output-side strength levels for different input-side strength values of rnmos, rpmos, and rcmos switches

Input strength	Output strength
Supply – drive	Pull – drive
Strong – drive	Pull – drive
Pull – drive	Weak – drive
Weak – drive	Medium – capacitive
Large – capacitive	Medium – capacitive
Medium – capacitive	Small – capacitive
Small – capacitive	Small – capacitive
High impedance	High impedance

10.2.3   pullup and pulldown

A MOS transistor functions as a resistive element when in the active state. Realization of resistance in this form takes less silicon area in the IC as compared to a resistance realized directly. **pullup** and **pulldown** represent such resistive elements. A typical instantiation here has the form

pullup (x);

Here the net x is pulled up to the **supply1** through a resistance. Similarly, the instantiation

pulldown(y);

pulls y down to the **supply0** level through a resistance. The **pullup** and **pulldown** primitives can be used as loads for switches or to connect the unused input ports to V_{CC} or GND, respectively. They can also form loads of switches in logic circuits. The default strengths for **pullup** and **pulldown** are **pull1** and **pull0** respectively. One can also specify strength values for the respective nets. For example,

pullup (strong1) (x)

specifies a resistive **pullup** of net x to **supply1**. One can also assign names to the **pullup** and **pulldown** primitives. Thus

pullup (strong1) rs(x)

represents an instantiation of **pullup** designated rs having strength **strong1**.

Difference between **tri** and **pullup** or **pulldown** is to be understood clearly. **pullup** is a functional element; it represents a resistive connection to **supply1**. In contrast **tri1** is a type of net; in the absence of an assignment, it remains connected to **supply1**. A similar difference exists between **pulldown** and **tri0**. The example below brings out the differences.

Example 10.1

Figure 10.3 shows two simple circuits that are apparently identical: Figure 10.3 (a) has the net o1 declared as tr1 and is pulled up in case it is left open. With the circuit in Figure 10.3(b), o2 is a wire type of net; it has a resistive element connecting it to **supply1**. Figure 10.4 shows a module incorporating both the circuits and a test bench for them. Note that the module instantiates the primitive **bufif1** for the controlled buffer (discussed in Chapter 4). The test bench has specific assignments to the two input signals which bring out the difference in contention resolution. For identical input signal values, the outputs o1 and o2 can differ in certain cases.

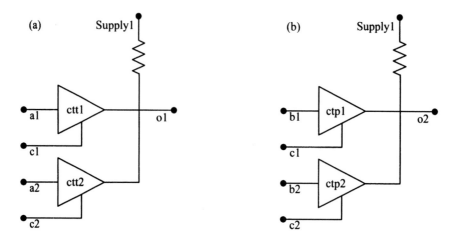

Figure 10.3 Two circuits to demonstrate the difference between `tri1` and `pullup`.

- At t = 0, all the inputs are at **x** state; o1 is also at **x** state: But o2=1 because of the **pullup**.
- At t = 1, all the switches are turned off; o1 and o2 are at 1 state.
- At t = 2, all the switches are on; the outputs follow the inputs and both of them are at 1 state.
- At t = 3, ctt1 & ctp1 are on; ctt2 & ctp2 are off; o1=a1=1 but o2=1 since its value is the result of contention resolution between b1 at **weak0** & the stronger pp at **pull1**.
- At t = 4, all switches are on; all inputs are at 0; o1=0 but o2=1 since the stronger pull1 prevails over the weaker 0's of b1 & b2.

```
module swt_aa (o1,o2, a1,a2,b1,b2,c1,c2);
output o1,o2; input a1,a2,b1,b2,c1,c2;
wire o2; tri1 o1;
bufif1 ctt1(o1,a1,c1), ctt2(o1,a2,c2);
bufif1 (weak1, weak0) ctp1(o2,b1,c1), ctp2(o2,b2,c2);
pullup pp(o2);
endmodule

module swt_aa_tb;
reg [5:0] rx; wire o1,o2, a1,a2,b1,b2,c1,c2;
assign {a1,a2,b1,b2,c1,c2} = rx;
swt_aa aa(o1,o2, a1,a2,b1,b2,c1,c2);
initial begin
```

continued

continued

```
#1      rx=6'o00;
#1      rx=6'o77;
#1      rx=6'o42;
#1      rx=6'o03;
#1      $stop;
     end
initial $monitor("time=%0d, c1=%b, o1=%b, a1=%b, a2=%b, c2=%b, o2=%b,
b1=%b, b2=%b",$time, c1,o1,a1,a2,c2,o2,b1,b2);
endmodule
```

#t	c1	o1	a1	a2	c2	o2	b1	b2
#0	x	x	x	x	x	1	x	x
#1	0	1	0	0	0	1	0	0
#2	1	1	1	1	1	1	1	1
#3	1	1	1	0	0	1	0	0
#4	1	0	0	0	1	1	0	0

Figure 10.4 Design module for the circuits of Figure 1.3 and its test bench.

Example 10.2 CMOS inverter

A CMOS inverter is formed by connecting an **nmos** and a **pmos** switch in series across the supply (see Figure 10.5). The output terminals are joined together to form the common output. Similarly, the input is used as the common control input to both the switches. Referring to the figure, we can see the following:

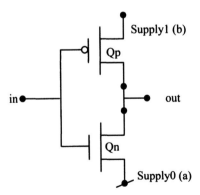

Figure 10.5 A CMOS inverter formed by connecting an NMOS and a PMOS set of transistors in series across the supply.

- When the input is low (0 V), transistor Q_n is off. But Q_p is on. **supply1** is connected to the output. Hence the output is high.
- When the input is high (5 V), transistor Q_p is off. But Q_n is on. **supply0** is connected to the output. Hence the output is low.

The inverter operation is clear from the above. The design description for the corresponding CMOS inverter is shown in Figure 10.6. The leads a and b are declared as nets – **supply0** and **supply1** respectively; *i.e.*, they are connected to ground and V_{CC} respectively. The two instantiations together describe the inverter operation.

Observations:

- Under steady-state operation of the CMOS inverter, only Q_p or Q_n is ON at a time. Hence the inverter does not draw any quiescent current from the supply. Current is drawn only to charge the internal capacitor associated with the transistors during the transition.
- The input and output sides of the switches refer to the signal flow directions and not that of the current flow. Thus for the NMOS switch under the ON condition, current flows from out to **supply0**. But the signal from a (at **supply0**) is made available at out.

Example 10.3 CMOS NOR gate

A CMOS nor gate with two inputs is shown in Figure 10.7. It employs four transistors.

- When only in1 is high, Q_{n1} is ON pulling out to **suppy0**. Output is zero. Q_{p2} is also on. But since in2 is low, Q_{p1} is off. Hence no current can be drawn from **supply1.**
- When only in2 is high, Q_{n2} is ON pulling out to **suppy0**. Output is zero. Q_{p1} is also on. But since in1 is low, Q_{p2} is off. Hence no current can be drawn from **supply1**.

```
module inv (in, out );
output out;
input in;
supply0 a;
supply1 b;
nmos (out, a, in );
pmos (out, b, in);
endmodule
```

Figure 10.6 design description of a CMOS inverter formed by connecting an NMOS transistor and a PMOS transistor in series.

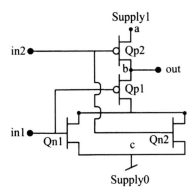

Figure 10.7 A 2 input CMOS NOR gate.

- When both in1 and in2 are low, Q_{n1} and Q_{n2} are OFF. Q_{p1} and Q_{p2} are ON and out is connected to **supply1**. But no current is drawn from the supply.

When both in1 and in2 are high, Q_{n1} and Q_{n2} are on. Out is grounded at c. Since Q_{p1} and Q_{p2} are off, no current is drawn from supply1. The design description for the NOR gate is shown in Figure 10.8. It has four instantiations – two of **pmos** and two of **nmos**, respectively.

```
module npnor_2(out, in1, in2 );
output out;
input in1, in2;
supply1 a;
supply0 c;
wire b;
pmos(b, a, in2), (out, b, in1);
nmos (out, c, in1), (out, c, in2) ;
endmodule
```

Figure 10.8 design description of a CMOS NOR gate.

Observations:

- A three-input NOR gate has three NMOS transistors in parallel on the ground side and three PMOS transistors in series on the V_{CC} side. Although the number of inputs can be increased in this manner, circuit operation is not satisfactory for more than two or three inputs [Bogart].
- NAND gate is formed by connecting the NMOS transistors in series on the ground side and the PMOS transistors in parallel on the V_{CC} side (NAND is the dual of NOR).

- Because NAND and NOR are universal gates, all other logic gates can be realized in terms of them.

Example 10.4 NMOS Inverter with Pull up Load

Figure 10.9 shows an NMOS inverter. Q_1 is the NMOS transistor. Q_2 properly biased, forms an active resistance and is the load on Q_1. The design description for the inverter is shown in Figure 10.10. When in = 0, Q_1 is OFF and out is pulled up and is at state 1. When in = 1, Q_1 is ON and out is pulled down to 0. A test bench and the results of simulating the test bench are also included in the figure.

Observations:

- When Q_1 is ON (in = 1), the gate has a standing current; contrast this with CMOS inverter in Example 10.2, where the quiescent current is always zero. The output is available as **strong0**.
- When Q_1 is OFF, the standing current is zero. But the output is available as **pull1**. Thus there is a difference in the strengths of the two states.
- If necessary, a different strength value can be assigned to **pullup**.

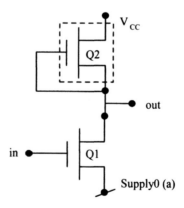

Figure 10.9 An NMOS inverter with an active pull up load.

```
module NMOSinv(out,in);
output out;input in;supply0 a;
pullup (out);
nmos(out,a,in);
endmodule
module tst_nm_in();
reg in;wire out;
NMOSinv nmv(out,in);
initial
in =1'b1;
always
#3 in =~in;
initial $monitor($time , " in = %b, output = %b ",in,out);
initial #30 $stop;
endmodule
```

#		
#	0 in = 1, output = 0	
#	3 in = 0, output = 1	
#	6 in = 1, output = 0	
#	9 in = 0, output = 1	
#	12 in = 1, output = 0	
#	15 in = 0, output = 1	
#	18 in = 1, output = 0	
#	21 in = 0, output = 1	
#	24 in = 1, output = 0	
#	27 in = 0, output = 1	

Figure 10.10 Design description of an NMOS inverter gate: A test bench for the inverter and the simulation results are also shown in the figure.

Example 10.5 An NMOS Three Input NOR Gate

Figure 10.11 shows a three-input NMOS NOR gate with Q4 – properly biased – forming a resistive pullup load. Output b is high when all the inputs – in1, in2 and in3 are low – keeping the respective mos transistors – Q1, Q2, and Q3 – off.   If any one of the three inputs goes high, the corresponding NMOS transistor turns ON and the output b is pulled down to zero. When output is in 1 state, it has strength **pull1**. When in the zero state, it has strength **strong0**. The design description for the gate is shown in Figure 10.12. Simulation results are given in Figure 10.13.

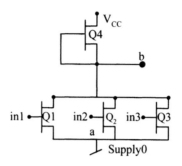

Figure 10.11 An NMOS NOR gate with active pull up.

```
module nor3NMOS(in1,in2,in3,b);
output b;
input in1,in2,in3;
supply0 a; wire b;
nmos(b,a,in1),(b,a,in2),(b,a,in3);
pullup(b);
endmodule

module tst_nor3NMOS();
reg in1,in2,in3;wire b;
nor3NMOS nn(in1,in2,in3,b);
initial
begin
in1=1'b1;in2=1'b1;in3=1'b1;
end
always #2 in1=~in1;
always #3 in2=~in2;
always #5 in3=~in3;
initial $monitor($time , "in1 = %b , in2 = %b , in3 = %b , output = %b
",in1,in2,in3,b);
initial #24 $stop;
endmodule
module  (b, in1, in2, in3 );
output b;
input in1, in2, in3;
supply0 a;
wire b;
nmos (b, a, in1), (b, a, in2), (b, a, in3) ;
pullup (b ) ;
endmodule
```

Figure 10.12 Design description of an NMOS NOR gate with active pull up.

#	0in1 = 1 , in2 = 1 , in3 = 1 , output = 0
#	2in1 = 0 , in2 = 1 , in3 = 1 , output = 0
#	3in1 = 0 , in2 = 0 , in3 = 1 , output = 0
#	4in1 = 1 , in2 = 0 , in3 = 1 , output = 0
#	5in1 = 1 , in2 = 0 , in3 = 0 , output = 0
#	6in1 = 0 , in2 = 1 , in3 = 0 , output = 0
#	8in1 = 1 , in2 = 1 , in3 = 0 , output = 0
#	9in1 = 1 , in2 = 0 , in3 = 0 , output = 0
#	10in1 = 0 , in2 = 0 , in3 = 1 , output = 0
#	12in1 = 1 , in2 = 1 , in3 = 1 , output = 0
#	14in1 = 0 , in2 = 1 , in3 = 1 , output = 0
#	15in1 = 0 , in2 = 0 , in3 = 0 , output = 1
#	16in1 = 1 , in2 = 0 , in3 = 0 , output = 0
#	18in1 = 0 , in2 = 1 , in3 = 0 , output = 0
#	20in1 = 1 , in2 = 1 , in3 = 1 , output = 0
#	21in1 = 1 , in2 = 0 , in3 = 1 , output = 0
#	22in1 = 0 , in2 = 0 , in3 = 1 , output = 0

Figure 10.13 Results of running the test bench in Figure 10.12.

Observations:

- When any of the inputs is high, the corresponding transistor is ON and the gate has a standing current. The standing current is zero only when all the three inputs are at zero state and Q_1, Q_2, and Q_3 are off. The standing current makes the power dissipation in the device much higher than that for its CMOS counterpart.

- Adding transistors in parallel can increase the number of inputs.

- NAND gate can be formed by connecting the NMOS transistors controlled by the inputs in series. However, NOR remains the preferred universal gate element with NMOS logic.

- One can use a **pmos**-type switch at the top with a **pulldown** type of load to the ground. It forms a PMOS inverter (see Figure 10.14). The different logic gates of PMOS technology can be built with it. Here again, due to the standing current, the power consumption of the device will be much higher than that of its CMOS counterpart.

- For any logic function the **nmos** or the **pmos** gate uses a much smaller number of transistors than does the CMOS gate. Despite this CMOS logic family stands out due to two reasons:-
  - Lowest power consumption
  - Uniformity in the element patterns

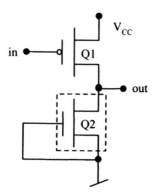

Figure 10.14 A PMOS inverter with active pull down load.

The advantages of CMOS technology often far outweigh the apparent complexity of the larger number of devices required on a per gate basis. Hence CMOS has proved to be much more popular.

10.3   CMOS SWITCH

A CMOS switch is formed by connecting a PMOS and an NMOS switch in parallel – the input leads are connected together on the one side and the output leads are connected together on the other side. Figure 10.15 shows the switch so formed. It has two control inputs:

- N_control turns ON the NMOS transistor and keeps it ON when it is in the 1 state.
- P_control turns ON the PMOS transistor and keeps it ON when it is in the 0 state.

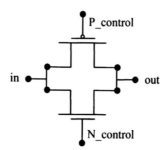

Figure 10.15 A CMOS switch formed by connecting a PMOS transistor and an NMOS transistor in parallel.

The CMOS switch is instantiated as shown below.
cmos csw (out, in, N_control, P_control);

Significance of the different terms is as follows:

- **cmos** : The keyword for the switch instantiation
- csw: Name assigned to the switch in the instantiation
- out: Name assigned to the output variable in the instantiation
- in: Name assigned to the input variable in the instantiation
- N_control: Name assigned to the control variable of the NMOS transistor in the instantiation
- P_control: Name assigned to the control variable of the PMOS transistor in the instantiation

Example 10.6 CMOS Switch – 1

Being a parallel combination of a PMOS and an NMOS switch, the CMOS switch can be realized by instantiating these to form a parallel switch. Design description of such a switch is shown in Figure 10.16 along with a test bench. The controls for the NMOS and the PMOS sides are separate in the primitive. The (partial) simulation results are shown in Figure 10.17.

```
module CMOSsw(out,in,n_ctr,p_ctr);
output out; input in,n_ctr,p_ctr;
nmos gn(out,in,n_ctr);
pmos gp(out,in,p_ctr);
endmodule

module tst_CMOSsw();
reg in,n_ctr,p_ctr; wire out;
CMOSsw cmsw(out,in,n_ctr,p_ctr);
initial begin in=1'b0;n_ctr=1'b1;p_ctr=~n_ctr; end
always #5 in =~in;
always begin #3 n_ctr=~n_ctr; #0p_ctr =~n_ctr; end
initial $monitor($time , "in = %b , n_ctr = %b , p_ctr = %b , output = %b
",in,n_ctr,p_ctr,out);
initial #39 $stop;
endmodule
```

Figure 10.16 Design description of a CMOS switch formed by paralleling a pair of NMOS and PMOS switches.

#	0in = 0 , n_ctr = 1 , p_ctr = 0 , output = 0
#	3in = 0 , n_ctr = 0 , p_ctr = 1 , output = z
#	5in = 1 , n_ctr = 0 , p_ctr = 1 , output = z
#	6in = 1 , n_ctr = 1 , p_ctr = 0 , output = 1
#	9in = 1 , n_ctr = 0 , p_ctr = 1 , output = z
#	10in = 0 , n_ctr = 0 , p_ctr = 1 , output = z
#	12in = 0 , n_ctr = 1 , p_ctr = 0 , output = 0

Figure 10.17 Partial results of simulating the test bench for the CMOS switch in Figure 10.16.

Example 10.7 CMOS Switch – 2

In normal use of a CMOS switch, the same control line drives the gates of the PMOS and the NMOS switches (as shown in Figure 10.18). With this change the switch becomes more compact for description as well. The module for the compact switch is shown in Figure 10.19; the figure also includes a test bench for it. The design module uses an instantiation of the NOT gate for generating P_control from con – the control input. The (partial) simulation results are in Figure 10.20.

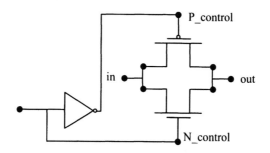

Figure 10.18 A CMOS switch with a single control line.

```
module CMOSsw1(out,in,con);
output out; input in,con; wire p_ctr;
not gn(p_ctr,con);
cmos gc(out,in,con,p-ctr);
endmodule
```

continued

continued

module tst_CMOSsw1();
reg in,con; wire out;
CMOSsw1 cmsw(out,in,con);
initial begin in=1'b0;con=1'b1; end
always #5 in =~in;
always #3 con=~con;
initial $monitor($time , "in = %b , con = %b , output = %b " ,in,con,out);
initial #40 $stop;
endmodule

Figure 10.19 Design description of a CMOS switch with a single control input.

#	0in = 0 , con = 1 , output = 0
#	3in = 0 , con = 0 , output = x
#	5in = 1 , con = 0 , output = x
#	6in = 1 , con = 1 , output = 1
#	9in = 1 , con = 0 , output = x
#	10in = 0 , con = 0 , output = x
#	12in = 0 , con = 1 , output = 0

Figure 10.20 Partial results of simulating the test bench for the CMOS switch in Figure 10. 19.

Example 10.8 A RAM Cell

Figure 10.21 shows a basic ram cell with facilities for writing data, storing data, and reading data. When switch sw2 is on, qb – the output of inverter g1 – forms the input to the inverter g2 and vice versa. The g1-g2 combination functions as a latch and freezes the last state entry before sw2 turns on. The step-by-step function of the cell is as follows (see the waveforms in Figure 10.22):

- When wsb (write/store) is high, switch sw1 is ON, and switch sw2 OFF. With sw1 on, input Din is connected to the input of gate g1 and remains so connected.
- When wsb goes low, din is isolated, since sw1 is OFF. But sw2 is ON and the data remains latched in the latch formed by g1-g2. In other words the data Din is stored in the RAM cell formed by g1-g2.
- When RD (Read) goes active (=1), the latched state is available as output Do. Reading is normally done when the latch is in the stored state.

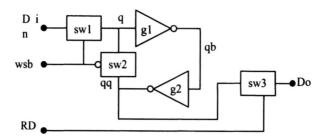

Figure 10.21 Basic RAM cell in block diagram form.

The design description for the ram cell as well as a test bench for it is given in Figure 10.23. It instantiates a **csw** module which is a basic CMOS switch with a single control line. If necessary, the **not** gate can be separately defined as a CMOS gate module and instantiated. Note that the output of gate **g2** – **qq**- has been declared as a **trireg** type of net. It is to ensure that the **q2** output is stored when **sw2** is OFF. It avoids any error during transition – that is, **sw2** turning off with a delay compared to that of **sw1**. The (partial) simulation results are in Figure 10.24. A full-fledged memory can be built using the ram cell. The memory address decoders are to form the enable signals to the write and read control signals here.

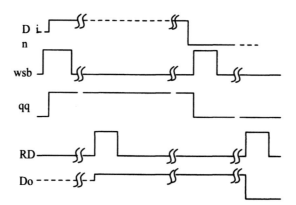

Figure 10.22 Waveforms of different signals in the operation of the basic RAM cell of Figure 10.21.

```
module ram_cell(do,din,wsb,rd);
output do; input din,wsb,rd; wire sb; wire q,qq; tri do;
csw sw1(q,din,wsb),sw2(q,qq,sb),sw3(do,q,rd);
not n1(sb,wsb),n2(qb,q),n3(qq,qb);
endmodule

module csw(out,in,n_ctr);
output out; input in,n_ctr; wire p_ctr;
assign p_ctr =~n_ctr;
cmos csw(out,in,n_ctr,p_ctr);
endmodule

module tst_ramcell();
reg din,wsb,rd; wire do;
ram_cell mc(do,din,wsb,rd);
initial begin din=1'b0;wsb=1'b0;rd=1'b0; end
always #10 din =~din;
always begin #3wsb=1'b1; #8wsb=1'b0; end
always begin #2 rd=1'b1; #5 rd =1'b0; end
initial $monitor ($time," rd= %b ,wsb = %b ,din = %b ,do = %b ",rd,wsb,din,do);
initial #40 $stop;
endmodule
```

Figure 10.23 Design description of a basic RAM cell.

```
#           0 rd= 0 ,wsb = 0 ,din = 0 ,do = z
#           2 rd= 1 ,wsb = 0 ,din = 0 ,do = x
#           3 rd= 1 ,wsb = 1 ,din = 0 ,do = 0
#           7 rd= 0 ,wsb = 1 ,din = 0 ,do = z
#           9 rd= 1 ,wsb = 1 ,din = 0 ,do = 0
#          10 rd= 1 ,wsb = 1 ,din = 1 ,do = 1
#          11 rd= 1 ,wsb = 0 ,din = 1 ,do = 1
#          14 rd= 0 ,wsb = 1 ,din = 1 ,do = z
#          16 rd= 1 ,wsb = 1 ,din = 1 ,do = 1
#          20 rd= 1 ,wsb = 1 ,din = 0 ,do = 0
#          21 rd= 0 ,wsb = 1 ,din = 0 ,do = z
```

Figure 10.24 Partial results of simulating the test bench for the CMOS switch in Figure 10.23.

Example 10.9 An Alternate RAM Cell Realization

Figure 10.25 shows an alternate and apparently simpler version of the ram cell (ram_1). The two inverters are connected permanently in a back-to-back fashion. Their output strength levels are **pull1** and **pull0**. Din can be of strength **strong**. Hence when the data write switch (sww) is turned ON, Din prevails and forces q to its own state. The condition is latched and remains so after switch sww is turned OFF. Another data can be written again by turning ON switch sww after making the new data available at Din. Data can be read out of the latch by turning on the switch – swr. It has the control line RD.

The module of the ram_1 cell is shown in Figure 10.26; the figure also includes a test bench. The design uses two instantiations of the **not** gate with strength **pull1** and **pull0**. The switches sww and swr are realized through instantiations of the CMOS switch modules csw. (Alternately, the same can be defined as a function inside the ram1 module and used as such.) Partial simulation results are shown in Figure 10.27. By adding address decoding and clock, the cell can be used as the basis for forming a full-fledged ram.

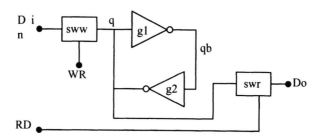

Figure 10.25 An alternate version of the RAM cell in block diagram form.

```
module ram1(do,din,wr,rd);
output do; input din,wr,rd; wire qb,q; tri do;
scw sww(q,din,wr),swr(do,q,rd);
not(pull1,pull0)n1(qb,q),n2(q,q);
endmodule

module scw(out,in,n_ctr);
output out; input in,n_ctr; wire p_ctr;
assign p_ctr =~n_ctr;
cmos sw(out,in,n_ctr,p_ctr);
endmodule
```

continued

continued

```
//test-bench
module tst_ram1();
reg din,wr,rd; wire do;
ram1 mm(do,din,wr,rd);
initial begin din=1'b0;wr=1'b0;rd=1'b0; end
always #10 din =~din;
always begin #3wr=1'b1; #8wr=1'b0; end
always  begin #2 rd=1'b1; #5 rd =1'b0; end
initial $monitor ($time," rd= %b ,wr = %b ,din = %b ,do = %b ",rd,wr,din,do);
initial #40 $stop;
endmodule
```

Figure 10.26 Design description of the RAM cell of Figure 10.24.

#	0 rd= 0 ,wr = 0 ,din = 0 ,do = z
#	2 rd= 1 ,wr = 0 ,din = 0 ,do = x
#	3 rd= 1 ,wr = 1 ,din = 0 ,do = 0
#	7 rd= 0 ,wr = 1 ,din = 0 ,do = z
#	9 rd= 1 ,wr = 1 ,din = 0 ,do = 0
#	10 rd= 1 ,wr = 1 ,din = 1 ,do = 1

Figure 10.27 Partial results of simulating the test bench for the CMOS switch in Figure 10.26.

Example 10.10 A Dynamic Shift Register

Figure 10.28 shows three successive stages of a dynamic shift register. It is operated through a two-phase clock system – $\phi1$ and $\phi2$. Each stage has a CMOS inverter. Successive stages are given input through CMOS switches (sw1, sw2, *etc.*). $\phi1$ and $\phi2$ are symmetric clock waveforms in anti-phase. Two successive stages together form one storage element.

- When $\phi2$ is ON AND $\phi1$ is OFF. Din is input to stage 1 through switch swd. sw1 and sw3 are OFF and sw2 is ON. State of stage 2 (attained when $\phi1$ was high last) is coupled as input to stage 3 through switch sw2, and stage 3 takes up the new state.
- In the next half cycle, $\phi1$ is ON and $\phi2$ is OFF. sw1 and sw3 are ON and sw2 is OFF. State of stage 1 (attained when $\phi2$ was high last) is coupled as input to stage 2 through switch sw1 and Do takes up the new state from stage3 through sw3.

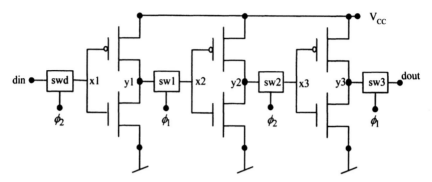

Figure 10.28 The basic functional unit of a dynamic shift register.

The data in the input line are latched and shifted right on successive clock cycles. The stages together act as a shift register stage. The design description of a shift-register module with a two-phase clock is shown in Figure 10.29 along with a test-bench for the same. The two-phase clock and switches are defined as functions. These are repeatedly called to realize the shift register. Figure 10.30 shows partial simulation results.

The shift register can be modified to suit a variety of needs:

- Dynamic logic incorporating NAND / NOR gates.
- Dynamic RAM with row and column select lines and refresh functions.
- A shift register to function as a right- or a left-shift-type shift register; a direction select bit can be used to alter the shift direction.

```
module shreg1(dout,din,phi1);
output dout;//tested ok on 22nd Non 2001
input din,phi1;
wire phi2;
trireg[3:0] x,y;
trireg dout;
assign phi2=~phi1;
cmos switch0(x[0],din,phi1,phi2), switch1(x[1],y[0],phi2,phi1),
switch2(x[2],y[1],phi1,phi2), switch3(x[3],y[2],phi2,phi1),
switch4(dout,y[3],phi1,phi2);
cell cc0(y[0],x[0]), cc1(y[1],x[1]), cc2(y[2],x[2]), cc3(y[3],x[3]);
endmodule

module cell(op,ip);
output op;
input ip;
```

continued

continued

```
supply1 pwr;
supply0 gnd;
nmos(op,gnd,ip);
pmos(op,pwr,ip);
endmodule

module tst_shreg1;
reg din,phi1;
wire dout;
shreg1 shr(dout,din,phi1);
initial {din,phi1}=2'B00;
always
begin
  #1      din=1'b1;  #2    din=1'b1;  #2    din=1'b0;
  #2      din=1'b0;  #2    din=1'b0;  #2    din=1'b1;
  #2      din=1'b1;
end
always # 2 phi1=~phi1;
initial $monitor($time," din= %b,   dout= %b,  phi1= %b", din,dout,phi1);
endmodule
```

Figure 10.29 Design description of the dynamic shift register of Figure 10.28.

```
#         0 din= 0,   dout= x,  phi1= 0
#         1 din= 1,   dout= x,  phi1= 0
#         2 din= 1,   dout= x,  phi1= 1
#         4 din= 1,   dout= x,  phi1= 0
#         5 din= 0,   dout= x,  phi1= 0
#         6 din= 0,   dout= x,  phi1= 1
#         8 din= 0,   dout= x,  phi1= 0
#        10 din= 0,   dout= 1,  phi1= 1
#        11 din= 1,   dout= 1,  phi1= 1
#        12 din= 1,   dout= 1,  phi1= 0
#        14 din= 1,   dout= 0,  phi1= 1
#        16 din= 1,   dout= 0,  phi1= 0
#        18 din= 0,   dout= 1,  phi1= 1
#        20 din= 0,   dout= 1,  phi1= 0
```

Figure 10.30 Partial results of running the test bench in Figure 10.29.

10.4   BI-DIRECTIONAL GATES

The gates discussed so far (**nmos**, **pmos**, **rnmos**, **rpmos**, **rcmos**) are all unidirectional gates. When turned ON, the gate establishes a connection and makes the signal at the input side available at the output side. Verilog has a set of primitives for bi-directional switches as well. They connect the nets on either side when ON and isolate them when OFF. The signal flow can be in either direction. None of the continuous-type assignments at higher levels dealt with so far has a functionality equivalent to the bi-directional gates. There are six types of bi-directional gates.

10.4.1   `tran` and `rtran`

The **tran** gate is a bi-directional gate of two ports. When instantiated, it connects the two ports directly. Thus the instantiation

`tran` (s1, s2);

connects the signal lines s1 and s2. Either line can be **input**, **inout** or **output**. **rtran** is the resistive counterpart of **tran**.

10.4.2   tranif1 and rtranif1

tranif1 is a bi-directional switch turned ON/OFF through a control line. It is in the ON-state when the control signal is at 1 (high) state. When the control line is at state 0 (low), the switch is in the OFF state. A typical instantiation has the form

`tranif1` (s1, s2, c);

Here c is the control line. If c=1, s1 and s2 are connected and signal transmission can be in either direction. **rtranif1** is the resistive counterpart of **tranif1**. It is instantiated in an identical manner.

10.4.3   tranif0 and rtranif0

tranif0 and **rtranif0** are again bi-directional switches. The switch is OFF if the control line is in the 1 (high) state, and it is ON when the control line is in the 0 (low) state. A typical instantiation has the form

`tranif0` (s1, s2, c);

With the above instantiation, if c = 0, s1 and s2 are connected and signal transmission can be in either direction. If c = 1, the switch is OFF and s1 and s2 are isolated from each other. **rtranif0** is the resistive counterpart of **tranif0**.

Observations:

- Any instantiation of a bi-directional switch of the above types can be given a name. But a name is not essential. It is true of the other switches also.
- With the bi-directional switches the signal on either side can be of **input**, **output**, or **inout** type. They can be nets or appearing as ports in the module. But the type declaration on the two sides has to be consistent.
- The connections to the bi-directional terminals of each of the bi-directional switches have to be scalars or individual bits of vectors and not vector themselves.
- In the above instantiation s1 can be an input port in a module. In that case, s2 has to be a net forming an input to another instantiated module or circuit block. s2 can be of **output** or **inout** type also. But it cannot be another input port.
  - s1 and s2 – both cannot be output ports.
  - s1 and s2 – both can be **inout** ports.
- With **tran**, **tranif1**, and **tranif0** bi-directional switches if the input signal has strength **supply1** (**supply0**), the output side signal has strength **strong1** (**strong0**). For all other strength values of the input signal, the strength value of the output side signal retains the strength of the input side signal.
- With **rtran**, **rtranif1** and **rtranif0** switches the output side signal strength is less than that of the input side signal. The strength reduction is on the lines shown in Table 10.4 for **rnmos**, **rpmos**, and **rcmos** switches.

Features of all the bi-directional switches are shown summarized in Table 10.5.

Example 10.11 Bus Switching

Figure 10.31 shows the circuit of a single-data line bus with the possibility of two-way data transfer; the module bus_tran in Figure 10.32 is the Verilog description of the circuit at the switch level. c is a **tran**-type switch with the possibility of connecting a and b. ar and br are registers which can be switched ON to the lines

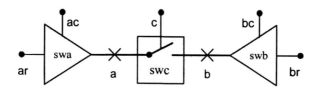

Figure 10.31 A circuit to demonstrate two-way signal transfer.

Table 10.5 Different bi-directional switches and their features

Type of Bi-directional switch	Typical instantiation	Condition to be ON	Remarks
2 port	**tran** (a, b);	Always ON (if instantiated)	Acts essentially as a buffer
	rtran (a, b);	– do –	Acts essentially as a buffer with reduction in the strength of the signal
3 port	**tranif1** (a, b, c);	ON if c = 1	Acts as a buffer if ON. Otherwise provides isolation
	tranif0 (a, b, c);	ON if c = 0	– do –
	rtranif1 (a, b, c);	ON if c = 1	Acts as a buffer if ON. Otherwise provides isolation; signal strength on the output side is lower than that on the input side
	rtranif0 (a, b, c);	ON if c = 0	– do –

```
module bus_tran(a,b,c);
inout  a,b; input c; wire a,b,c;
tranif1  gg (a,b,c);
endmodule

module bus_tst;
reg ar,br,ac,bc,c;wire a,b;
bufif1 swa(a,ar,ac), swb(b,br,bc);
bus_tran bs(a,b,c);
initial    begin
           $display("t\tar\tac\ta\tc\tb\tbc\tbr");
  #1       {ar,ac,c,bc,br}=5'b01100; repeat(3) #1 ar=~ar;
  #1       {ar,ac,c,bc,br}=5'b00110; repeat(3) #2 br=~br;
  #1       {ar,ac,c,bc,br}=5'b11010; repeat(3) #1 ar=~ar;
           repeat(3) #2 br=~br;
  #1       $stop;
           end
initial $monitor("%0d\t%b\t%b\t%b\t%b\t%b\t%b\t%b",$time,ar,ac,a,c,b,bc,br);
endmodule
```

Figure 10.32 Design and test modules for the circuit of Figure 10.31.

a and b. Two-way signal transmission is demonstrated through the test bench in the figure; the simulation results reproduced in Figure 10.33 bring out the following:

- Up to 4 ns, switch swa is ON, swb is OFF, and swc is ON. Data in ar – ar toggles 3 times and is available on a and b.
- During 5 ns to 11 ns, switch swa is OFF, swb is ON, and swc is ON. Data in br – br toggles 3 times and is available on b and a.
- During 1 2ns to 21 ns, switch swc is ON, swa and swb are OFF; a follows ar while b follows br.

#t	ar	ac	a	c	b	bc	br
#0	x	x	x	x	x	x	x
#1	0	1	0	1	0	0	0
#2	1	1	1	1	1	0	0
#3	0	1	0	1	0	0	0
#4	1	1	1	1	1	0	0
#5	0	0	0	1	0	1	0
#7	0	0	1	1	1	1	1
#9	0	0	0	1	0	1	0
#11	0	0	1	1	1	1	1
#12	1	1	1	0	0	1	0
#13	0	1	0	0	0	1	0
#14	1	1	1	0	0	1	0
#15	0	1	0	0	0	1	0
#17	0	1	0	0	1	1	1
#19	0	1	0	0	0	1	0
#21	0	1	0	0	1	1	1

Figure 10.33 Simulation results with the test bench for the circuit in Figure 10.31.

Example 10.12 Another RAM Cell

Figure 10.34 shows a single RAM cell. It can be instantiated in vector form to form a full-fledged ram. a_d is the decoded address line. When active, it turns on the bi-directional switch g3 and establishes a two-way connection between net ddd and net q. g1 and g2 together form a latch in feedback fashion. When g3 is OFF, the latch stores the state it was last in. It is connected to ddd through g3 by activating a_d for writing and reading. The design description for the RAM is shown in Figure 10.35. The simulation results are (partially) reproduced in Figure 10.36. The following are possible after such selection and connection:

- When wr = 1, **cmos** gate g4 turns ON; the data at the input port di (with strength **strong0 / strong1**) are connected to q through ddd. It forces the latch to its state – since q has strength **pull0 / pull1** only – di prevails here. This constitutes the write operation.
- When rd = 1, **cmos** gate g5 turns ON. The net ddd is connected to the output net do. The data stored in the latch are made available at the output port do. This constitutes the read operation.

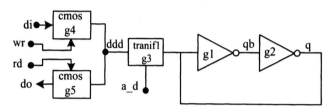

Figure 10.34 Circuit of a RAM cell,in block diagram form.

```
module ram_cell1(do,di,wr,rd,a_d);
output do; input di,wr,rd,a_d; wire ddd,q,qb,wrb,rdb;
not(rdb,rd),(wrb,wr);
not(pull1,pull0)(q,qb),(qb,q);
tranif1 g3(ddd,q,a_d);
cmos g4(ddd,di,wr,wrb),g5(do,ddd,rd,rdb);
endmodule

//test bench
module tst_ramcell1();
reg din,wr,rd,a_d; wire do;
ram_cell1 rmc1(do,din,wr,rd,a_d);
initial begin a_d=1'b0;din=1'b0;wr=1'b0;rd=1'b0; end
always #3a_d=1'b1;
always #10 din =~din;
always begin #3wr=1'b1; #8 wr=1'b0; end
always begin #2 rd=1'b1; #5 rd =1'b0; end
initial $monitor ($time," rd= %b ,wr = %b ,din = %b ,a_d = %b ,do = %b
",rd,wr,din,a_d,do);
initial #40 $stop;
endmodule
```

Figure 10.35 Design description of the RAM cell of Figure 10.34.

```
#                0 rd= 0 ,wr = 0 ,din = 0 ,a_d = 0 ,do = z
#                2 rd= 1 ,wr = 0 ,din = 0 ,a_d = 0 ,do = z
#                3 rd= 1 ,wr = 1 ,din = 0 ,a_d = 1 ,do = 0
#                7 rd= 0 ,wr = 1 ,din = 0 ,a_d = 1 ,do = z
#                9 rd= 1 ,wr = 1 ,din = 0 ,a_d = 1 ,do = 0
#               10 rd= 1 ,wr = 1 ,din = 1 ,a_d = 1 ,do = 1
#               11 rd= 1 ,wr = 0 ,din = 1 ,a_d = 1 ,do = 1
#               14 rd= 0 ,wr = 1 ,din = 1 ,a_d = 1 ,do = z
#               16 rd= 1 ,wr = 1 ,din = 1 ,a_d = 1 ,do = 1
#               20 rd= 1 ,wr = 1 ,din = 0 ,a_d = 1 ,do = 0
#               21 rd= 0 ,wr = 1 ,din = 0 ,a_d = 1 ,do = z
```

Figure 10.36 Partial results of simulating the test bench for the CMOS switch in Figure 10.35.

10.5   TIME DELAYS WITH SWITCH PRIMITIVES

Propagation delays can be specified for switch primitives on the same lines as was done with the gate primitives in Chapter 5. For example, an NMOS switch instantiated as

nmos g1 (out, in, ctrl);

has no delay associated with it. The instantiation

nmos (delay1) g2 (out, in, ctrl);

has delay1 as the delay for the output to rise, fall, and turn OFF. The instantiation

nmos (delay_r, delay_f) g3 (out, in, ctrl);

has delay_r as the rise-time for the output. delay_f is the fall-time for the output. The turn-off time is zero. The instantiation

nmos (delay_r, delay_f, delay_o) g4 (out, in, ctrl);

has delay_r as the rise-time for the output. delay_f is the fall-time for the output delay_o is the time to turn OFF when the control signal ctrl goes from 0 to 1. Delays can be assigned to the other uni-directional gates (**rcmos, pmos, rpmos, cmos,** and **rcmos**) in a similar manner. Bi-directional switches do not delay transmission – their rise- and fall-times are zero. They can have only turn-on and turn-off delays associated with them. **tran** has no delay associated with it.

tranif1 (delay_r, delay_f) g5 (out, in, ctrl);

represents an instantiation of the controlled bi-directional switch. When control changes from 0 to 1, the switch turns on with a delay of delay_r. When control changes from 1 to 0, the switch turns off with a delay of delay_f.

transif1 (delay0) g2 (out, in, ctrl);

represents an instantiation with delay0 as the delay for the switch to turn on when control changes from 0 to 1, with the same delay for it to turn off when control changes from 1 to 0. When a delay value is not specified in an instantiation, the turn-on and turn-off are considered to be ideal that is, instantaneous. Delay values similar to the above illustrations can be associated with **rtranif1, tranif0**, and **rtranif0** as well.

10.6   INSTANTIATIONS WITH STRENGTHS AND DELAYS

In the most general form of instantiation, strength values and delay values can be combined. For example, the instantiation

nmos (strong1, strong0) (delay_r, delay_f, delay_o) gg (s1, s2, ctrl) ;

means the following:

- It has strength **strong0** when in the low state and strength **strong1when** in the high state.
- When output changes state from low to high, it has a delay time of delay_r.
- When the output changes state from high to low, it has a delay time of delay_f.
- When output turns-off it has a turn-off delay time of delay _o.

rnmos, pmos, and **rpmos** switches too can be instantiated in the general form in the same manner. The general instantiation for the bi-directional gates too can be done similarly.

10.7   STRENGTH CONTENTION WITH TRIREG NETS

As was explained in Chapter 5, nets declared as **trireg** can have capacitive storage. Such storage can be assigned one of three strengths – **large, medium,** or **small**. Driving such a net from different sources can lead to contention; the relative strength levels of the sources also have a say in the signal level taken by the net. The contention resolution is brought out here through an illustrative example. A similar procedure of analysis can be followed in other cases as well.

Example 10.13

Figure 10.37 shows a circuit where a set of switches connect nets in series to a signal source. Strengths have been assigned to the nets and a test bench to bring out contention shown in Figure 10.38. The thicker line representation of net a3 in Figure 10.37 signifies that the capacitive storage strength of net a3 is stronger than that of net a2. The progress of simulation is depicted in Figure 10.39 showing the switch status and corresponding signal values at different times. Simulation results are shown in Figure 10.40. One can see that whenever a2 and a3 are connected (but isolated from a1), the stronger a3 prevails.

Observations:

- When a net is connected to a single signal source through intervening switches and capacitive nets, the source decides the value of the signal on the net.
- When 2 capacitive nets are connected, in case of a contention the stronger one prevails.
- When a signal source and a capacitive net drive another net, in case of a contention the signal value is dictated by the stronger of the two (see Table 5.5).

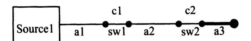

Figure 10.37 A simple circuit to demonstrate contention resolution with **trireg** nets.

```
module demo_1;
trireg(large)a3; trireg(small)a2; wire a1; reg c1,c2,b;
buf(strong1,strong0) source1(a1,b);
tranif1 sw1(a2,a1,c1), sw2(a3,a2,c2);
initial begin
        $display("t\ta1\tc1\ta2\tc2\ta3");
    #0  {c1,c2,b}=3'b111; #1 {c1,c2,b}=3'b011; #1 {c1,c2,b}=3'b001;
    #1  {c1,c2,b}=3'b000; #1 {c1,c2,b}=3'b100; #1 {c1,c2,b}=3'b000;
    #1  {c1,c2,b}=3'b010; #1 {c1,c2,b}=3'b000; #1 {c1,c2,b}=3'b100;
    #1  {c1,c2,b}=3'b000; #1 {c1,c2,b}=3'b010; #1 {c1,c2,b}=3'b000;
    #1  {c1,c2,b}=3'b001; #1 {c1,c2,b}=3'b101; #1 {c1,c2,b}=3'b111;
    #1  $stop;
    end
initial $monitor("%0d\t%b\t%b\t%b\t%b\t%b",$time,a1,c1,a2,c2,a3);
endmodule
```

Figure 10.38 A test bench for the circuit in Figure 10.37.

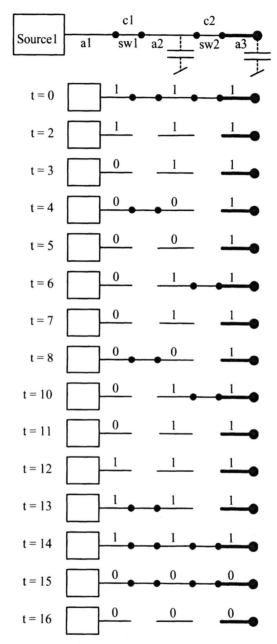

Figure 10.39 Changes in signal values at different times in Example 10.13 as the status of switches changes.

#t	a1	c1	a2	c	a3
#0	1	1	1	1	1
#1	1	0	1	1	1
#2	1	0	1	0	1
#3	0	0	1	0	1
#4	0	1	0	0	1
#5	0	0	0	0	1
#6	0	0	1	1	1
#7	0	0	1	0	1
#8	0	1	0	0	1
#9	0	0	0	0	1
#10	0	0	1	1	1
#11	0	0	1	0	1
#12	1	0	1	0	1
#13	1	1	1	0	1
#14	1	1	1	1	1

Figure 10.40 Results of the simulation-run with the test bench in Figure 10.38.

10.8   EXERCISES

1. Implement NAND, AND, OR GATES using MOS switches; test it with a suitable test-bench.
2. Implement a pseudo-NMOS 4-input NOR logic gate. Write a test bench and test it.
3. Implement a dynamic logic NAND gate for 4 inputs; the pullup is to be a precharge transistor, and the pulldown is to be an evaluation transistor, with the output being precharged in precharge phase of the clock. The output should be available during the evaluation phase. Write a test bench and test the switch level dynamic gate.
4. Implement a 4-to-1 MUX using CMOS transmission gates.
5. Build a dynamic 2-to-4 NOR gate based decoder and a dynamic 2-to-4 NAND gate-based decoder using NMOS switches and PMOS switches.
6. Implement a one-bit full adder using CMOS logic and test it using a test bench.
7. Implement a 4-bit look-ahead adder using CMOS logic and test it with a test bench.
8. Implement a 4-bit barrel shifter using NMOS switches.

9. Form an edge-triggered flip-flop; using it, form an 8-bit port as shown in Figure 10.40. Form a latch and modify it to provide two flags – data input flag (DIF) and data output flag (DOF). Normally, Wr and Rd are low; old state is retained. If Wr goes high, Di bits are loaded into the port at the next clock pulse. DIF flag is set. DOF is at zero state. If RD goes high, Do bits are loaded into the port at the next clock pulse. DOF Flag is set. DIF is at zero state. Design the port module; test it with a test bench.

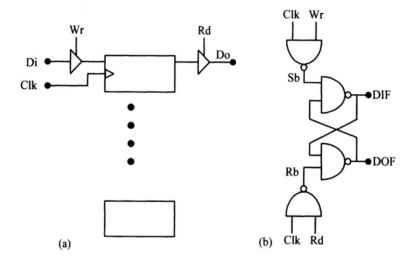

Figure 10.40 Figure for Exercise 9.

11

SYSTEM TASKS, FUNCTIONS, AND COMPILER DIRECTIVES

11.1 INTRODUCTION

A number of facilities in Verilog relate to the management of simulation; starting and stopping of simulation, selectively monitoring the activities, testing the design for timing constraints, *etc.,* are amongst them. Although a variety of such constructs is available in Verilog for such activities [IEEE], we discuss the leading ones and illustrate their use through representative example.

11.2 PARAMETERS

Often designers keep debugged modules for reuse. Such modules call for flexibility on two counts:

- They should be adaptable to designs conforming to different technologies. Timing parameters used for testing should be flexible.
- They should have a scalable feature; that is, bus width, register size, *etc.,* should be flexible.

The parameter constructs facilitate such flexibility. Constants signifying timing values, ranges of variables, wires, *etc.,* can be specified in terms of assigned names. Such assigned names are called parameters. The parameter values can be specified and changed to suit the design environment or test environment. Such changes are effected and frozen at instantiation. The assigned values cannot change during testing or synthesis. In this respect a parameter is different from a net or a variable.

Two types of parameters are of use in modules:

- Parameters related to timings, time delays, rise and fall times, *etc.,* are technology-specific and used during simulation. Parameter values can be assigned or overridden with the keyword "**specparam**" preceding the assignments.

• Parameters related to design, bus width, and register size are of a different category. They are related to the size or dimension of a specific design; they are technology-independent. Assignment or overriding is with assignments following the keyword "**defparam**".

The two types of parameters are treated differently in Verilog. The former type is discussed here and the latter type is discussed in Section 11.4.

11.2.1   Timing-Related Parameters

The constructs associated with parameters are discussed here through specific design or test modules.

Example 11.1

The half-adder module of in Figure 4.24 is reconsidered here in Figure 11.1. Gate delays of the type discussed in Chapter 5 have been added to all output transitions of the sum bit (**s**) as well as the carry bit (**ca**). Simulation results are partially reproduced in Figure 11.2. The following observations are in order here:

• a=0 and b=0 at start of simulation. Because of the transition times, the outputs remain indecisive at the **x** state.

```
module ha_1(s,ca,a,b);
input a,b; output s,ca;
xor #(1,2) (s,a,b);
and #(3,4) (ca,a,b);
endmodule

//test-bench
module tstha_1();
reg a,b; wire s,ca;
ha_1 hh(s,ca,a,b);
initial begin a=0;b=0; end
always begin #5 a=1;b=0; #5 a=0;b=1; #5 a=1;b=1; #5 a=0;b=0; end
initial $monitor($time , " a = %b , b = %b ,out carry = %b , outsum = %b "
,a,b,ca,s);
initial #30 $stop;
endmodule
```

Figure 11.1 Module of a half-adder with delays assigned to the output transitions; a test bench is also included in the figure.

```
#              0  a = 0 , b = 0 ,out carry = x , outsum = x
#              2  a = 0 , b = 0 ,out carry = x , outsum = 0
#              4  a = 0 , b = 0 ,out carry = 0 , outsum = 0
#              5  a = 1 , b = 0 ,out carry = 0 , outsum = 0
#              6  a = 1 , b = 0 ,out carry = 0 , outsum = 1
#             10  a = 0 , b = 1 ,out carry = 0 , outsum = 1
#             15  a = 1 , b = 1 ,out carry = 0 , outsum = 1
#             17  a = 1 , b = 1 ,out carry = 0 , outsum = 0
#             18  a = 1 , b = 1 ,out carry = 1 , outsum = 0
```

Figure 11.2 Partial results of simulating the test bench in Figure 11.1.

- The sum bit falls down to 0 state with the specified delay of 2 ns. The carry bit falls down to 0 state with its specified delay of 4 ns.
- a=1 and b=0 at 5 ns. The sum bit rises to the 1 state at 6 ns (with the specified delay of 1 ns).
- a=0 and b=1 at 10 ns. But the sum and carry bits remain unchanged.
- a=b=1 at 15 ns. The sum bit falls down to 0 state at 17 ns (with the specified fall delay of 2 ns). The carry bit rises to the 1 state at 18 ns (with the specified rise time delay of 3 ns).
- Subsequent output transitions too can be explained in a similar manner.

11.2.2   Parameter Declarations and Assignments

Declaration of parameters in a design as well as assignments to them can be effected using the keyword "**Parameter**." A declaration has the form

parameter alpha = a, beta = b;

where

- **parameter** is the keyword,
- alpha and beta are the names assigned to two parameters and
- a, b are values assigned to alpha and beta, respectively.

In general a and b can be constant expressions. The parameter values can be overridden during instantiation but cannot be changed during the run-time. If a parameter assignment is made through the keyword "**localparam**," its value cannot be overridden.

Observations:

- As mentioned earlier, **parameter**s are constants which can be altered during compilation but not during runtime.
- A **Parameter** can be signed or unsigned in nature; it can be an integer or a real number.
- Its nature – signed or not, real or integral type as well as range – can be specified at the time of declaration or decided by default based on assignment.

    Examples

    parameter a = 3;   // a is a positive integer
    parameter b = - 3; // b is a signed integer
    parameter c = 3.0, d = 3.0e2; //c and d are unsigned real numbers.

    In all the above cases the **parameter** type and range are decided by default.

    parameter integer e = 3; /* e is declared to be an integer type of parameter and assigned the value 3. */
    parameter real f = 3.0; /* f is declared to be a real unsigned real number and assigned the value 3. */

    In the last two cases the **parameter** type is declared explicitly and remains so.

- Whenever a **parameter** value is overridden during instantiation (as in some of the cases discussed below), type, signed/unsigned, *etc.*, remain unchanged.

Example 11.2

The half-adder module in Figure 11.1 has been modified and shown in Figure 11.3. The rise and fall times of the primitive gate instantiation are assigned identifier names. Specific numeric values are assigned to them through a separate parameter declaration statement. The numerical values assigned are the same as those assigned in Example 11.1 above. The simulation results are identical to those in Figure 11.2.

The scheme of Figure 11.3 has an apparent advantage compared to that of Figure 11.1. Different rise and fall times, time delays, *etc.*, need not be fully specified in the design or its test bench. Values can be assigned separately through parameter declaration as done here. Numeric values can be changed by assigning the required values to the parameters afresh: It avoids the unpleasant task of scanning the module file and changing the numerical values all through.

```
module ha_2(s,ca,a,b);
input a,b; output s,ca;
parameter dl1r=1,dl2f=2,dl3r=3,dl4f=4;
xor #(dl1r,dl2f) (s,a,b);
and #(dl3r,dl4f) (ca,a,b);
endmodule

//test-bench
module tstha_2();
reg a,b; wire s,ca;
ha_2 hh(s,ca,a,b);
initial begin a=0;b=0; end
always begin #5 a=1;b=0; #5 a=0;b=1; #5 a=1;b=1; #5 a=0;b=0; end
initial $monitor($time , " a = %b , b = %b ,out carry = %b , outsum = %b "
,a,b,ca,s);
initial #30 $stop;
endmodule
```

Figure 11.3 The half-adder module of Figure 11.1 with the time delays assigned through parameter declarations.

Example 11.3

Figure 11.4 shows the half-adder module with the test bench being modified. The rise and fall times are specified separately in the test bench. They override the values specified in the half-adder module itself. The time delay values are specified within the instantiation statement. Four numbers are specified there; they override the first four parameters declared in the module instantiated and in the same order. Specifically, the numbers 4, 3, 2, and 1 are assigned to the parameters dl1r, dl2f, dl3, and dl4f, respectively. The simulation results are given in Figure 11.5. The quantities representing delayed response are shown in bold italics. Thus the change in a at 5 ns causes the sum bit to get set with a delay of dlir (4 ns here)–that is, at 9 ns. As pointed out earlier, the 4 ns delay value for dl1r has been specified in the test bench at instantiation, and it overrides the value of 1 ns assigned in the module definition. Subsequent changes to s and c too can be explained in a similar manner. The overriding illustrated here can be done separately for each instantiation in a module or in different modules.

```
module ha_2(s,ca,a,b);
input a,b; output s,ca;
parameter dl1r=1,dl2f=2,dl3r=3,dl4f=4;
xor #(dl1r,dl2f) (s,a,b);
and #(dl3r,dl4f) (ca,a,b);
endmodule

//test-bench
module tstha_3();
reg a,b; wire s,ca;
ha_2 #(4,3,2,1) hh(s,ca,a,b);
initial begin a=0;b=0; end
always begin #5 a=1;b=0; #5 a=0;b=1; #5 a=1;b=1; #5 a=0;b=0; end
initial  $monitor($time , "  a = %b , b = %b ,out carry = %b , outsum = %b  "
,a,b,ca,s);
initial #30 $stop;
endmodule
```

Figure 11.4 The half-adder module of Figure 11.3 with the time delay values reassigned from the test bench.

#	0 a = 0 , b = 0 ,out carry = x , outsum = x
#	1 a = 0 , b = 0 ,out carry = **0** , outsum = x
#	3 a = 0 , b = 0 ,out carry = 0 , outsum = **0**
#	5 a = 1 , b = 0 ,out carry = 0 , outsum = 0
#	9 a = 1 , b = 0 ,out carry = 0 , outsum = **1**
#	10 a = 0 , b = 1 ,out carry = 0 , outsum = 1
#	15 a = 1 , b = 1 ,out carry = 0 , outsum = 1
#	17 a = 1 , b = 1 ,out carry = **1** , outsum = 1
#	18 a = 1 , b = 1 ,out carry = 1 , outsum = **0**
#	20 a = 0 , b = 0 ,out carry = 1 , outsum = 0
#	21 a = 0 , b = 0 ,out carry = **0** , outsum = 0
#	25 a = 1 , b = 0 ,out carry = 0 , outsum = 0
#	29 a = 1 , b = 0 ,out carry = 0 , outsum = **1**

Figure 11.5 Results of simulating the test bench in Figure 11.4.

Example 11.4

Figure 11.6 shows the half adder module considered above and its test bench with one change in the test bench. The module instantiation has three numbers representing three time delays. They override the first three parameters (dl1r, dl2f, and dl3r, respectively) as declared in the instantiation. All other parameters (only dl4f here) remain unchanged. The simulation results are given in Figure 11.7. The numerals pertaining to the delayed changes are shown in bold italics in the figure. The fall time of ca for its 0 to 1 transition (specified by dl4f) in the instantiated module can be seen to be unchanged at 4 ns – as is evident from the line representing the values of variables at 24 ns in Figure 11.7.

```
module ha_2(s,ca,a,b);
input a,b; output s,ca;
parameter dl1r=1,dl2f=2,dl3r=3,dl4f=4;
xor #(dl1r,dl2f) (s,a,b);
and #(dl3r,dl4f) (ca,a,b);
endmodule

//test-bench
module tstha_4();
reg a,b; wire s,ca;
ha_2 #(4,3,2) hh(s,ca,a,b);
initial begin a=0;b=0; end
always begin #5 a=1;b=0; #5 a=0;b=1; #5 a=1;b=1; #5 a=0;b=0; end
initial $monitor($time , " a = %b , b = %b ,out carry = %b , outsum = %b "
,a,b,ca,s);
initial #30 $stop;
endmodule
```

Figure 11.6 The half-adder module with only some of the time delays assigned afresh from the test bench.

#		
#	0 a = 0 , b = 0 ,out carry = x , outsum = x
#	*3* a = 0 , b = 0 ,out carry = x , outsum = *0*
#	*4* a = 0 , b = 0 ,out carry = *0* , outsum = 0
#	5 a = 1 , b = 0 ,out carry = 0 , outsum = 0
#	*9* a = 1 , b = 0 ,out carry = 0 , outsum = *1*
#	10 a = 0 , b = 1 ,out carry = 0 , outsum = 1
#	15 a = 1 , b = 1 ,out carry = 0 , outsum = 1
#	*17* a = 1 , b = 1 ,out carry = *1* , outsum = 1
#	*18* a = 1 , b = 1 ,out carry = 1 ,.outsum = *0*
#	20 a = 0 , b = 0 ,out carry = 1 , outsum = 0
#	24 a = 0 , b = 0 ,out carry = 0 , outsum = 0
#	25 a = 1 , b = 0 ,out carry = 0 , outsum = 0
#	*29* a = 1 , b = 0 ,out carry = 0 , outsum = *1*

Figure 11.7 Results of simulating the test bench in Figure 11.6.

The numbers specified in the test bench will be automatically assigned to the parameters in the instantiated module – in the same order as they are defined in the parameter statement. With such an implicit approach one cannot do an assignment to a selected set of parameters. For example, dl4f cannot be assigned a different value directly.

Example 11.5

The test bench in the module of Figure 11.4 has been modified and the modified module shown in Figure 11.8. The parameters dl1r, dl2f, dl3r, and dl4f are assigned values through the **defparam** statement. Each parameter, whose value has to be overridden, has to be specified hierarchically. One can also follow the approach here to assign values to parameters in different instantiated modules. Such values can be assigned to all the desired parameters at one place in the manner done here through a **defparam** construct. Simulation results are identical to those of Figure 11.5.

```
module ha_2(s,ca,a,b);
input a,b; output s,ca;
parameter dl1r=1,dl2f=2,dl3r=3,dl4f=4;
xor #(dl1r,dl2f) (s,a,b);
and #(dl3r,dl4f) (ca,a,b);
endmodule

//test-bench
module tstha_5();
reg a,b; wire s,ca;
defparam hh.dl1r=4,hh.dl2f=3,hh.dl3r=2,hh.dl4f=1;
ha_2 hh(s,ca,a,b);
initial begin a=0;b=0; end
always begin #5 a=1;b=0; #5 a=0;b=1; #5 a=1;b=1; #5 a=0;b=0; end
initial $monitor($time , " a = %b , b = %b ,out carry = %b , outsum = %b "
,a,b,ca,s);
initial #30 $stop;
endmodule
```

Figure 11.8 Use of **defparam** for assignment of values to specific parameters.

Example 11.6

The half-adder module of Figure 11.4 has been reproduced in Figure 11.9 with one change; the parameter assignments are done with constant expressions on the right ride. Note that the parameters appearing in a constant expression have to be defined (value assigned) prior to such use. The expressions here are such that the numerical values of dl1r, dl2f, dl3r, and dl4f are the same as those in Example 11.2. The simulation results too are the same.

```
module ha_6(s,ca,a,b);
input a,b; output s,ca;
parameter dl1r=1,dl2f=dl1r+1,dl3r=3,dl4f=dl2f*2;
xor #(dl1r,dl2f) (s,a,b);
and #(dl3r,dl4f) (ca,a,b);
endmodule

//test-bench
module tstha_6();
reg a,b; wire s,ca;
ha_6 hh(s,ca,a,b);
initial begin a=0;b=0; end
always begin #5 a=1;b=0; #5 a=0;b=1; #5 a=1;b=1; #5 a=0;b=0; end
initial $monitor($time , " a = %b , b = %b ,out carry = %b , outsum = %b "
,a,b,ca,s);
initial #30 $stop;
endmodule
```

Figure 11.9 Illustration of the use of constant expressions for parameter assignments.

11.2.3   Type Declarations for Parameters

Examples 11.2 to 11.6 above do not have any type declaration statements for the parameters dl1r, dl2f, dl3r, and dl4f. However, integer value assignments are made to each of them; implicitly they are taken as integers by the simulator. But in general one can use constant expressions on the right-hand side of the assignments. With the module of Figure 11.9, consider the parameter assignment statement

parameter dl1r =1, dl2f =dl1r + 1, dl3r =3 , dl4f = dl2f*2;

As mentioned earlier, all four parameters are automatically taken as integers by the simulator. If the above statement is modified as

parameter dl1r =1, dl2f =dl1r + 1.0, dl3r =3 , dl4f = dl2f*2;

the parameter types will be radically different. dl1r and dl3r will be treated as integers but dl2f and hence dl4f will be treated as real. However, the numerical values assigned will remain unaltered and hence the simulation results too will be the same.

11.3   PATH DELAYS

The time delays discussed so far (from Chapter 5 onwards) are all delays associated with individual operations or activities in a module. They refer to basic circuit elements in a design – at the microlevel itself. These are called "distributed delays" in LRM. Verilog has the provision to specify and check delays associated with total paths – from any input to any output of a module. Such paths and delays are at the chip or system level. They are referred to as "module path delays." Constructs available make room for specifying their paths and assigning delay values to them – separately or together.

11.3.1   Specify Blocks

Module paths are specified and values assigned to their delays through **specify** blocks. They are used to specify rise time, fall time, path delays pulse widths, and the like. A "**specify**" block can have the form shown in Figure 11.10.

```
specify
        specparam rise_time = 5, fall_time = 6;
        (a =>b) = (rise_time, fall_time);
        (c => d) = (6, 7);
endspecify
```

Figure 11.10 Structure of a **specparam** block

The block starts with the keyword "**specify**" and ends with the keyword "**endspecify**". **Specify** blocks can appear anywhere within a module. The block can have two types of statements:

- One type starts with the keyword **specparam** and assigns numerical values to timing parameters declared elsewhere. The **specparam** statements can appear within a module or within a specify block. (In earlier versions of the LRM its presence was restricted to the specify blocks.) The right sides of the assignments can be constants or constant expressions involving such parameters already assigned.
- The second type specifies paths and assigns values to time delays to them. Details of different possibilities for such paths are discussed later.

A **specify** block can have only the above types of assignments. Circuit function assignments, assignments to module parameters, *etc.*, are not permitted within it.

11.3.2   Module Paths

Module paths can be specified in different ways inside a specify block. The simplest has the form

A*>B

Here "A" is the source and "B" the destination. The source can be an input or an **inout** port. The destination can be output or an **inout** port. The symbol combination "*>" specifies the path from the source to the destination. It encompasses all the possible paths from A to B. If A and B are scalars, it signifies a single path.

- If A is a vector and B is a scalar, it signifies all the paths from every bit of A to the scalar B. Thus if A is a 4-bit-wide vector, 4 paths are specified.
- If A is a scalar and B is a vector, it signifies all the paths from A to every bit of the vector B. Thus if B is an 8-bit vector, it signifies all 8 possible paths.
- If both A and B are vectors, it signifies all the possible paths from every bit of the vector A to every bit of the vector B; thus if A is a 4-bit vector and B is an 8-bit vector, it signifies $4 \times 8 = 32$ possible paths; a total of 32 delay values (all being equal to each other) are implied here.

Figure 11.11(a) illustrates a case of all possible paths from a 2-bit vector A to another 2-bit vector B; the specification implies 4 paths. A statement of the type

C => D

signifies only all the parallel paths. Here C and D have to be vectors of the same size. The path specified signifies transmission from every bit of vector C to the corresponding bit of vector D. In this sense the path description is more restrictive than that of A*>B above. Figure 11.11(b) illustrates a case of all possible parallel paths from a 2-bit vector C to another 2-bit vector D; the specification implies a total of 2 paths only.

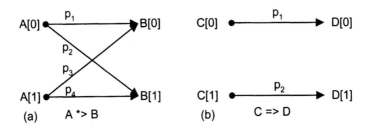

Figure 11.11 Illustration of the difference between the operators *> and =>.

Example 11.7

The module in Figure 11.12 specifies path delays from the input pins a and b to the output pins s and ca. The delay values are assigned within a specify block. The assignment

(a,b *> s) =1;

implies that

- The propagation delay from input a to output s is 1 ns and
- The propagation delay from input b to the output s is also 1 ns.
- Further the delay value is 1 ns for the change in the state of s from 0 to 1 as well as from 1 to 0.

Similarly the statement

(a,b *> ca) = 2;

implies that the delay from a to ca as well as that from b to ca is 2 ns; further, it holds for any transition in ca. The simulation results are reproduced in Figure 11.13. The values specific to the delayed changes are shown in bold italics. The

```
module ha_7(s,ca,a,b);
input a,b; output s,ca;
specify
        (a,b*>s)=1;
        (a,b*>ca)=2;
endspecify
xor  (s,a,b);
and  (ca,a,b);
endmodule

//test-bench
module tstha_7();
reg a,b; wire s,ca;
ha_7 hh(s,ca,a,b);
initial begin a=0;b=0; end
always begin #5 a=1;b=0; #5 a=0;b=1; #5 a=1;b=1; #5 a=0;b=0; end
initial $monitor($time , "  a = %b , b = %b ,out carry = %b , outsum = %b  "
,a,b,ca,s);
initial #30 $stop;
endmodule
```

Figure 11.12 A module to demonstrate use of path delay assignments.

```
#              0  a = 0 , b = 0 ,out carry = x , outsum = x
#              1  a = 0 , b = 0 ,out carry = x , outsum = 0
#              2  a = 0 , b = 0 ,out carry = 0 , outsum = 0
#              5  a = 1 , b = 0 ,out carry = 0 , outsum = 0
#              6  a = 1 , b = 0 ,out carry = 0 , outsum = 1
#             10  a = 0 , b = 1 ,out carry = 0 , outsum = 1
#             15  a = 1 , b = 1 ,out carry = 0 , outsum = 1
#             16  a = 1 , b = 1 ,out carry = 0 , outsum = 0
#             17  a = 1 , b = 1 ,out carry = 1 , outsum = 0
#             20  a = 0 , b = 0 ,out carry = 1 , outsum = 0
#             22  a = 0 , b = 0 ,out carry = 0 , outsum = 0
#             25  a = 1 , b = 0 ,out carry = 0 , outsum = 0
#             26  a = 1 , b = 0 ,out carry = 0 , outsum = 1
```

Figure 11.13 Simulation results of the test bench in Figure 11.12.

results can be seen to be consistent with the delay specifications. If the propagation delay values are the same in all the cases, the same could have been specified as

```
specify
        (a, b *> s, ca) = 1;
endspecify
```

Example 11.8

The module of Figure 11.12 has been slightly modified and reproduced in Figure 11.14. The delay values are specified as parameters and the parameters assigned values through respective **specparam** statements. Further, the **specparam** statement

specparam dl2 = dl1 +1;

uses a constant expression involving previously specified parameter values on the right side.

The delay paths and the values assigned to them are identical to those in Example 11.7 above. The simulation results too are identical to those in Figure 11.13.

```
module ha_8(s,ca,a,b);
input a,b; output s,ca;
specify
        specparam dl1=1;
        specparam dl2=dl1+1;
        (a,b*>s)=dl1;
        (a,b*>ca)=dl2;
endspecify
xor  (s,a,b);
and  (ca,a,b);
endmodule

//test-bench
module tstha_8();
reg a,b; wire s,ca;
ha_8 hh(s,ca,a,b);
initial begin a=0;b=0; end
always begin #5 a=1;b=0; #5 a=0;b=1; #5 a=1;b=1; #5 a=0;b=0; end
initial  $monitor($time , "  a = %b , b = %b ,out carry = %b , outsum = %b  "
,a,b,ca,s);
initial #30 $stop;
endmodule
```

Figure 11.14 Illustration of **specparam** with path delays.

Example 11.9

In the half-adder module of Figure 11.15 the rise and fall times at the output have been specified separately; effectively the specifications are the same as those for Example 11.1; but here they are at the chip level in contrast to those in Example 11.1, where they are at the gate level. The simulation results are shown in Figure 11.16; the values that pertain to the delayed response are shown in bold italics in the figure.

```
module ha_9(s,ca,a,b);
input a,b; output s,ca;
specify
        (a,b*>s) = (1,2);
        (a,b*>ca) = (3,4);
endspecify
xor  (s,a,b);
and  (ca,a,b);
endmodule
```

continued

continued

```
//test-bench
module tstha_9();
reg a,b; wire s,ca;
ha_9 hh(s,ca,a,b);
initial begin a=0;b=0; end
always begin #5 a=1;b=0; #5 a=0;b=1; #5 a=1;b=1; #5 a=0;b=0; end
initial $monitor($time , "  a = %b , b = %b ,out carry = %b , outsum = %b  "
,a,b,ca,s);
initial #30 $stop;
endmodule
```

Figure 11.15 Use of specify block to specify out rise and fall times separately for pin-to-pin delays.

```
#            0  a = 0 , b = 0 ,out carry = x , outsum = x
#            2  a = 0 , b = 0 ,out carry = x , outsum = 0
#            4  a = 0 , b = 0 ,out carry = 0 , outsum = 0
#            5  a = 1 , b = 0 ,out carry = 0 , outsum = 0
#            6  a = 1 , b = 0 ,out carry = 0 , outsum = 1
#           10  a = 0 , b = 1 ,out carry = 0 , outsum = 1
#           15  a = 1 , b = 1 ,out carry = 0 , outsum = 1
#           17  a = 1 , b = 1 ,out carry = 0 , outsum = 0
#           18  a = 1 , b = 1 ,out carry = 1 , outsum = 0
#           20  a = 0 , b = 0 ,out carry = 1 , outsum = 0
#           24  a = 0 , b = 0 ,out carry = 0 , outsum = 0
#           25  a = 1 , b = 0 ,out carry = 0 , outsum = 0
#           26  a = 1 , b = 0 ,out carry = 0 , outsum = 1
```

Figure 11.16 Simulation results of the test bench in Figure 11.15.

Example 11.10

Figure 11.17 shows the module of Figure 6.20 modified with an additional group delay specification. The block

```
specify
        (a => d) = 1;
endspecify
```

```
module alu_1 (d, co, a, b, f,cci);
output [3:0] d; output co; wire[3:0]d; input cci; input [3 : 0 ] a, b;
input [1 : 0] f; //F IS A 2 BIT FUNCTION SELECT INPUT
specify
        (a=>d)=1;
endspecify
assign {co,d}=(f==2'b00)?(a+b+cci):((f==2'b01)?(a-b):((f==2'b10)?
            {1'bz,a^b}:{1'bz,~a}));
endmodule

//test-bench
module tst_alu1();
reg [3:0]a,b; reg[1:0] f; reg cci; wire[3:0]d; wire co;
alu_1 aa(d,co,a,b,f,cci);
initial begin cci=1'b0; f=2'b00;a=4'b0;b=4'h0; #30 $stop; end
always   begin
  #2     cci =1'b0;f=2'b00;a=4'h1;b=4'h0; #2 cci =1'b1;f=2'b00;a=4'h8;b=4'hf;
  #2     cci =1'b1;f=2'b01;a=4'h2;b=4'h1; #2 cci =1'b0;f=2'b01;a=4'h3;b=4'h7;
  #2     cci =1'b1;f=2'b01;a=4'h3;b=4'h3; #2 cci =1'b1;f=2'b10;a=4'h3;b=4'h3;
  #2     cci =1'b1;f=2'b11;a=4'hf;b=4'hc;
         end
initial $monitor($time, " cci = %b , a= %b ,b = %b ,f = %b ,d =%b ,co= %b ",cci
,a,b,f,d,co);
endmodule
```

Figure 11.17 Illustration of the use of group delay with an ALU module.

```
# 0 cci = 0 , a= 0000 ,b = 0000 ,f = 00 ,d =xxxx ,co= 0
# 1 cci = 0 , a= 0000 ,b = 0000 ,f = 00 ,d =0000 ,co= 0
# 2 cci = 0 , a= 0001 ,b = 0000 ,f = 00 ,d =0000 ,co= 0
# 3 cci = 0 , a= 0001 ,b = 0000 ,f = 00 ,d =0001 ,co= 0
# 4 cci = 1 , a= 1000 ,b = 1111 ,f = 00 ,d =0001 ,co= 1
# 5 cci = 1 , a= 1000 ,b = 1111 ,f = 00 ,d =1000 ,co= 1
# 6 cci = 1 , a= 0010 ,b = 0001 ,f = 01 ,d =1001 ,co= 0
# 7 cci = 1 , a= 0010 ,b = 0001 ,f = 01 ,d =0001 ,co= 0
# 8 cci = 0 , a= 0011 ,b = 0111 ,f = 01 ,d =1101 ,co= 1
# 9 cci = 0 , a= 0011 ,b = 0111 ,f = 01 ,d =1100 ,co= 1
```

Figure 11.18 Partial simulation results of the test bench in Figure 11.17.

signifies a group delay. It implies that any change in any bit of vector a propagates to the corresponding bit of vector d with a delay of 1 ns. The delay is the same for rise or fall in the bits of vector d. Partial results of simulation are shown in Figure 11.18. The values related to the delayed response are shown in bold italics in the figure. The following points are noteworthy here:

- No propagation delay has been specified for the changes in the input vector b or input ci affecting the outputs d or co. Hence all such transitions are instantaneous.
- The propagation delay from a to d has been described as a parallel path delay. Thus any change in a bit of vector a propagates to the corresponding bit of vector d with a delay of 1 ns; but the propagation to the other bits of d is without any delay. Thus the delays associated with the carry bit are zero: those with the sum bits are 1 ns each. Addition operation has been specified up to 6 ns in the test bench (since f = 0 up to 6 ns). At time 4 ns the input values are

a=1000
b=1111 and
ci = 1

The corresponding output values are
d = 1000 and
co =1.

One can see that co attains the final value without any time delay; but every bit of d attains the final value with a delay of 1 ns. The delays considered here are hypothetical and hence need neither be realistic nor consistent with practical circuits.

Example 11.11

The module of Figure 11.17 has been modified and shown in Figure 11.19. Propagation delays have been specified for the changes in the input vector a as well as the vector b affecting the output vector d. All delays affect in a parallel manner. Thus a change in a[2] will transmit to d[2] with a 1 ns delay. But if it affects d[3], the same propagates at the same time step (instantaneously); changes in b vector too affects d in a similar manner. Partial results of simulation are reproduced in Figure 11.20; the values that relate to the delayed changes are shown in italics; they can be seen to conform to the parallel delay specifications.

```
module alu_2 (d, co, a, b, f, cci);
output [3:0] d; output co; wire[3:0]d; input cci; input [3 : 0 ] a, b;
input [1 : 0] f;
specify
        (a,b=>d)=1;
endspecify
assign {co,d}=(f==2'b00)?(a+b+cci):((f==2'b01)?(a-b):((f==2'b10)?
        {1'bz, a^b}:{1'bz,~a})));
endmodule

//test-bench
module tst_alu2();
reg [3:0]a,b; reg[1:0] f; reg cci; wire[3:0]d; wire co;
alu_2 aa(d,co,a,b,f,cci);
initial begin cci=1'b0; f=2'b00;a=4'b0;b=4'h0; #30 $stop; end
always  begin
   #2    cci =1'b0;f=2'b00;a=4'h1;b=4'h0; #2 cci =1'b1;f=2'b00;a=4'h8;b=4'hf;
   #2    cci =1'b1;f=2'b01;a=4'h2;b=4'h1; #2 cci =1'b0;f=2'b01;a=4'h3;b=4'h7;
   #2    cci =1'b1;f=2'b01;a=4'h3;b=4'h3; #2 cci =1'b1;f=2'b10;a=4'h3;b=4'h3;
   #2    cci =1'b1;f=2'b11;a=4'hf;b=4'hc;
        end
initial $monitor($time, " cci = %b , a= %b ,b = %b ,f = %b ,d =%b ,co= %b ",cci
,a,b,f,d,co);
endmodule
```

Figure 11.19 Illustration of assignment of multiple group delays through a specify block.

```
#               0 cci = 0 , a= 0000 ,b = 0000 ,f = 00 ,d =xxxx ,co= 0
#               1 cci = 0 , a= 0000 ,b = 0000 ,f = 00 ,d =0000 ,co= 0
#               2 cci = 0 , a= 0001 ,b = 0000 ,f = 00 ,d =0000 ,co= 0
#               3 cci = 0 , a= 0001 ,b = 0000 ,f = 00 ,d =0001 ,co= 0
#               4 cci = 1 , a= 1000 ,b = 1111 ,f = 00 ,d =0001 ,co= 1
#               5 cci = 1 , a= 1000 ,b = 1111 ,f = 00 ,d =1000 ,co= 1
#               6 cci = 1 , a= 0010 ,b = 0001 ,f = 01 ,d =1000 ,co= 0
#               7 cci = 1 , a= 0010 ,b = 0001 ,f = 01 ,d =0001 ,co= 0
#               8 cci = 0 , a= 0011 ,b = 0111 ,f = 01 ,d =0001 ,co= 1
#               9 cci = 0 , a= 0011 ,b = 0111 ,f = 01 ,d =1100 ,co= 1
```

Figure 11.20 Simulation results with the test bench in Figure 11.19.

Example 11.12

The module in Figure 11.19 has been modified as shown in Figure 11.21. The statement (within the specify block)

(a , b => d) =(1 , 2);

implies that

- All parallel transmission from the input pins of a and b vectors have propagation delays.
- The propagation delay for the rise of d is 1 ns while that for the fall of d is 2 ns.
- Propagation to the noncorresponding bits of output vector d is effected without any delay.

Figure 11.22 shows the partial simulation results. The values in bold italics in the figure relate to the delayed changes.

```
module alu_3 (d, co, a, b, f,cci);
output [3:0] d; output co; wire[3:0]d; input cci;
input [3 : 0 ] a, b; input [1 : 0] f;//F IS A 2 BIT FUNCTION SELECT INPUT
specify
        (a,b=>d)=(1,2);
endspecify
assign {co,d}=(f==2'b00)?(a+b+cci):((f==2'b01)?(a-b):((f==2'b10)?
        {1'bz,a^b}:{1'bz,~a}));
endmodule

//test-bench
module tst_alu3();
reg [3:0]a,b; reg[1:0] f; reg cci; wire[3:0]d; wire co;
alu_3 aa(d,co,a,b,f,cci);
initial begin cci=1'b0; f=2'b00;a=4'b0;b=4'h0; #30 $stop; end
always  begin
   #3     cci =1'b0;f=2'b00;a=4'h1;b=4'h0; #3 cci =1'b1;f=2'b00;a=4'h8;b=4'hf;
   #3     cci =1'b1;f=2'b01;a=4'h2;b=4'h1; #3 cci =1'b0;f=2'b01;a=4'h3;b=4'h7;
   #3     cci =1'b1;f=2'b01;a=4'h3;b=4'h3; #3 cci =1'b1;f=2'b10;a=4'h3;b=4'h3;
   #3     cci =1'b1;f=2'b11;a=4'hf;b=4'hc;
          end
initial $monitor($time, " cci = %b , a= %b ,b = %b ,f = %b ,d =%b ,co= %b ",cci
,a,b,f,d,co);
endmodule
```

Figure 11.21 Module to illustrate assignment of different group delays for rise and fall times using a specify block.

#	0 cci = 0 , a= 0000 ,b = 0000 ,f = 00 ,d =xxxx ,co= 0
#	2 cci = 0 , a= 0000 ,b = 0000 ,f = 00 ,d =**0000** ,co= 0
#	3 cci = 0 , a= 0001 ,b = 0000 ,f = 00 ,d =0000 ,co= 0
#	4 cci = 0 , a= 0001 ,b = 0000 ,f = 00 ,d =**0001** ,co= 0
#	6 cci = 1 , a= 1000 ,b = 1111 ,f = 00 ,d =0001 ,co= 1
#	7 cci = 1 , a= 1000 ,b = 1111 ,f = 00 ,d =**1001** ,co= 1
#	8 cci = 1 , a= 1000 ,b = 1111 ,f = 00 ,d =**1000** ,co= 1
#	9 cci = 1 , a= 0010 ,b = 0001 ,f = 01 ,d =1000 ,co= 0
#	10 cci = 1 , a= 0010 ,b = 0001 ,f = 01 ,d =**1001** ,co= 0
#	11 cci = 1 , a= 0010 ,b = 0001 ,f = 01 ,d =**0001** ,co= 0
#	12 cci = 0 , a= 0011 ,b = 0111 ,f = 01 ,d =0001 ,co= 1
#	13 cci = 0 , a= 0011 ,b = 0111 ,f = 01 ,d =**1101** ,co= 1
#	14 cci = 0 , a= 0011 ,b = 0111 ,f = 01 ,d =**1100** ,co= 1

Figure 11.22 Results of simulating the test bench in Figure 11.21.

Example 11.13

The module of Figure 11.23 has a set of two propagation delay specifications in the specify block – the first one specifies a parallel group delay from the input vectors to the output vector as in the previous example. The second

(a , b , cci *> co) = 1;

implies that any transmission in any of the pins of ports a or b or the pin cci propagates to co with a delay of 1 ns. It is the same for rise as well as fall in the status of the output pin. Figure 11.24 shows part of the simulation results; the values in italics pertain to the delayed response.

```
module alu_4 (d, co, a, b, f,cci);
output [3:0] d; output co; wire[3:0]d; input cci; input [3 : 0 ] a, b;
input [1 : 0] f; //F
specify
        (a,b=>d)=(1,2);
        (a,b,cci*>co)=1;
endspecify
assign {co,d}=(f==2'b00)?(a+b+cci):((f==2'b01)?(a-b):((f==2'b10)?
        {1'bz,a^b}:{1'bz,~a}));
endmodule
```

continued

continued

```
//test-bench
module tst_alu4();
reg [3:0]a,b; reg[1:0] f; reg cci; wire[3:0]d; wire co;
alu_4 aa(d,co,a,b,f,cci);
initial begin cci=1'b0; f=2'b00;a=4'b0;b=4'h0; #30 $stop; end
always    begin
  #3      cci =1'b0;f=2'b00;a=4'h1;b=4'h0; #3 cci =1'b1;f=2'b00;a=4'h8;b=4'hf;
  #3      cci =1'b1;f=2'b01;a=4'h2;b=4'h1; #3 cci =1'b0;f=2'b01;a=4'h3;b=4'h7;
  #3      cci =1'b1;f=2'b01;a=4'h3;b=4'h3; #3 cci =1'b1;f=2'b10;a=4'h3;b=4'h3;
  #3      cci =1'b1;f=2'b11;a=4'hf;b=4'hc;
          end
initial $monitor($time, " cci = %b , a= %b ,b = %b ,f = %b ,d =%b ,co= %b ",cci
,a,b,f,d,co);
endmodule
```

Figure 11.23 A module to illustrate combining of assignments of individual and group delays of the pin-to-pin type.

```
#              0 cci = 0 , a= 0000 ,b = 0000 ,f = 00 ,d =xxxx ,co= x
#              1 cci = 0 , a= 0000 ,b = 0000 ,f = 00 ,d =xxxx ,co= 0
#              2 cci = 0 , a= 0000 ,b = 0000 ,f = 00 ,d =0000 ,co= 0
#              3 cci = 0 , a= 0001 ,b = 0000 ,f = 00 ,d =0000 ,co= 0
#              4 cci = 0 , a= 0001 ,b = 0000 ,f = 00 ,d =0001 ,co= 0
#              6 cci = 1 , a= 1000 ,b = 1111 ,f = 00 ,d =0001 ,co= 0
#              7 cci = 1 , a= 1000 ,b = 1111 ,f = 00 ,d =1001 ,co= 1
#              8 cci = 1 , a= 1000 ,b = 1111 ,f = 00 ,d =1000 ,co= 1
#              9 cci = 1 , a= 0010 ,b = 0001 ,f = 01 ,d =1000 ,co= 1
#             10 cci = 1 , a= 0010 ,b = 0001 ,f = 01 ,d =1001 ,co= 0
#             11 cci = 1 , a= 0010 ,b = 0001 ,f = 01 ,d =0001 ,co= 0
```

Figure 11.24 Results of simulating the test bench in Figure 11.23.

11.3.3   Conditional Pin-to-Pin Delays

The pin to pin path of a signal may change depending on the value of another signal; in turn the number of circuit elements in the alternate path may differ. Conditional selection and assignment of path delays facilitates simulation in such cases.

Example 11.14

The specify block in the module of Figure 11.25 is

```
specify
        if(f==2'b00)(a=>d)=1;
        if(f >2'b00)(a=>d)=2;
            (b,cci*>co)=1;
endspecify
```

It has three propagation statements. The statement

b ,cci *> c0 =1;

is similar to the corresponding one in the previous example. It implies that all transitions to co – if due to changes in any pin of ports a and b or the pin cci – take place with a delay of 1 ns. But the propagation delays associated with changes in the output port d are dependent on the defined functions. For the case

```
module alu_5 (d, co, a, b, f,cci);
output [3:0] d; output co; wire[3:0]d; input cci; input [3 : 0 ] a, b;
input [1 : 0] f;
specify
        if(f==2'b00)(a=>d)=1;
        if(f >2'b00)(a=>d)=2;
            (b,cci*>co)=1;
endspecify
assign {co,d}=(f==2'b00)?(a+b+cci):((f==2'b01)?(a-b):((f==2'b10)?
            {1'bz,a^b}:{1'bz,~a}));

//test-bench
module tst_alu5();
reg [3:0]a,b; reg[1:0] f; reg cci; wire[3:0]d; wire co;
alu_5 aa(d,co,a,b,f,cci);
initial begin cci=1'b0; f=2'b00;a=4'b0;b=4'h0; #30 $stop; end
always   begin
    #3     cci =1'b0;f=2'b00;a=4'h1;b=4'h0; #3 cci =1'b1;f=2'b00;a=4'h8;b=4'hf;
    #3     cci =1'b1;f=2'b01;a=4'h2;b=4'h1; #3 cci =1'b0;f=2'b01;a=4'h3;b=4'h7;
    #3     cci =1'b1;f=2'b01;a=4'h3;b=4'h3; #3 cci =1'b1;f=2'b10;a=4'h3;b=4'h3;
    #3     cci =1'b1;f=2'b11;a=4'hf;b=4'hc;
                end
initial $monitor($time, " cci = %b , a= %b ,b = %b ,f = %b ,d =%b ,co= %b ",cci
,a,b,f,d,co);
endmodule
```

Figure 11.25 Illustration of conditional assignments for delay values through a **specify** block.

#	0 cci = 0 , a= 0000 ,b = 0000 ,f = 00 ,d =xxxx ,co= x
#	**1** cci = 0 , a= 0000 ,b = 0000 ,f = 00 ,d =**0000** ,co= **0**
#	3 cci = 0 , a= 0001 ,b = 0000 ,f = 00 ,d =0000 ,co= 0
#	**4** cci = 0 , a= 0001 ,b = 0000 ,f = 00 ,d =**0001** ,co= 0
#	6 cci = 1 , a= 1000 ,b = 1111 ,f = 00 ,d =0001 ,co= 0
#	**7** cci = 1 , a= 1000 ,b = 1111 ,f = 00 ,d =**1000** ,co= **1**
#	9 cci = 1 , a= 0010 ,b = 0001 ,f = 01 ,d =1001 ,co= 1
#	**10** cci = 1 , a= 0010 ,b = 0001 ,f = 01 ,d =1001 ,co= **0**
#	**11** cci = 1 , a= 0010 ,b = 0001 ,f = 01 ,d =**0001** ,co= 0

Figure 11.26 Results of simulating the test bench in Figure 11.25.

of addition (f=2'00) the propagation delay is 1 ns. For all other types of functions it is 2 ns. Similar conditional propagation delays can be defined separately for each of the functions of the ALU. Figure 11.26 shows the simulation results partially; the values pertaining to the delayed response are in bold italics in the figure.

Observations:

A simple condition was used in Example 11.14 to illustrate conditional assignment to delay values. In a general case a conditional expression can be more involved with different logical operations performed in tandem, with the following restrictions:

- The expression can involve any logical reduction operation.
- All bit-wise logical operations can be used.
- If a conditional expression evaluates to multiple bits, the least significant bit decides the delay.

Example 11.15

The half-adder of Figure 11.3 has been slightly modified and is shown in Figure 11.27. The propagation delays for rise and fall are kept the same here for simplicity. However, the test bench has two instantiations of the module. The propagation delays are assigned one set of values for the instantiation h1 and another for the instantiation h2. The alternate assignments are made through a **defparam** statement. The access to the parameters is by suitably specifying the hierarchy. Note that if the parameters had been specified through a specify block, and specparam assignment, such an alteration at the time of instantiation, is not feasible. The simulation results are reproduced partially in Figure 11.28; the figures in bold italics relate to the delayed changes.

```
module ha_a(s,ca,a,b);
input a,b; output s,ca; parameter dl1r=1,dl3r=3;
xor #(dl1r) (s,a,b);
and #(dl3r) (ca,a,b);
endmodule

//test-bench
module tstha_a();
reg a,b; wire s,ca;
ha_a h1(s1,ca1,a,b);
ha_a h2(s2,ca2,a,b);
defparam
        h1.dl1r=2,
        h1.dl3r=1,
        h2.dl1r=2,
        h2.dl3r=2;
initial begin a=0;b=0; end
always begin #5 a=1;b=0; #5 a=0;b=1; #5 a=1;b=1; #5 a=0;b=0; end
initial  $monitor($time , "  a = %b , b = %b ,ca1 = %b , s1 = %b,ca2 = %b , s2 =
%b " ,a,b,ca1,s1,ca2,s2);
initial #30 $stop;
endmodule
```

Figure 11.27 Illustration of Multiple instantiations with assignment of different time delays.

```
#            0 a = 0 , b = 0 ,ca1 = x , s1 = x,ca2 = x , s2 = x
#            1 a = 0 , b = 0 ,ca1 = 0 , s1 = x,ca2 = x , s2 = x
#            2 a = 0 , b = 0 ,ca1 = 0 , s1 = 0,ca2 = 0 , s2 = 0
#            5 a = 1 , b = 0 ,ca1 = 0 , s1 = 0,ca2 = 0 , s2 = 0
#            7 a = 1 , b = 0 ,ca1 = 0 , s1 = 1,ca2 = 0 , s2 = 1
#           10 a = 0 , b = 1 ,ca1 = 0 , s1 = 1,ca2 = 0 , s2 = 1
#           15 a = 1 , b = 1 ,ca1 = 0 , s1 = 1,ca2 = 0 , s2 = 1
#           16 a = 1 , b = 1 ,ca1 = 1 , s1 = 1,ca2 = 0 , s2 = 1
#           17 a = 1 , b = 1 ,ca1 = 1 , s1 = 0,ca2 = 1 , s2 = 0
```

Figure 11.28 Results of simulating the test bench in Figure 11.27.

Example 11.16

The half-adder module of Figure 11.15 is modified and shown in Figure 11.29. The specify block has the rise- and fall-time values at output specified. The "minimum, typical, maximum" format has been used here for the time delay values specified. The test bench uses typical delay values (2 ns and 4 ns for **s** and 3 ns and 7 ns for **ca**, respectively) by default. The simulation results are shown in Figure 11.30: The values representing delayed response are in bold italics. Testing with minimum or maximum delay values can be carried out in the normal manner.

```
module ha_c(s,ca,a,b);
input a,b; output s,ca;
specify
        (a,b*>s)=(1:2:3, 2:4:6);
        (a,b*>ca)=(1:3:5, 5:7:9);
endspecify
xor  (s,a,b);
and  (ca,a,b);
endmodule
//test-bench
module tstha_c();
reg a,b; wire s,ca;
ha_c hh(s,ca,a,b);
initial begin a=0;b=0;  #100 $stop; end
always   begin #15 a=1;b=0; #15 a=0;b=1; #15 a=1;b=1; #15 a=0;b=0; end
initial $monitor($time , "  a = %b , b = %b ,out carry = %b , outsum = %b  "
,a,b,ca,s);
endmodule
```

Figure 11.29 Illustration of specifying minimum, typical, and maximum values for path delays in a specify block.

#	0 a = 0 , b = 0 ,out carry = x , outsum = x
#	**4** a = 0 , b = 0 ,out carry = x , outsum = **0**
#	**7** a = 0 , b = 0 ,out carry = **0** , outsum = 0
#	15 a = 1 , b = 0 ,out carry = 0 , outsum = 0
#	**17** a = 1 , b = 0 ,out carry = 0 , outsum = **1**
#	30 a = 0 , b = 1 ,out carry = 0 , outsum = 1
#	45 a = 1 , b = 1 ,out carry = 0 , outsum = 1
#	**48** a = 1 , b = 1 ,out carry = **1** , outsum = 1
#	**49** a = 1 , b = 1 ,out carry = 1 , outsum = **0**
#	60 a = 0 , b = 0 ,out carry = 1 , outsum = 0
#	**67** a = 0 , b = 0 ,out carry = **0** , outsum = 0
#	75 a = 1 , b = 0 ,out carry = 0 , outsum = 0
#	**77** a = 1 , b = 0 ,out carry = 0 , outsum = **1**
#	90 a = 0 , b = 1 ,out carry = 0 , outsum = 1

Figure 11.30 Results of simulating the test bench in Figure 11.29.

11.3.4   Edge-Sensitive Paths

Behavior level modules can have signal paths activated following an edge in a different signal. Verilog has the provision to specify such delays during simulation. They can be specified in a variety of ways. The path may get activated following a positive edge or a negative edge in a signal. The path delay may be specified for rise or fall in the output or for positive or negative polarity transitions separately. The delay assignment can be made conditional on an expression; such a path specification is an "edge sensitive state dependent path".

Example 11.17

The D flip-flop module of Figure 7.27 has been modified and is shown in Figure 11.31. The specify block specifies the delay from di to do following a negative edge of clock. The simulation results are partially reproduced in Figure 11.32; the flip-flop is to latch the input data di at the negative-going edges of the clock – that is, at the 6th ns, 12th ns, *etc.* The latching is delayed by 1 ns as demanded by the specified delay and takes effect at the 7th ns, 13th ns, *etc.*

```
module dff_p(do,di,clk);
output do; input di,clk;
specify
        (negedge clk *>(do:di)) =1;
endspecify
reg do;
initial do=1'b0;
always@(negedge clk) do=di;
endmodule

//test-bench
module tst_dff_pbeh();
reg di,clk; wire do;
dff_p d1(do,di,clk);
initial begin clk=0;di=1'b0; #35 $stop; end
always #3clk=~clk;
always #5 di=~di;
initial $monitor($time,"clk=%b,di=%b,do=%b",clk,di,do);
endmodule
```

Figure 11.31 A module to illustrate edge-sensitive path delay and its test bench.

#	0clk=0,di=0,do=x
#	1clk=0,di=0,do=0
#	3clk=1,di=0,do=0
#	5clk=1,di=1,do=0
#	6clk=0,di=1,do=0
#	7clk=0,di=1,do=1
#	9clk=1,di=1,do=1
#	10clk=1,di=0,do=1
#	12clk=0,di=0,do=1
#	13clk=0,di=0,do=0

Figure 11.32 Partial results of simulating the test bench in Figure 11.31.

Example 11.18

The module in Figure 11.33 is a slightly modified version of that in Figure 8.20. The specify block specifies the input to output delay following a positive edge of clock; further it is effective only when clr and pr are inactive. The path specified here is an "edge-sensitive state-dependent path". Partial simulation results are in Figure 11.34.

```
module dff_aa(q,qb,di,clk,clr,pr);
output q,qb; input di,clk,clr,pr;
reg q;
assign qb=~q;
specify
        if(!clr && !pr) (posedge clk *> (q:di))=1;
endspecify
always@(posedge clk)
 begin if(clr)q = 1'b0; else if(pr) q = 1'b1; else  q=di;  end
endmodule

//test-bench
module dff_aa_tst();
reg di,clk,clr,pr; wire q,qb;
dff_aa dd(q,qb,di,clk,clr,pr);
initial begin clr=1'b1;pr=1'b0;clk=1'b0;di=1'b0;  #100 $stop;
end
always #3 clk=~clk;
always  begin
        # 4  di =~di; # 3  di =~di; # 3  di =~di; # 6  di =~di;
```

continued

continued

```
        # 6  di =~di; # 3  di =~di; # 2  di =~di;
        end
initial begin #5  pr=1'b1; #5  pr=1'b0; #35 pr=1'b1; #25 pr=1'b1; end
initial #25 clr=1'b0;
initial $monitor( $time , "clk = %b , clr = %b , pr = %b , di = %b , q =
%b ", clk,clr,pr,di,q);
endmodule
```

Figure 11.33 A module to illustrate delay assignment for an edge-sensitive state-dependent path.

t	clk	clr	pr	di	q		t	clk	clr	pr	di	q
0	0	1	0	0	x		25	0	0	0	0	0
3	1	1	0	0	0		27	1	0	0	1	0
4	1	1	0	1	0		28	1	0	0	1	1
5	1	1	1	1	0		30	0	0	0	1	1
6	0	1	1	1	0		31	0	0	0	0	1
7	0	1	1	0	0		33	1	0	0	0	1
9	1	1	1	0	0		34	1	0	0	1	0
10	1	1	0	1	0		36	0	0	0	1	0
12	0	1	0	1	0		37	0	0	0	0	0
15	1	1	0	1	0		39	1	0	0	0	0
16	1	1	0	0	0		42	0	0	0	0	0
18	0	1	0	0	0		43	0	0	0	1	0
21	1	1	0	0	0		45	1	0	1	1	1
22	1	1	0	1	0		48	0	0	1	1	1
24	0	1	0	1	0							

Figure 11.34 Partial results of simulating the test bench in Figure 11.33.

Observations:

- Until the 25th nanosecond, the clr input is active and the flip-flop remains reset. The pr signal is high from the 5th to the 10th ns; but since the clr has priority, the flip-flop remains reset. The delay specified is not relevant.
- After the 45th ns, pr is active and the flip-flop remains set. Clk and di are not relevant.
- Only in the 25th to the 45th ns interval the flip-flop responds to di at the positive-going edge of the clock; it happens at the 28th and 34rd ns. Specifically, 27th ns and 33rd ns represent positive going edges of the clock. Changes in di preceding them get reflected as changes in do with a delay of 1 ns–that is, at 28th ns and 34rd ns, respectively.

11.3.5   Pulse Filtering and its Control

All transitions on an input pin with less than a specified module path delay are termed "pulses." Normally, when a module path delay is specified, all pulses are ignored; that is, the simulator does not take cognizance of such narrow transitions. However, response to such narrow pulses can be specified through **specparam** in a specify block. A statement

specparam PATHPULSE$ (x , y) = (a, b);

implies the following concerning the module pulse path from x to y:

- Ignore all pulses of width less than a ns. a is referred to as the "rejection limit" for the pulse path.
- Take cognizance of all the pulses wider than b ns. Note that the specification has relevance only if the delay value for the pulse path (specified in the specify block) is larger than b.
- For all pulses of width value between a and b, the output is in error and in **x** state.

The **PATHPULSE$** specification is governed by the following:

- It has to appear within a specify block as a **specparam** assignment as shown above.
- It specifies the limits for the path pulse-error limit as well as reject limit for the specified path.
- A statement as

    specparam PATHPULSE$ = (a, b);

    implies that a and b are the error and reject limits for the pulse widths for all the paths specified within the specify block; the simulator checks for the pulse width and if it is between a and b values, the output goes to **x** state.

- A set of statements

    Specparam PATHPULSE$ (x, y) = (a, b);
    Specparam PATHPULSE$ = (c, d);

    implies that for the path from x to y a and b are the error and reject limits, respectively; further, for all other pulse paths within the specify block, the limits for error and rejection are c and d, respectively. If only one limit is specified as

    Specparam PATHPULSE$ =a;

    a is taken as the error limit as well as reject limit for the concerned paths.

Example 11.19

The module in Figure 11.35 is the half-adder module of earlier examples. A pin-to-pin delay of 4 ns is specified from a and b inputs to the sum bit s and the carry bit ca. Further a PATHPULSE limit of 3 ns is specified; hence any pulse of width less than 3 ns will be ignored by the simulator. Simulation results are shown in Figure 11.36. The following may be noted in this connection:

- During the interval of 8 ns – 10 ns the input a is at zero. It represents a pulse input. It is ignored and the sum bit does not revert to zero during the corresponding delayed interval of 12 ns – 14 ns. Similar response is repeated for the change in a to zero during the interval 28 ns – 30 ns.
- During the interval 14 ns – 15 ns, input b goes high. The same pulse, being narrower than the specified limit of 3 ns, is ignored. Neither the sum bit s nor the carry bit ca is affected.
- At 34 ns, b goes to 1 and remains so up to 37 ns; it is treated as a pulse and ignored by the simulator.

```
module ha_pt(s,ca,a,b);
input a,b; output s,ca;
specify
        (a,b*>s,ca) =4;
        specparam pathpulse$ = 3;
endspecify
xor  (s,a,b);
and  (ca,a,b);
endmodule

//test-bench
module tstha_pt();
reg a,b; wire s,ca;
ha_pt hh(s,ca,a,b);
initial begin a=0;b=0; #50 $stop;end
initial begin #4 a=1;b=0; #4 a=0;b=0; #2 a=1;b=0; #4 a=1;b=1; #1 a=1;b=0;
        #4 a=1;b=1; #4 a=1;b=0; #1 a=1;b=0; #4 a=0;b=0; #2 a=1;b=0;
        #4 a=1;b=1; #3 a=1;b=0; #4 a=1;b=1; #4 a=1;b=0; #1 a=1;b=0;
    end
initial  $monitor($realtime , " a = %b , b = %b ,out carry = %b , outsum = %b "
,a,b,ca,s);
endmodule
```

Figure 11.35 A module to illustrate the use of the PATHPULSE limit.

```
# 0  a = 0 , b = 0 ,out carry = x , outsum = x
# 4  a = 1 , b = 0 ,out carry = 0 , outsum = 0
# 8  a = 0 , b = 0 ,out carry = 0 , outsum = 1
# 10  a = 1 , b = 0 ,out carry = 0 , outsum = 1
# 14  a = 1 , b = 1 ,out carry = 0 , outsum = 1
# 15  a = 1 , b = 0 ,out carry = 0 , outsum = 1
# 19  a = 1 , b = 1 ,out carry = 0 , outsum = 1
# 23  a = 1 , b = 0 ,out carry = 1 , outsum = 0
# 27  a = 1 , b = 0 ,out carry = 0 , outsum = 1
# 28  a = 0 , b = 0 ,out carry = 0 , outsum = 1
# 30  a = 1 , b = 0 ,out carry = 0 , outsum = 1
# 34  a = 1 , b = 1 ,out carry = 0 , outsum = 1
# 37  a = 1 , b = 0 ,out carry = 0 , outsum = 1
# 41  a = 1 , b = 1 ,out carry = 0 , outsum = 1
# 45  a = 1 , b = 0 ,out carry = 1 , outsum = 0
# 49  a = 1 , b = 0 ,out carry = 0 , outsum = 1
```

Figure 11.36 Results of simulating the test bench in Figure 11.35.

Example 11.20

The module in Figure 11.37 is a slightly modified version of that in Figure 11.35. Here 2 ns is specified as the error limit and 3 ns as the rejection limit for all module path pulses. The test bench remains unchanged. The following can be observed with the simulation results shown in Figure 11.38:

- In the interval 8ns – 10ns as well as the interval 28 ns –30 ns the input a goes down and remains at the 0 state. These represent pulse widths less than the reject limit but more than the error limit. Hence an error is indicated and the output goes to **x** state in the corresponding intervals 12 ns – 14 ns and 32 ns – 34 ns, respectively. Ca remains unaltered as expected.
- In the interval 14 ns –15 ns, b is at the 1 state. Because the pulse width is less than the error limit, the pulse is ignored. Neither s nor ca responds to it.

```
module ha_ptt(s,ca,a,b);
input a,b; output s,ca;
specify
        (a,b*>s,ca) =4;
        specparam PATHPULSE$ = (2,3);
endspecify
xor  (s,a,b);
and  (ca,a,b);
```

continued

continued

```
endmodule
//test-bench
module tstha_ptt();
reg a,b; wire s,ca;
ha_ptt hh(s,ca,a,b);
initial begin a=0;b=0; #50 $stop;end
initial    begin
           #4 a=1;b=0; #4 a=0;b=0; #2 a=1;b=0; #4 a=1;b=1; #1 a=1;b=0;
           #4 a=1;b=1; #4 a=1;b=0; #1 a=1;b=0; #4 a=0;b=0; #2 a=1;b=0;
           #4 a=1;b=1; #3 a=1;b=0; #4 a=1;b=1; #4 a=1;b=0; #1 a=1;b=0;
           end
initial  $monitor($realtime , " a = %b , b = %b ,out carry = %b , outsum = %b "
,a,b,ca,s);
endmodule
```

Figure 11.37 Module to illustrate error limit and rejection limit with PATHPULSE.

```
# 0  a = 0 , b = 0 ,out carry = x , outsum = x
# 4  a = 1 , b = 0 ,out carry = 0 , outsum = 0
# 8  a = 0 , b = 0 ,out carry = 0 , outsum = 1
# ** Warning: D:/chap11/chap_11/ha_ptt.v.txt(5): path pulse error on net
tstha_ptt.s
#    Time: 10 ns Iteration: 1  Instance: /tstha_ptt/hh
# 10  a = 1 , b = 0 ,out carry = 0 , outsum = 1
# 12  a = 1 , b = 0 ,out carry = 0 , outsum = x
# 14  a = 1 , b = 1 ,out carry = 0 , outsum = 1
# 15  a = 1 , b = 0 ,out carry = 0 , outsum = 1
# 19  a = 1 , b = 1 ,out carry = 0 , outsum = 1
# 23  a = 1 , b = 0 ,out carry = 1 , outsum = 0
# 27  a = 1 , b = 0 ,out carry = 0 , outsum = 1
# 28  a = 0 , b = 0 ,out carry = 0 , outsum = 1
# ** Warning: D:/chap11/chap_11/ha_ptt.v.txt(5): path pulse error on net
tstha_ptt.s
#    Time: 30 ns Iteration: 1  Instance: /tstha_ptt/hh
# 30  a = 1 , b = 0 ,out carry = 0 , outsum = 1
# 32  a = 1 , b = 0 ,out carry = 0 , outsum = x
# 34  a = 1 , b = 1 ,out carry = 0 , outsum = 1
# 37  a = 1 , b = 0 ,out carry = 0 , outsum = 1
# 38  a = 1 , b = 0 ,out carry = 1 , outsum = 0
# 41  a = 1 , b = 1 ,out carry = 0 , outsum = 1
# 45  a = 1 , b = 0 ,out carry = 1 , outsum = 0
# 49  a = 1 , b = 0 ,out carry = 0 , outsum = 1
```

Figure 11.38 Results of simulating the test bench in Figure 11.37.

The module in Figure 11.37 was modified and the PULSEPATH$ assignment is removed. The default value of 4 ns is the reject as well as the error limit here. Simulation results obtained with the modified module are shown in Figure 11.39. One can see that all pulses of width less than 4 ns (in the intervals 8 ns – 10 ns and 28 ns – 30 ns for a; 14 ns – 15 ns and 34 ns – 37 ns for b) are ignored by the design module.

```
# 0  a = 0 , b = 0 ,out carry = x , outsum = x
# 4  a = 1 , b = 0 ,out carry = 0 , outsum = 0
# 8  a = 0 , b = 0 ,out carry = 0 , outsum = 1
# 10 a = 1 , b = 0 ,out carry = 0 , outsum = 1
# 14 a = 1 , b = 1 ,out carry = 0 , outsum = 1
# 15 a = 1 , b = 0 ,out carry = 0 , outsum = 1
# 19 a = 1 , b = 1 ,out carry = 0 , outsum = 1
# 23 a = 1 , b = 0 ,out carry = 1 , outsum = 0
# 27 a = 1 , b = 0 ,out carry = 0 , outsum = 1
# 28 a = 0 , b = 0 ,out carry = 0 , outsum = 1
# 30 a = 1 , b = 0 ,out carry = 0 , outsum = 1
# 34 a = 1 , b = 1 ,out carry = 0 , outsum = 1
# 37 a = 1 , b = 0 ,out carry = 0 , outsum = 1
# 41 a = 1 , b = 1 ,out carry = 0 , outsum = 1
# 45 a = 1 , b = 0 ,out carry = 1 , outsum = 0
# 49 a = 1 , b = 0 ,out carry = 0 , outsum = 1
```

Figure 11.39 Results of simulating the test bench in Figure 11.37 with the PATHPULSE specification in the module ha_ptt deleted.

11.4   MODULE PARAMETERS

Module parameters are associated with size of bus, register, memory, ALU, and so on. They can be specified within the concerned module but their value can be altered during instantiation. The alterations can be brought about through assignments made with **defparam**. Such **defparam** assignments can appear anywhere in a module.

The rules of assigning values for the module parameters, deciding their size, type, *etc.*, are all similar to those of **specify** parameters discussed in Section 11.2.

Example 11.21

The module of Figure 11.23 has been modified and shown in Figure 11.40. The parameter msb specifies the ALU size — consistently in the input and the output vectors of the ALU. The size assignment has been made separately through the assignment statement

parameter msb = 3;

With the test bench in Figure 11.23 the simulation results are identical to those of Figure 11.24. The ALU size can be scaled up to any value by reassigning a value to msb during instantiation.

```
module alu_6 (d, co, a, b, f,cci);
parameter msb=3;
output [msb:0] d; output co; wire[msb:0]d;
input cci; input [msb : 0 ] a, b; nput [1 : 0] f;
specify
        (a,b=>d)=(1,2);
        (a,b,cci*>co)=1;
endspecify
assign {co,d}=(f==2'b00)?(a+b+cci):((f==2'b01)?(a-b):((f==2'b10)?
        {1'bz,a^b}:{1'bz,~a}));
endmodule
```

Figure 11.40 The ALU module in Figure 11.23 with its size declared as a parameter.

Example 11.22

Figure 11.41 shows a design where the ALU module of Figure 11.40 has been retained and the test bench of Figure 11.23 included; the test bench has been altered whenever the parameter msb is assigned a different value (=7) which overrides the assignment in the instantiation. The simulation results are shown in Figure 11.42 from the 15th ns.

```
module alu_6 (d, co, a, b, f,cci);
parameter msb=3;
output [msb:0] d; output co; wire[msb:0]d; input cci;
input [msb : 0 ] a, b; input [1 : 0] f;
specify  (a,b=>d)=(1,2);  (a,b,cci*>co)=1; endspecify
assign {co,d}=(f==2'b00)?(a+b+cci):((f==2'b01)?(a-b):((f==2'b10)?
        {1'bz,a^b}:{1'bz,~a}));
endmodule
//test-bench
module tst_alu7();
defparam aa.msb=7; parameter nl=7;
reg [nl:0]a,b; reg[1:0] f; reg cci; wire[nl:0]d; wire co;
```

continued

continued

```
alu_6 aa(d,co,a,b,f,cci);
initial begin cci=1'b0;  f=2'b00;a=8'h00;b=8'h00;  #30 $stop;end
always   begin
#3 cci =1'b0;f=2'b00;a=8'h01;b=8'h00;  #3 cci =1'b1;f=2'b00;a=8'h08;b=8'h0f;
#3 cci =1'b1;f=2'b01;a=8'h02;b=8'h01;  #3 cci =1'b0;f=2'b01;a=8'h23;b=8'h27;
#3 cci =1'b1;f=2'b01;a=8'h23;b=8'h23;  #3 cci =1'b1;f=2'b10;a=8'h23;b=4'h23;
#3 cci =1'b1;f=2'b11;a=8'h2f;b=8'h2c;
        end
initial $monitor($time, " cci = %b , a= %b ,b = %b ,f = %b ,d =%b ,co= %b ",cci
,a,b,f,d,co);
endmodule
```

Figure 11.41 An ALU module with its size being redefined during instantiation

```
# 15 cci = 1 , a= 00100011 ,b = 00100011 ,f = 01 ,d =11111100 ,co= 1
# 16 cci = 1 , a= 00100011 ,b = 00100011 ,f = 01 ,d =11111100 ,co= 0
# 17 cci = 1 , a= 00100011 ,b = 00100011 ,f = 01 ,d =00000000 ,co= 0
# 18 cci = 1 , a= 00100011 ,b = 00000011 ,f = 10 ,d =00000000 ,co= 0
# 19 cci = 1 , a= 00100011 ,b = 00000011 ,f = 10 ,d =00100000 ,co= z
# 21 cci = 1 , a= 00101111 ,b = 00101100 ,f = 11 ,d =00100000 ,co= z
# 22 cci = 1 , a= 00101111 ,b = 00101100 ,f = 11 ,d =11110000 ,co= z
# 23 cci = 1 , a= 00101111 ,b = 00101100 ,f = 11 ,d =11010000 ,co= z
# 24 cci = 0 , a= 00000001 ,b = 00000000 ,f = 00 ,d =11010000 ,co= z
# 25 cci = 0 , a= 00000001 ,b = 00000000 ,f = 00 ,d =11010001 ,co= 0
# 26 cci = 0 , a= 00000001 ,b = 00000000 ,f = 00 ,d =00000001 ,co= 0
# 27 cci = 1 , a= 00001000 ,b = 00001111 ,f = 00 ,d =00000001 ,co= 0
# 28 cci = 1 , a= 00001000 ,b = 00001111 ,f = 00 ,d =00011001 ,co= 0
# 29 cci = 1 , a= 00001000 ,b = 00001111 ,f = 00 ,d =00011000 ,co= 0
```

Figure 11.42 Results of simulating the test bench in Figure 11.41.

11.5   SYSTEM TASKS AND FUNCTIONS

Verilog has a number of System Tasks and Functions defined in the LRM. They are for taking output from simulation, control simulation, debugging design modules, testing modules for specifications, *etc.* A "$" sign preceding a word or a word group signifies a system task or a system function. Some of the system tasks and functions have been extensively used in the earlier chapters. Some others with the potential for common use are described and illustrated here. The complete list is available in the LRM.

11.5.1　Output Tasks

A number of system tasks are available to output values of variables and selected messages, *etc.*, on the monitor. Out of these **$monitor** and **$display** tasks have been extensively used in the preceding chapters. These and related tasks are discussed below.

11.5.2　Display Tasks

The **$display** task, whenever encountered, displays the arguments in the desired format; and the display advances to a new line. **$write** task carries out the desired display but does not advance to the new line. For both the format is identical to that of **scanf** and **printf** in C language [Gottfried]. The features are briefly outlined here:

- The arguments are displayed in the same order as they appear in the display statement.
- The arguments can be variables, an expression involving variables, or quoted strings.
- The strings are output as such except the escape sequences. An escape sequence starts with the character \ or the character %.
- "\" signifies one of a set of special characters in Table 11.1.
- "%m" signifies that the hierarchical name of the particular argument is to be displayed (see Example 11.23).
- "%" followed by a character – as given in Table 11.2 – specifies the format for display of the following argument or an aspect of the following argument.
- If the format for the display of an argument is not specified, a default format is assumed. It is binary for **$displayb** and **$writeb**, octal for **$displayo** and **$writeo**, decimal for **$displayd** and **$writed**, hex for **$displayh** and **$writeh**.
- If any argument is in the form of an expression, it is evaluated and the result displayed or written; it is sized automatically. With decimal numbers the leading zeros are suppressed. Insertion of a "0" character (zero digit) between the "%" symbol and the radix overrides the automatic sizing.

Table 11.1 Escape sequences

Sequence	Implication
\n	Display to advance to a new line.
\t	Insert a tab in the display.
\\	Insert a '\' character in the display.
\"	Insert the double quote character '"' in the display.
\aaa	Insert an ASCII character specified by the octal number "aaa", in the display.
%%	Inset the character '%' in the string displayed

Table 11.2 Format for display of arguments

Character combination	Implication
%h or %H	Display in hex format
%d or %D	Display in decimal format
%o or %O	Display in octal format
%b or %B	Display in binary format
%c or %C	Display in ASCII character format
%l or %L	Display library binding information
%v or %V	Display net signal strength
%s or %S	Display as a string
%t or %T	Display in current time format
%u or %U	Unformatted 2-value data
%z or %Z	Unformatted 4-value data
%f or %F	Display real in decimal format
%g or %G	Display real in exponential or decimal format, whichever is shorter

Example 11.23

The module in Figure 11.31 has been modified and shown in Figure 11.43. A **$display** (`"%m"`) has been added to the test bench as well as to the design module itself. Partial simulation results are also included in the figure. The task displays the hierarchical name of the module it is in. Thus when encountered in the test-bench, the hierarchical name of the test bench – namely "tst_dff_p_b.d1"– is displayed.

The task is useful to identify the *"parentage"* of the module when a design has a number of instantiations and values are not clearly traceable to sources. Note that the task does not require an argument to be tagged to it.

```
module dff_p(do,di,clk);
output do; input di,clk;
specify
        (negedge clk *>(do:di)) =1;
endspecify
reg do;
initial do=1'b0;
always@(negedge clk) do=di;
initial    $display ("%m");
endmodule
```

continued

continued

```
//test-bench
module tst_dff_p_b();
reg di,clk; wire do;
dff_p d1(do,di,clk);
initial begin clk=0;di=1'b0; #35 $stop;end
always #3clk=~clk;
always #5 di=~di;
initial    $display ("%m");
initial    $monitor($time,"clk=%b,di=%b,do=%b",clk,di,do);
endmodule
```

Simulation results (shown partially)
```
# tst_dff_p_b.d1
# tst_dff_p_b
#               0clk=0,di=0,do=x
#               1clk=0,di=0,do=0
#               3clk=1,di=0,do=0
#               5clk=1,di=1,do=0
```

Figure 11.43 A module and its test bench to illustrate the use of "%m" in the display task: The simulation results are also shown partially.

Example 11.24

The 4-to-16 decoder considered in Chapter 4 has repeated nested instantiations. The module listing is reproduced in Figure 11.44. The test bench is omitted; a "$display ("%m")" statement is included in the 2-to-4 decoder module (dec2_4a_). Whenever it is instantiated, its hierarchical name is displayed. The simulation results are reproduced in Figure 11.45. The 3-to-8 decoder module is instantiated twice (as g3 and g4) in the 4-to-16 decoder module. In turn, these (g3 and g4) instantiate the 2-to-4 decoders twice (as g1 and g2). The hierarchical names are displayed in the simulation run as can be seen from Figure 11.45.

```
module dec3_8a(pp,q,enn);
output[7:0]pp; input[2:0]q; input enn; wire qq; wire[7:0]p;
not(qq,q[2]);
dec2_4a g1(.a(p[3:0]),.b(q[1:0]),.en(qq));
dec2_4a g2(.a(p[7:4]),.b(q[1:0]),.en(q[2]));
and g3_8_7(pp[7],p[7],enn), g3_8_6(pp[6],p[6],enn),
g3_8_5(pp[5],p[5],enn), g3_8_4(pp[4],p[4],enn), g3_8_3(pp[3],p[3],enn),
g3_8_2(pp[2],p[2],enn), g3_8_1(pp[1],p[1],enn), g3_8_0(pp[0],p[0],enn);
endmodule
```

continued

continued

module dec2_4a (a,b,en);
output [3:0] a; input [1:0]b; input en; wire [1:0]bb;
not(bb[1],b[1]),(bb[0],b[0]);
and(a[0],en,bb[1],bb[0]),(a[1],en,bb[1],b[0]),
(a[2],en,b[1],bb[0]),(a[3],en,b[1],b[0]));
initial $display ("%m");
endmodule

test-bench
module dec4_16_tba;
wire[15:0]m;
wire l,m,n;
reg[3:0]n;
dec4_16a gg(m,n);
endmodule

Figure 11.44 A 4-to-16 decoder module with a "**$display ("%m");**" statement inserted to display hierarchy.

```
//# dec4_16_tba.gg.g3.g1
//# dec4_16_tba.gg.g3.g2
//# dec4_16_tba.gg.g4.g1
//# dec4_16_tba.gg.g4.g2
```

Figure 11.45 Results of simulating the module in Figure 11.44.

Example 11.25 Display of Strength

Figure 11.46 shows the module of Figure 5.33 along with its test bench. Through the **$monitor** task the strength of the output variable o is displayed. Simulation results are shown in Figure 11.47. The strength of o is consistent with the signal status in each case:

- Whenever i1 = i2 = 0, o =0 and has a strength of **pull0** (represented as **pu0**).
- Whenever i1 = 0 and i2 = 1, o = x and has a strength of **pullx** (represented as **pux**).
- Whenever i1 = 1 and i2 = 0, strong1 dominates over **pull0**; output is at 1 state and of strength **strong1** (represented as **st1**).
- Whenever i1 = 1 and i2 = 1, o = 1; **strong1** dominates and decides the strength (represented as **st1**).

```
module strng_1(o,i1,i2);
input i1,i2; output o; //wire o
buf(strong1 ,pull0)g1(o,i1);
buf(pull1,pull0)g2(o,i2);
endmodule

//TEST BENCH
module tst_strng_1;
reg i1,i2;
strng_1 cc(o,i1,i2);
initial begin i1 =0;i2 =0; #40 $stop; end
always begin #4 i1 = 0;i2 = 1; #4 i1 =1; i2 =0;  #4 i1 =1 ;i2 = 1; end
initial $monitor($time   ," i1 = %b  ,i2 = %b ,o = %b(strength of o = %v) "
,i1,i2,o,o);
endmodule
```

Figure 11.46 A module set to illustrate display of strength levels.

```
#              0 i1 = 0  ,i2 = 0 ,o = 0(strength of o = Pu0)
#              4 i1 = 0  ,i2 = 1 ,o = x(strength of o = PuX)
#              8 i1 = 1  ,i2 = 0 ,o = 1(strength of o = St1)
#             12 i1 = 1  ,i2 = 1 ,o = 1(strength of o = St1)
#             16 i1 = 0  ,i2 = 1 ,o = x(strength of o = PuX)
#             20 i1 = 1  ,i2 = 0 ,o = 1(strength of o = St1)
#             24 i1 = 1  ,i2 = 1 ,o = 1(strength of o = St1)
#             28 i1 = 0  ,i2 = 1 ,o = x(strength of o = PuX)
#             32 i1 = 1  ,i2 = 0 ,o = 1(strength of o = St1)
#             36 i1 = 1  ,i2 = 1 ,o = 1(strength of o = St1)
```

Figure 11.47 Results of simulating the test bench in Figure 11.46.

11.5.3 $strobe Task

When a variable or a set of variables is sampled and its value displayed, the **$strobe** task can be used; it senses the value of the specified variables and displays them.   The form of specifying arguments is identical to that of the **$display** task.   The **$strobe** task is executed as the last activity in the concerned time step.   It is useful to check for specific activities and debug modules.

Example 11.26

The module of Figure 7.27 is reproduced in Figure 11.48 with the addition of a **$strobe** command. Simulation results are shown (partially) in Figure 11.49. The **$monitor** task ensures that all specified items are displayed when any of them changes. The **$strobe** task is activated at the specified time of 9 ns and the values of concerned arguments are displayed.

```
module dff_c(do,di,clk);
output do; input di,clk;
specify
        (negedge clk *>(do:di)) =1;
endspecify
reg do;
initial do=1'b0;
always@(negedge clk) do=di;
endmodule

//test-bench
module tst_dff_cbeh();
reg di,clk; wire do;
dff_c d1(do,di,clk);
initial  begin clk=0;di=1'b0; #35 $stop; end
always
#3clk=~clk;
always   #5 di=~di;
initial $monitor($time,"clk=%b,di=%b,do=%b",clk,di,do);
initial #9 $strobe ("at time %t, di=%b, do=%b",$time, di, do);
endmodule
```

Figure 11.48 A module set to illustrate the use of **$strobe** task.

#	0clk=0,di=0,do=x
#	1clk=0,di=0,do=0
#	3clk=1,di=0,do=0
#	5clk=1,di=1,do=0
#	6clk=0,di=1,do=0
#	7clk=0,di=1,do=1
# at time	9, di=1, do=1
#	9clk=1,di=1,do=1
#	10clk=1,di=0,do=1
#	12clk=0,di=0,do=1

Figure 11.49 Partial results of simulating the test bench in Figure 11.48.

11.5.4 $monitor Task

The **$monitor** task has been used extensively in the examples so far. The form of specifying arguments is identical to that of the **$display** task.

Observations:

- Only one **$monitor** task can be active at any time.
- **$monitor** task is activated and displays the arguments specified whenever any of the arguments changes. This excludes **$time**, **$stime**, and **$realtime** tasks.

$monitoroff and **$monitoron** are two additional tasks allied to the **$monitor** task; they are useful to enable and disable the monitoring activity. **$monitoroff** turns off the monitoring at the specified time, while **$monitoron** turns it on at the specified time.

Example 11.27

Figure 11.50 shows the half-adder module considered earlier; **$monitoroff** and **$monitoron** tasks have been included in the test bench. Monitoring is turned off at 30 ns, turned on at 60 ns and again turned off at 90 ns. The simulation results are shown in Figure 11.51. Monitoring activity can be seen to continue up to the 26th ns. At the next time step of any change in the module variables– that is, at the 30th ns–it stops; it resumes at 60 ns and continues up to 86 ns.

```
module ha_1(s,ca,a,b);
input a,b; output s,ca;
xor #(1,2) (s,a,b);
and #(3,4) (ca,a,b);
endmodule

//test-bench
module tstha_e();
reg a,b; wire s,ca;
ha_1 hh(s,ca,a,b);
initial begin a=0;b=0; #100 $stop; end
always    begin #5 a=1;b=0; #5 a=0;b=1; #5 a=1;b=1; #5 a=0;b=0; end
initial $monitor($time , "  a = %b , b = %b ,out carry = %b , outsum = %b  "
,a,b,ca,s);
initial begin        #30 $monitoroff; #30 $monitoron; #30 $monitoroff; end
endmodule
```

Figure 11.50 A module set to illustrate the use of **$monitoroff** and **$monitoron** tasks.

```
#       0  a = 0 , b = 0 ,out carry = x , outsum = x
#       2  a = 0 , b = 0 ,out carry = x , outsum = 0
#       4  a = 0 , b = 0 ,out carry = 0 , outsum = 0
#       5  a = 1 , b = 0 ,out carry = 0 , outsum = 0
#       6  a = 1 , b = 0 ,out carry = 0 , outsum = 1
#      10  a = 0 , b = 1 ,out carry = 0 , outsum = 1
#      15  a = 1 , b = 1 ,out carry = 0 , outsum = 1
#      17  a = 1 , b = 1 ,out carry = 0 , outsum = 0
#      18  a = 1 , b = 1 ,out carry = 1 , outsum = 0
#      20  a = 0 , b = 0 ,out carry = 1 , outsum = 0
#      24  a = 0 , b = 0 ,out carry = 0 , outsum = 0
#      25  a = 1 , b = 0 ,out carry = 0 , outsum = 0
#      26  a = 1 , b = 0 ,out carry = 0 , outsum = 1
#      60  a = 0 , b = 0 ,out carry = 1 , outsum = 0
#      64  a = 0 , b = 0 ,out carry = 0 , outsum = 0
#      65  a = 1 , b = 0 ,out carry = 0 , outsum = 0
#      66  a = 1 , b = 0 ,out carry = 0 , outsum = 1
#      70  a = 0 , b = 1 ,out carry = 0 , outsum = 1
#      75  a = 1 , b = 1 ,out carry = 0 , outsum = 1
#      77  a = 1 , b = 1 ,out carry = 0 , outsum = 0
#      78  a = 1 , b = 1 ,out carry = 1 , outsum = 0
#      80  a = 0 , b = 0 ,out carry = 1 , outsum = 0
#      84  a = 0 , b = 0 ,out carry = 0 , outsum = 0
#      85  a = 1 , b = 0 ,out carry = 0 , outsum = 0
#      86  a = 1 , b = 0 ,out carry = 0 , outsum = 1
```

Figure 11.51 Results of simulating the test bench in Figure 11.50.

11.5.5 $stop and $finish Tasks

The **$stop** task suspends simulation. The compiled design remains active; simulation can be resumed through commands available in the simulator. In contrast **$finish** stops simulation, closes the simulation environment, and reverts to the operating system.

11.5.6 $random Function

A set of random number generator functions are available as system functions. One can start with a seed number (optional) and generate a random number repeatedly. Such random number sequences can be fruitfully used for testing.

Example 11.28

The module of Figure 11.17 is reproduced in Figure 11.52 – with a modification in the test bench. The values assigned to the input vectors a and b are decided by the successive output values of the $random function. The first of the lot is decided by the seed number (4 here). The simulation results are reproduced in Figure 11.53.

```
module alu_8 (d, co, a, b, f,cci);
output [3:0] d; output co; wire[3:0]d; input cci;
input [3 : 0 ] a, b; input [1 : 0] f;
assign {co,d}=(f==2'b00)?(a+b+cci):((f==2'b01)?(a-b):((f==2'b10)?
          {1'bz,a^b}:{1'bz,~a}));
endmodule
//test-bench
module tst_alu8();
reg [3:0]a,b; reg[1:0] f; reg cci; wire[3:0]d; wire co;
alu_8 aa(d,co,a,b,f,cci);
initial begin cci=1'b0; f=2'b00;a=4'b0;b=4'h0; #30 $stop; end
always   begin
#2 cci =1'b0;f=2'b00;{a,b}=$random(4);
#2 cci =1'b1;f=2'b00;{a,b}=$random; #2 cci =1'b1;f=2'b01;{a,b}=$random;
#2 cci =1'b0;f=2'b01;{a,b}=$random; #2 cci =1'b1;f=2'b01;{a,b}=$random;
#2 cci =1'b1;f=2'b10;{a,b}=$random; #2 cci =1'b1;f=2'b11;{a,b}=$random;
        end
initial $monitor($time, " cci = %b , a= %b ,b = %b ,f = %b ,d =%b ,co= %b ",cci
,a,b,f,d,co);
endmodule
```

Figure 11.52 A module to illustrate the use of the system function $random.

```
#          0 cci = 0 , a= 0000 ,b = 0000 ,f = 00 ,d =0000 ,co= 0
#          4 cci = 1 , a= 0010 ,b = 0100 ,f = 00 ,d =0111 ,co= 0
#          6 cci = 1 , a= 1000 ,b = 0001 ,f = 01 ,d =0111 ,co= 0
#          8 cci = 0 , a= 0000 ,b = 1001 ,f = 01 ,d =0111 ,co= 1
#         10 cci = 1 , a= 0110 ,b = 0011 ,f = 01 ,d =0011 ,co= 0
#         12 cci = 1 , a= 0000 ,b = 1101 ,f = 10 ,d =1101 ,co= z
#         14 cci = 1 , a= 1000 ,b = 1101 ,f = 11 ,d =0111 ,co= z
#         16 cci = 0 , a= 0000 ,b = 0000 ,f = 00 ,d =0000 ,co= 0
#         18 cci = 1 , a= 0110 ,b = 0101 ,f = 00 ,d =1100 ,co= 0
#         20 cci = 1 , a= 0001 ,b = 0010 ,f = 01 ,d =1111 ,co= 1
#         22 cci = 0 , a= 0000 ,b = 0001 ,f = 01 ,d =1111 ,co= 1
#         24 cci = 1 , a= 0000 ,b = 1101 ,f = 01 ,d =0011 ,co= 1
#         26 cci = 1 , a= 0111 ,b = 0110 ,f = 10 ,d =0001 ,co= z
#         28 cci = 1 , a= 0011 ,b = 1101 ,f = 11 ,d =1100 ,co= z
```

Figure 11.53 Results of simulating the test bench in Figure 11.52.

Observations:

- If the seed is not changed with every simulation, the same sequence of random numbers is generated.
- If the seed is changed, the values in the random number sequence too change.
- If the seed is not specified, the **$random** function uses a default seed and generates the random number.
- Only the lowest 8 bits of the random number generated are used to assign values to **a** and **b** here.

11.6   FILE-BASED TASKS AND FUNCTIONS

LRM has the provision to accommodate and integrate design and test modules kept in different files. It makes room for structuring the design in an elegant manner and developing it with a "cross-functional" approach. Different facilities are specified in the LRM. That to output results to a file is discussed here as a specific case.

To carry out any file-based task, the file has to be opened, reading, writing, *etc.*, completed and the file closed. The keywords for all file-based tasks start with the letter f to distinguish them from the other tasks. A typical sequence of activities to write to a file can be as shown in Table 11.3.

Observations:

- The listing lines used need not be contiguous but have to be in the same sequence.
- All the system tasks to output information can be used to output to a file. **$display**, **$strobe**, **$monitor**, *etc.*, are of this category. The

Table 11.3 A typical (partial) sequence of a file-based operation

Line in module listing	Significance
Integer fileno;	fileno is declared as an **integer**
.....	
fileno = **$fopen**("info.txt");	A file with a name 'info.txt' is opened. The value of fileno signifies the same
.....	
$fdisplay(fileno, "string", arguments);	The arguments are displayed as specified
....	
$fclose(fileno);	The file is closed
.....	

respective keywords to output to the file are **$fdisplay**, **$fstrobe**, **$fmonitor**, respectively.

- The first field of the task statement is an argument – the file descriptor. The subsequent fields are identical to the corresponding nonfile tasks.
- The specified file will be opened and sustained in the directory of the executable file of the simulator.

Example 11.29

The half-adder module is reproduced in Figure 11.54 along with an associated test bench. The test bench uses a file "ha_f_rslt.txt." The file is opened and assigned the name "info." Later the **$fmonitor** task writes values of specified variables into the opened file. On completion of simulation, the file is closed automatically. One could also have closed the file beforehand through the "**$close**(info)" task. The contents of file ha_f_rslt.txt are reproduced in Figure 11.55.

```
module ha_1(s,ca,a,b);
input a,b; output s,ca;
xor #(1,2) (s,a,b);
and #(3,4) (ca,a,b);
endmodule

//test-bench
module tstha_f();
integer info; reg a,b; wire s,ca;
ha_1 hh(s,ca,a,b);
initial    begin
           a=0;b=0;
           info=$fopen("ha_f_rslt.txt");
           end
always begin #5 a=1;b=0; #5 a=0;b=1; #5 a=1;b=1; #5 a=0;b=0; end
initial  $fmonitor(info,$time , " a = %b , b = %b ,out carry = %b , outsum = %b "
,a,b,ca,s);
initial begin      #30 $display(info);      #0 $stop; end
endmodule
```

Figure 11.54 A module set to illustrate writing into a file.

```
//              0  a = 0 , b = 0 ,out carry = x , outsum = x
//              2  a = 0 , b = 0 ,out carry = x , outsum = 0
//              4  a = 0 , b = 0 ,out carry = 0 , outsum = 0
//              5  a = 1 , b = 0 ,out carry = 0 , outsum = 0
//              6  a = 1 , b = 0 ,out carry = 0 , outsum = 1
//             10  a = 0 , b = 1 ,out carry = 0 , outsum = 1
//             15  a = 1 , b = 1 ,out carry = 0 , outsum = 1
//             17  a = 1 , b = 1 ,out carry = 0 , outsum = 0
//             18  a = 1 , b = 1 ,out carry = 1 , outsum = 0
//             20  a = 0 , b = 0 ,out carry = 1 , outsum = 0
//             24  a = 0 , b = 0 ,out carry = 0 , outsum = 0
//             25  a = 1 , b = 0 ,out carry = 0 , outsum = 0
//             26  a = 1 , b = 0 ,out carry = 0 , outsum = 1
```

Figure 11.55 Contents of the file 'ha_f_rslt.txt' after the test bench in Figure 11.54 is simulated.

11.7   COMPILER DIRECTIVES

A number of compiler directives are available in Verilog. They allow for macros, inclusion of files, and timescale-related parameters for simulation. All compiler directives are preceded by the '`' (accent grave) character. Representative compiler directives are discussed here with illustrations.

11.7.1   `define Directive

The `define directive is for macro substitution. It substitutes the macro by a defined text. Hence a macro name can be used in place of such a group of characters in the listing wherever the group is to appear. Subsequently, the macro name can be substituted during compilation by the actual text. The `define directive is used to define and associate the desired text with the macro name.

The `define compiler directive can also be used to substitute a number by a macro name. It allows for deciding bus-width, specific delay values, *etc.*, at compilation time.

Example 11.30

The ALU module in Figure 11.56 is a modified version of that considered earlier. Three macro-names – namely add, subtract, and exor – are used in the module listing. The `define directives assign values to them. Note that despite the replacement the compiled file will remain unaltered.

module alu_a (d, co, a, b, f,cci);

`define add 2'b00
`define subtract 2'b01
`define exor 2'b10

output [3:0] d; utput co; wire[3:0]d;
input cci; input [3 : 0] a, b; input [1 : 0] f;
assign {co,d}=(f==`add)?(a+b+cci):((f==`subtract)?(a-b):((f==`exor)?
{1'bz,a^b}:{1'bz,~a}));
endmodule

Figure 11.56 A module to illustrate the use of the `define directive.

11.7.2   Time-Related Tasks

A set of compiler directives and system tasks relate to the running time of simulation as well as the delays in the concerned modules. A wide range of timescales as well as precision levels are available for selection during simulation.

`timescale

The `timescale compiler directive allows the time scale to be specified for the design. When a `timescale directive is encountered in a file, the same is valid for all subsequent modules within the file. The `timescale directive has two components: Figure 11.57 shows its form. A few examples are given below:

* `timescale 1 ms/100 μs
   implies that in the following design all the time values specified are in ms and they have a precision of 100 μs. Thus
   3, 3.0, 3.022 are all interpreted as 3 ms;
   3.1, 3.12,3.199 are all interpreted as 3.1 ms; and
   0.1, 0.12 are interpreted as 100 μs .

* `timescale 10 ms/100 μs
   implies that in the following design all the timescales are specified as multiples of 10 ms with a precision of 100 μs . Thus
   3 and 3.0 are interpreted as 30 ms;
   3.022 is interpreted as 30.2 ms;
   3.1 is interpreted as 31 ms;
   3.12 is interpreted as 31.2 ms;
   3.199 is interpreted as 31.9 ms;
   0.1 is interpreted as 1 ms and
   0.12 is interpreted as 1.2 ms .

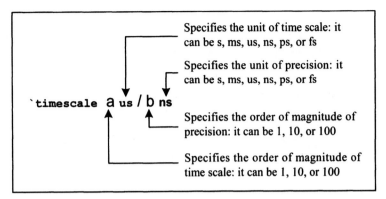

Figure 11.57 Form of specifying timescale: s, ms, us, ns, ps, and fs stand for seconds, milliseconds, microseconds, picoseconds and femtoseconds, respectively.

* `` `timescale `` 1 ms/1 ms
  implies that in the following design all the time values specified are in ms and they have a precision of 1 ms. Thus
  3, 3.0, 3.022, 3.1, 3.12, 3.199 are all interpreted as 3 ms and
  0.1, 0.12 are interpreted as 0 ms .

$timeformat

The timescale and the format for display can be changed during simulation with the help of $timeformat task. The syntax for the task is explained in Figure 11.58. Whenever "μs" (microsecond) is to be specified for defining or changing time scale, it is specified as "us." Conventions for all other timescale values (s, ms, ns, ps, and fs) remain unaltered.

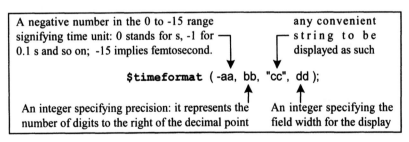

Figure 11.58 Syntax for $timeformat.

Simulation Time

Simulation time value can be obtained, displayed or used in specific expressions; a limited amount of flexibility is available here: –

- $time returns the value of simulation time as an integer.
- $realtime returns the value of simulation time as a real number.

Default Timescale

If the time scale values are not specified in the source file, simulation is carried out with the default values specified in the tool used for simulation. The default value of time unit is taken as nanosecond in this book.

Example 11.31

Figure 11.59 shows two illustrative modules and a test bench instantiating both of them. For all the modules the time unit is set at 1 µs and precision at 100 ns. The simulation results are in Figure 11.60.

```
`timescale 1us /100ns
module show_1;
reg  ai, bi; wire ao, bo;
show_2 aa(ao, ai);
show_3 bb(bo, bi);
initial $timeformat(-3, 5, "ms", 12);
initial $monitor("%m has  ai=%b,ao=%b,bi=%b,bo=%b, at time
%t",ai,ao,bi,bo,$realtime);
always     begin
                  #3{ai,bi} =2'b00; #3{ai,bi} =2'b01;
                  #3{ai,bi} =2'b10; #3{ai,bi} =2'b11;
           end
initial #12 $stop;
endmodule

`timescale 1us / 100ns
module show_3(bo,bi);
output bo; input bi; wire bo, bi;
not #1.2 (bo,bi);
endmodule

`timescale 1us / 100ns
module show_2(ao,ai);
output ao; input ai; wire ao, ai;
not #2 (ao,ai);
endmodule
```

Figure 11.59 A simple set of modules to illustrate the functioning of `timescale compiler directive.

# show_1 has ai=x,ao=x,bi=x,bo=x, at time	0.00000ms
# show_1 has ai=0,ao=x,bi=0,bo=x, at time	0.00300ms
# show_1 has ai=0,ao=x,bi=0,bo=1, at time	0.00420ms
# show_1 has ai=0,ao=1,bi=0,bo=1, at time	0.00500ms
# show_1 has ai=0,ao=1,bi=1,bo=1, at time	0.00600ms
# show_1 has ai=0,ao=1,bi=1,bo=0, at time	0.00720ms
# show_1 has ai=1,ao=1,bi=0,bo=0, at time	0.00900ms

Figure 11.60 Results of simulating the module set in Figure 11.59.

Observations:

- All propagation to bo from bi take place with a delay of 1.2 µs as specified.
- All propagation to ao from ai take place with a delay of 2 µs as specified.
- The display format specifies time to be displayed in milliseconds with a precision of 5 decimal places (second field of argument 5). Further, "ms" is to be displayed after the time display to signify that the time is displayed in milliseconds

Example 11.32

The module in Figure 11.59 has been modified and shown in Figure 11.61. "-3" in the first field of time format has been changed to "-4." It implies that the time displayed is in 100 µs units. Simulation results are shown in Figure 11.62. A number 0.042 there signifies $0.042 \times 10^{-4} = 4.2$ µs, since the time unit is specified as -4 in the $timeformat task.


```
`timescale 1us /100ns
module show_a;
reg  ai, bi; wire ao, bo;
show_b aa(ao, ai);
show_c bb(bo, bi);
initial $timeformat(-4, 5, " ", 12);
initial $monitor("%m has ai=%b,ao=%b,bi=%b,bo=%b, at time
%t",ai,ao,bi,bo,$realtime);
always begin
         #3        {ai,bi} =2'b00; #3        {ai,bi} =2'b01;
         #3        {ai,bi} =2'b10; #3        {ai,bi} =2'b11;
         end
```

continued

continued

initial #12 $stop;
endmodule
`timescale 1us / 100ns
module show_b(ao,ai);
output ao; input ai; wire ao, ai;
not #2 (ao,ai);
endmodule

`timescale 1us / 100ns
module show_c(bo,bi);
output bo;
input bi;
wire bo, bi;
not #1.2 (bo,bi);
endmodule

Figure 11.61 A modified version of the set of modules in Figure 11.59.

```
# show_a has  ai=x,ao=x,bi=x,bo=x, at time   0.00000
# show_a has  ai=0,ao=x,bi=0,bo=x, at time   0.03000
# show_a has  ai=0,ao=x,bi=0,bo=1, at time   0.04200
# show_a has  ai=0,ao=1,bi=0,bo=1, at time   0.05000
# show_a has  ai=0,ao=1,bi=1,bo=1, at time   0.06000
# show_a has  ai=0,ao=1,bi=1,bo=0, at time   0.07200
# show_a has  ai=1,ao=1,bi=0,bo=0, at time   0.09000
```

Figure 11.62 Results of simulating the set of modules in Figure 11.61.

Example 11.33

The module in Figure 11.59 is repeated in Figure 11.63: $realtime is replaced by $time. Simulation results are in Figure 11.64. Time units displayed here are in integers in contrast to those in Figure 11.61 where they are real numbers. Further, the integers displayed are in "ms" (the "-3" field signifies this), shown with 5-digit precision. Thus the delay of 1.2 µs for the transition in bo appears as only a 1 µs delay (The lines ending with 0.00400 ms, 0.00700, ms *etc.*, signify this.)

```
`timescale 1us /100ns
module show_aatb;
reg  ai, bi; wire ao, bo;
show_aa aa1(ao, ai);
show_bb bb1(bo, bi);
initial $timeformat(-3, 5, "ms", 12);
initial $monitor("%m has  ai=%b,ao=%b,bi=%b,bo=%b, at time
%t",ai,ao,bi,bo,$time);
always    begin
          #3        {ai,bi} =2'b00; #3         {ai,bi} =2'b01;
          #3        {ai,bi} =2'b10; #3         {ai,bi} =2'b11;
          end
initial #12 $stop;
endmodule

`timescale 1us / 100ns
module show_bb(bo,bi);
output bo; input bi; wire bo, bi;
not #1.2 (bo,bi);
endmodule

`timescale 1us / 100ns
module show_aa(ao,ai);
output ao; input ai; wire ao, ai;
not #2 (ao,ai);
endmodule
```

Figure 11.63 A modified version of the set of modules in Figure 11.59.

```
# show_aatb has  ai=x,ao=x,bi=x,bo=x, at time   0.00000ms
# show_aatb has  ai=0,ao=x,bi=0,bo=x, at time   0.00300ms
# show_aatb has  ai=0,ao=x,bi=0,bo=1, at time   0.00400ms
# show_aatb has  ai=0,ao=1,bi=0,bo=1, at time   0.00500ms
# show_aatb has  ai=0,ao=1,bi=1,bo=1, at time   0.00600ms
# show_aatb has  ai=0,ao=1,bi=1,bo=0, at time   0.00700ms
# show_aatb has  ai=1,ao=1,bi=0,bo=0, at time   0.00900ms
# show_aatb has  ai=1,ao=1,bi=0,bo=1, at time   0.01000ms
# show_aatb has  ai=1,ao=0,bi=0,bo=1, at time   0.01100ms
```

Figure 11.64 Results of simulating the set of modules in Figure 11.63.

Example 11.34

Figure 11.65 shows another modification of Figure 11.59; the compiler directive

`timescale 1 us / 100 ns

preceding the module "show_3" has been replaced by the directive

`timescale 1 us / 1 us

here. The two designs are identical in all other respects. Here the time step is in μs with a precision of 1 μs itself. A delay specified as 1.2 μs is taken as 1 μs by the simulator. The simulation results in Figure 11.66 confirm this.

```
`timescale 1us /100ns
module show_bbb;
reg  ai, bi; wire ao, bo;
show_2 aa(ao, ai);
show_3 bb(bo, bi);
initial $timeformat(-3, 5, "ms", 12);
initial $monitor("%m has  ai=%b,ao=%b,bi=%b,bo=%b, at time
%t",ai,ao,bi,bo,$realtime);
always    begin
        #3        {ai,bi} =2'b00; #3{ai,bi} =2'b01;
        #3        {ai,bi} =2'b10; #3{ai,bi} =2'b11;
        end
initial #12 $stop;
endmodule

`timescale 1us / 1us
module show_3(bo,bi);
output bo; input bi; wire bo, bi;
not #1.2 (bo,bi);
endmodule

`timescale 1us / 100ns
module show_2(ao,ai);
output ao;
input ai;
wire ao, ai;
not #2 (ao,ai);
endmodule
```

Figure 11.65 Another modified version of Figure 11.59.

```
# show_bbb has  ai=x,ao=x,bi=x,bo=x, at time   0.00000ms
# show_bbb has  ai=0,ao=x,bi=0,bo=x, at time   0.00300ms
# show_bbb has  ai=0,ao=x,bi=0,bo=1, at time   0.00400ms
# show_bbb has  ai=0,ao=1,bi=0,bo=1, at time   0.00500ms
# show_bbb has  ai=0,ao=1,bi=1,bo=1, at time   0.00600ms
# show_bbb has  ai=0,ao=1,bi=1,bo=0, at time   0.00700ms
# show_bbb has  ai=1,ao=1,bi=0,bo=0, at time   0.00900ms
# show_bbb has  ai=1,ao=1,bi=0,bo=1, at time   0.01000ms
# show_bbb has  ai=1,ao=0,bi=0,bo=1, at time   0.01100ms
```

Figure 11.66 Results of simulating the set of modules in Figure 11.65.

11.8 HIERARCHICAL ACCESS

A Verilog design will normally have a module or two at the apex level. A number of modules and UDPs will be instantiated within it. They can have other instantiations within them. They can also have functions and tasks defined in them and invoked repeatedly. In addition, **begin–end** and **fork–join** blocks too may be present. All these represent identified functional blocks in a design. Despite the variety here, one should have access to every variable, net as well as named identity in a design. The access can be to sample and display the values, to change specific parameters or disable selected blocks. Verilog has the provision to access each such item in a unique and hierarchical manner. Due to its importance, one has to understand the mode of deciding the hierarchical name and accessing each item (Such accessing has been dealt with in passing in Sections 4.5.1 and 11.5.2.) We discuss it in more detail here through illustrative examples.

Example 11.35

Figure 11.67 shows a module and its simulation results. The function fad in the module adds two integers a and b and returns the sum. The function has been called twice – once within the block alpha and the second time within the block beta. Each time the two numbers as well as the sum are displayed. Figure 11.68 shows the hierarchy of the blocks and the lineage for the variables concerned. The function fad has been called within the block alpha. The variables within alpha are accessed there as "fad.a," "fad.b," and "fad.fad," and their values are displayed. Similarly, they are called within block beta and displayed in the same manner.

```
module hier_a;
integer aa, bb, cc, pp, qq, rr;
initial
begin: alpha
        aa = 2; bb = 3;
        cc = fad(aa,bb);
        $display("fad.a = %0d, fad.b = %0d, fad.fad = %0d", fad.a,fad.b,fad.fad);
end
initial
begin: beta
        pp = 4;qq =6;
        rr = fad(pp,qq);
        $display("fad.a = %0d, fad.b = %0d, fad.fad = %0d", fad.a,fad.b,fad.fad);
end
function integer fad;
input [7:0] a, b;
fad = a + b;
endfunction
endmodule

# fad.a = 2, fad.b = 3, fad.fad = 5
# fad.a = 4, fad.b = 6, fad.fad = 10
```

Figure 11.67 A simple module to illustrate hierarchy and its simulation results.

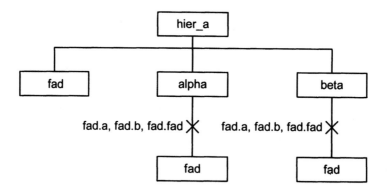

Figure 11.68 Hierarchy of the blocks and module instantiation in Example 11.35.

Example 11.36

Figure 11.69 shows the module of Example 11.35 with a **$display** statement added within the function definition itself. Simulation results are also included in the figure. The function fad has been called twice; both the times the values of the variables a, b, and fad are accessed and displayed. Further, each time the same quantities have been accessed using hierarchical names and displayed again.

```
module hier_b;
integer aa, bb, cc, pp, qq, rr;
initial
begin
        aa = 2; bb = 3;
        cc = fad(aa,bb);
        $display("fad.a = %0d, fad.b = %0d, fad.fad = %0d", fad.a,fad.b,fad.fad);
end
initial
begin
        pp = 4;qq =6;
        rr = fad(pp,qq);
        $display("fad.a = %0d, fad.b = %0d, fad.fad = %0d", fad.a,fad.b,fad.fad);
end
function integer fad;
input [7:0] a, b;
begin
        fad = a + b;
        $display("a = %0d, b = %0d, fad = %0d", a,b,fad);
end
endfunction
endmodule

# a = 2, b = 3, fad = 5
# fad.a = 2, fad.b = 3, fad.fad = 5
# a = 4, b = 6, fad = 10
# fad.a = 4, fad.b = 6, fad.fad = 10
```

Figure 11.69 A modified version of the module in Figure 11.67 and the simulation results.

Example 11.37

An additional display statement has been added to the module in Example 11.36 and shown in Figure 11.70. The variables aa and bb in the module have been accessed from within the function fad. Such "parallel" accessing from one block to another at the same level of hierarchy (see Figure 11.71) is possible in Verilog.

```
module hier_c;
integer aa, bb, cc, pp, qq, rr;
initial
begin
        aa = 2; bb = 3;
        cc = fad(aa,bb);
        $display("fad.a = %0d, fad.b = %0d, fad.fad = %0d", fad.a,fad.b,fad.fad);
end
initial
begin

        pp = 4;qq =6;
        rr = fad(pp,qq);
        $display("fad.a = %0d, fad.b = %0d, fad.fad = %0d", fad.a,fad.b,fad.fad);
end
function integer fad;
input [7:0] a, b;
begin
        fad = a + b;
        $display("hier_c.aa = %0d, hier_c.bb = %0d", hier_c.aa,hier_c.bb);
        $display("a = %0d, b = %0d, fad = %0d", a,b,fad);
end
endfunction
endmodule

# hier_c.aa = 2, hier_c.bb = 3
# a = 2, b = 3, fad = 5
# fad.a = 2, fad.b = 3, fad.fad = 5
# hier_c.aa = 2, hier_c.bb = 3
# a = 4, b = 6, fad = 10
# fad.a = 4, fad.b = 6, fad.fad = 10
```

Figure 11.70 The module in Figure 11.69 modified to illustrate "parallel" accessing.

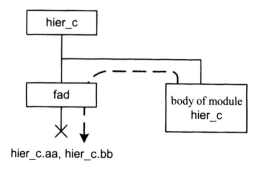

Figure 11.71 Parallel hierarchical accessing.

Example 11.38

The module in Figure 11.72 is similar to that in Example 11.36. A task – tad – defines the addition operation. It has been invoked to carry out the addition operation. Variables for display have been specified hierarchically. Simulation results are appended to the module in the figure.

```
module hier_d;
integer aa, bb, cc, pp, qq, rr;
initial
begin
        aa = 2; bb = 3;
        tad(aa,bb,cc);
        $display("tad.a = %0d, tad.b = %0d, tad.c = %0d", tad.a,tad.b,tad.c);
end
initial
begin
        pp = 4;qq =6;
        tad(pp,qq,rr);
        $display("tad.a = %0d, tad.b = %0d, tad.c = %0d", tad.a,tad.b,tad.c);
end
task tad;
input a, b;
output c;
integer a,b,c;
c = a + b;
endtask
endmodule

# tad.a = 2, tad.b = 3, tad.c = 5
# tad.a = 4, tad.b = 6, tad.c = 10
```

Figure 11.72 The module for Example 11.38 along its simulation results.

Example 11.39

The module in Figure 11.73 is a modified version of that in Figure 11.72. It is similar to the module in Figure 11.69; tasks have been defined and used here instead of functions.

```
module hier_e;
integer aa, bb, cc, pp, qq, rr;
initial
begin
        aa = 2; bb = 3;
        tad(aa,bb,cc);
        $display("tad.a = %0d, tad.b = %0d, tad.c = %0d", tad.a,tad.b,tad.c);
end
initial
begin
        pp = 4;qq =6;
        tad(pp,qq,rr);
        $display("tad.a = %0d, tad.b = %0d, tad.c = %0d", tad.a,tad.b,tad.c);
end
task tad;
input a, b;
output c;
integer a,b,c;
begin
        c = a + b;
        $display("a = %0d, b = %0d, c = %0d",a,b,c);
end
endtask
endmodule
```

```
# a = 2, b = 3, c = 5
# tad.a = 2, tad.b = 3, tad.c = 5
# a = 4, b = 6, c = 10
# tad.a = 4, tad.b = 6, tad.c = 10
```

Figure 11.73 The module for Example 11.39 along with simulation results.

Example 11.40

The module in Figure 11.73 has been modified and shown in Figure 11.74. The variables in the "parallel" module hier_f have been accessed from within the task "tad," hierarchically specifying the lineage. The accessing is similar to that in Figure 11.71 carried out in a parallel manner.

```
module hier_f;
integer aa, bb, cc, pp, qq, rr;
initial
begin
        aa = 2; bb = 3;
        tad(aa,bb,cc);
        $display("tad.a = %0d, tad.b = %0d, tad.c = %0d", tad.a,tad.b,tad.c);
end
initial
begin
        pp = 4;qq =6;
        tad(pp,qq,rr);
        $display("tad.a = %0d, tad.b = %0d, tad.c = %0d", tad.a,tad.b,tad.c);
end
task tad;
input a, b;
output c;
integer a,b,c;
begin
        c = a + b;
        $display("a = %0d, b = %0d, c = %0d",a,b,c);
        $display("hier_f.aa = %0d, hier_f.bb = %0d",hier_f.aa, hier_f.bb);
end
endtask
endmodule
```

```
# a = 2, b = 3, c = 5
# hier_f.aa = 2, hier_f.bb = 3
# tad.a = 2, tad.b = 3, tad.c = 5
# a = 4, b = 6, c = 10
# hier_f.aa = 2, hier_f.bb = 3
# tad.a = 4, tad.b = 6, tad.c = 10
```

Figure 11.74 The module for Example 11.40 along with simulation results.

Example 11.41

Figure 11.75 shows a module to add two octal numbers. It is done to further illustrate the features of hierarchy. ha is a half-adder module in the figure. It has been instantiated twice within the full-adder module fa. These two modules have been instantiated in the main module hier_l to carry out the addition of the octal numbers. hier_l has been instantiated in hier_ltst – the test bench for it. The

```
module hier_l(cc2,s2,s1,s0,a2,a1,a0,b2,b1,b0);
input a2,a1,a0,b2,b1,b0;
output cc2,s2,s1,s0;
wire cc1,cc0;
ha aaa(s0,cc0,a0,b0);
fa faa1(s1,cc1,a1,b1,cc0);
fa faa2(s2,cc2,a2,b2,cc1);
//location 3
endmodule
module ha(s,c,a,b);
input a,b;
output s,c;
assign {c,s}={a&b,a^b};
//location 5
endmodule

module fa(sf,cf,af,bf,ci);
input af,bf,ci;
output sf,cf;
wire sf1,cc1,cc2;
ha fha1(sf1,cc1,af,bf);
ha fha2(sf,cc2,sf1,ci);
//location 4
or rr(cf,cc1,cc2);
endmodule

module hier_ltst;
reg a2,a1,a0,b2,b1,b0;
wire cc2,s2,s1,s0;
hier_l ddd(cc2,s2,s1,s0,a2,a1,a0,b2,b1,b0);
initial
begin
#0      {a2,a1,a0,b2,b1,b0}=6'o34;
        $monitor("na = %0o, nb = %0o, ns =
%0o",{a2,a1,a0},{b2,b1,b0},{cc2,s2,s1,s0});
#2      $stop;
end
//location 1
initial #1 $display("sum = %b%b",ddd.faa2.fha2.c,ddd.faa2.fha2.s);
endmodule
#na = 3, nb = 4, ns = 7
#sum = 01
```

Figure 11.75 A module to illustrate hierarchy and its simulation results.

simulation results are also appended to the figure. Figure 11.76 shows the scheme of instantiations of the modules. The hierarchy of instantiated modules is shown in Figure 11.77. By way of illustration, the Sum and Carry bits of the half-adder instantiation **fha2** within the full-adder instantiation **faa2** have been selected. The module at the top is the test-bench **hier_ltst**. The hierarchical addresses of these two bits, as "looked" from the module hier_ltst, are

ddd.faa2.fha2.c

and

ddd.faa2.fha2.s.

They have been accessed and their values displayed at the end in the figure as

sum = 01.

The same variables can be accessed at "location3" and displayed. The hierarchical addresses to be used are

faa2.fha2.c

and

faa2.fha2.s.

"location4" is in the full-adder module **fa**. If accessed from there, the hierarchical names to be used are

fha2.c and
fha2.s.

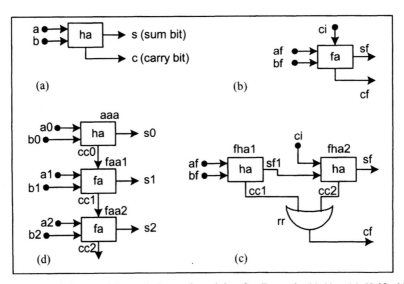

Figure 11.76 Scheme of instantiations of modules for Example 11.41. (a) Half-adder. (b) Full-adder. (c) Full-adder. (d) Octal adder.

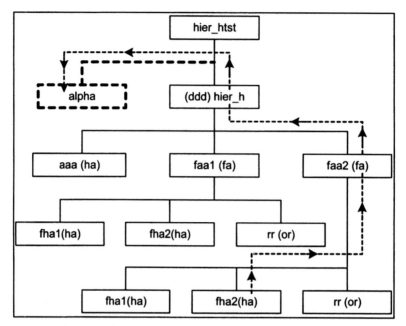

Figure 11.77 Hierarchy of instantiated modules in Example 11.41: The dashed lines pertain to Example 11.42.

If the display statement is inserted at "location4," the values of c and s will be displayed wherever fa is instantiated. Referring to Figure 11.75, fa is instantiated twice – first as faa1 and second as faa2. Hence these values will be displayed twice.

Consider "location5" in the figure within the Half-adder module. The variables can be accessed here directly with the names assigned to them – as "c" and "s." Their values will be displayed whenever the module is instantiated.

Example 11.42

The module hier_l has been altered slightly and shown in Figure 11.78. The display statement has been inserted within a "**begin–end**" block called "alpha." As a block it is parallel to the instantiation ddd; the variables accessed for display have been fully specified here as

hier_ntst.ddd. faa2.fha2.c and
hier_ntst.ddd.faa2.fha2.s.

The access path has been indicated separately in Figure 11.77. However, it suffices to specify the variables as

ddd. faa2.fha2.c and
ddd.faa2.fha2.s.

```
module hier_n(cc2,s2,s1,s0,a2,a1,a0,b2,b1,b0);
input a2,a1,a0,b2,b1,b0;
output cc2,s2,s1,s0;
wire cc1,cc0;
ha aaa(s0,cc0,a0,b0);
fa faa1(s1,cc1,a1,b1,cc0);
fa faa2(s2,cc2,a2,b2,cc1);
endmodule
module ha(s,c,a,b);
input a,b;
output s,c;
assign {c,s}={a&b,a^b};
endmodule

module fa(sf,cf,af,bf,ci);
input af,bf,ci;
output sf,cf;
wire sf1,cc1,cc2;
ha fha1(sf1,cc1,af,bf);
ha fha2(sf,cc2,sf1,ci);
or rr(cf,cc1,cc2);
endmodule

module hier_ntst;
reg a2,a1,a0,b2,b1,b0;
wire cc2,s2,s1,s0;
hier_n ddd(cc2,s2,s1,s0,a2,a1,a0,b2,b1,b0);
initial
begin
#0      {a2,a1,a0,b2,b1,b0}=6'o34;
        $monitor("na = %0o, nb = %0o, ns =
%0o",{a2,a1,a0},{b2,b1,b0},{cc2,s2,s1,s0});
#2      $stop;
end
initial
begin: alpha
//location1
#1      $display("sum = %b%b",hier_ntst.ddd.faa2.fha2.c,ddd.faa2.fha2.s);
end
endmodule
```

Figure 11.78 A modified version of the module in Figure 11.75.

Observations:

- Every entity in a design has a unique hierarchical name. Any entity can be accessed from a location in a module, if the location is in the hierarchical path.
- An entity can be accessed from above or below in the hierarchy, if it can be fully specified from the accessed location. As an example, consider the module designated "a" in Figure 11.79. It has two blocks b and c within it. Block d is within block b and block e is within block c. We can see the following from the figure:
  - Module a can access quantities in any of the other instantiated modules or blocks down the hierarchy.
  - Instantiated module b can access quantities within block d or module a.
  - Block d can access quantities in b or a.
  - Quantities in parallel blocks can be accessed. For example, block b can access those in block c as well as those in block e.
  - Access upwards beyond the parallel level is not possible. Thus an item within the block c or e cannot be accessed from the block d.
- Functions can be called hierarchically. Tasks too can be invoked hierarchically.
- Automatic tasks or functions cannot be accessed hierarchically.

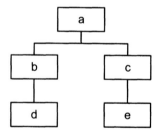

Figure 11.79 A hierarchical scheme of instantiations and blocks.

11.9   GENERAL OBSERVATIONS

Facilities of the type discussed in this chapter enhance the flexibility of one's approach to simulation. With their mastery, simulation and presentation of results can be made elegant. A great deal of useful information can be generated and used as input to refine design.

11.10   EXERCISES

1.   Consider the module "demux" of Example 8.1.

a. Define a parameter del in the module and assign a value of 0.5 ns to it. Obtain simulation results.

b. Reassign the value to del as 0.8 from the test bench. Obtain simulation results.

c. If a signal line is selected, the delay is to be 0.5 ns. If not it is to be 1 ns. Do the conditional assignments in the module. Obtain simulation results.

d. Specify pin-to-pin delay from b to a of 1.2 ns. Let er and rf be the error and rejection limits. Vary them in steps of 0.2 ns from 0.8 to 2.2 ns. Obtain simulation results and comment on the same.

2.   Modify the OR gate realization of Example 8.14 to realize an AND gate. Have a module parameter to decide the number of inputs and assign a value of 4 to it. Change the value from the test bench to 8. Obtain simulation results.

3.   Redo Example 8.15 by defining the input and output sizes to be module parameters. How are the two sizes related?

a. Assign a value of 12 to the input parameter from the test bench and obtain simulation results. Repeat the same with a value of 16.

b. Assign input values using **$random** system function repeatedly with input size of 16 bits. Run the simulation for 20 successive values of the **$random** function output.

4.   Modify the mod-n counter of Example 8.2 with n declared as a module parameter. Assign different values to n from the test-bench. Obtain simulation results.

5.   Consider Example 8.10 for memory loading; obtain a sequence of 8-bit random numbers and load them into the memory. Change the **for** loop in the module suitably.

6.   Consider the different examples in Chapters 4, 5, 6, and 7 where time delays are present. Alter them by defining **specify parameter** and redo the simulation.

7.   Consider the different examples in the Chapters 4, 5, 6, and 7 where register sizes, input sizes, output sizes, or bus sizes are present. Alter them by defining module parameters and redo the simulation.

8.   Complete Example 11.41 with proper insertions at the specified locations.

12

QUEUES, PLAS, AND FSMS

12.1   INTRODUCTION

Queues of the FIFO and the LIFO types form key blocks of many designs. They are used in many applications as buffers and for storage [Heuring & Jordan]. Verilog has a set of system tasks to set up and use FIFO and LIFO types of queues; adding to the queue and removing items from the queue are accomplished through others. The tasks are discussed here and illustrated through examples.

Synthesized circuits of the illustrative examples in the book are all realized with FPGAs. PLAs form a limited and more compact family. They are in wide use at least by a segment of designers. The constructs in Verilog to model them are explained through illustrative examples.

The long enduring importance of Finite State Machines (FSMs) is inherent due to their basic nature. Verilog constructs are used to model and simulate FSMs.

12.2   QUEUES

Queues can be modeled and their status checked with the help of a few tasks dedicated for the purpose; writing into a queue and reading from it are accomplished through others. The statements to invoke the related tasks and the arguments in each case are shown in Figure 12.1. Explanation of each task follows.

12.2.1 $q_initialize

The **$q_initialize** task is to initialize a new queue. All four arguments in the task invoking statement are variables of the integer type. The first is the identifier; it has the role of a queue address. The second specifies the type of queue – a FIFO or a LIFO. Only two types are possible. The third argument specifies the

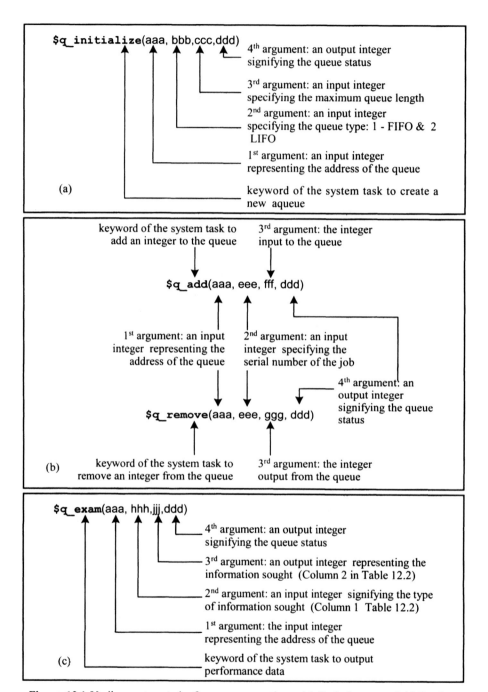

Figure 12.1 Verilog system tasks for queue operations: (a) Task for queue initialization. (b) Tasks for adding and removing from the queue. (c) Task to get queue statistics.

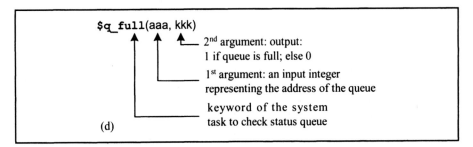

Figure 12.1 Verilog system tasks for queue operations (*continued*): (d) Task to check whether the queue is full or not.

allocated length of the queue. The fourth argument returns the status for all other operations. Any queue is to be initialized before carrying out an activity with it.

12.2.2 $q_add

$q_add is the system task to add a new entry to a queue. The first argument specifies the queue to which the entry is to be made. The second is the job serial number. The third is to be entered into the queue as an integer. The fourth returns a value signifying the q status. The value and the associated status it represents are given in the Table 12.1.

12.2.3 $q_remove

The task is to remove an entry from the queue (signifying servicing of the queue). The first argument signifies the queue to be accessed. The second is the serial number of the job. The third returns the value of the variable read from the queue. The fourth is an integer representing the queue status – as given in Table 12.1.

12.2.4 $q_exam

The task is to elicit updated statistical information about the queue operation. The first argument is the queue ID and the second is the code for requested information. The information sought is returned as the third argument. The fourth is the status of the queue. Respective codes are as in Table.12.1. Table 12.2 lists the types of information that can be obtained from the specified queue. The first column gives the code for the requested information and the second description of the quantity returned.

Table 12.1 Status of queue

Code value returned	Meaning
0	OK
1	Queue is full
2	Queue address not specified
3	Queue is empty
4	Queue type not available
5	Length specified is negative
6	Queue address already exists
7	Inadequate memory

Table 12.2 Information from the queue

Code value	Item returned
1	Current queue length
2	Mean inter arrival time
3	Maximum queue length
4	Lowest wait time
5	Highest wait time
6	Average wait time

12.2.5 $q_full

The task checks the specified queue for possible available space for entry. The first argument is the queue identification number; the second argument – returned from the queue – signifies its status: It is 1 if the queue is full and 0 otherwise. Queue-related activities are illustrated through a set of examples.

Example 12.1

A FIFO type of queue is set up with a length of 10 locations. The module is shown in Figure 12.2, and the simulation results in Figure 12.3. A random number in the range 0 to 100 is generated and successive ones added to the queue. After the 10th entry status 1 is returned signifying "buffer full."

```
module mulqua();
reg clk;
integer in, out, tim,iden,typ,len,status,jid ;
initial    begin
           iden=8;  typ=1;   len =10; jid=1;   clk=1'b0;
           $display ("time\t\tiden\ttyp\tlen\ttim\tstatus\tjid");
           $q_initialize(iden,typ,len,status);
           end
always #3 clk=~clk;
always@(posedge clk)      begin
tim={$random}%100;
$q_add(iden,jid,tim,status);
```

continued

continued

$display("%0d\t\t%0d\t%0d\t%0d\t%0d\t%0d\t%0d",$time,iden,typ,len,tim,status,
jid);
jid=jid+1;
                                    end
initial #70 $stop;
endmodule

Figure 12.2 A module to illustrate queue generation: A test bench is also included in the figure.

# time	iden	typ	len	tim	status	jid
# 3	8	1	10	48	0	1
# 9	8	1	10	97	0	2
# 15	8	1	10	57	0	3
# 21	8	1	10	87	0	4
# 27	8	1	10	57	0	5
# 33	8	1	10	57	0	6
# 39	8	1	10	25	0	7
# 45	8	1	10	82	0	8
# 51	8	1	10	61	0	9
# 57	8	1	10	29	0	10
# 63	8	1	10	18	1	11
# 69	8	1	10	97	1	12

Figure 12.3 Results of simulating the module set in Figure 12.2.

Example 12.2

Figure 12.4 shows a module for sequential entry into a queue followed by removal of the entries. A random number generates positive numbers in the range 0 to 100; 7 of them are entered into the queue. Subsequently, they are flushed out. The simulation results are reproduced in Table 12.3.

Table 12.3 Simulation results of the module set in Figure 12.4

#1	a	98	2	a	1	3	a	11	4	a	41	5	a	57	6	a	57	7	a	29
#1	r	98	2	r	1	3	r	11	4	r	41	5	r	57	6	r	57	7	r	29

```
module dem_qb;
integer alpha, beta ,gama,i,n;
initial begin
        $q_initialize (1,1,10,alpha); //start a fifo queue no:1
        n=$random(22);
        for(i=1;i<8;i=i+1)            begin
                beta=50 + $random%50;
                $q_add (1,i,beta,alpha);
                $write ("%0d  a  %0d  ",i,beta);
                                      end
    $display;

        for(i=1;i<8;i=i+1)            begin
                $q_remove (1,i,gama,alpha);
                $write ("%0d  r  %0d  ",i,gama);
                                      end

        $display;
        end
endmodule
```

Figure 12.4 A module to illustrate the formation of a typical queue.

Example 12.3

The module in Figure 12.5 is a modified version of that in Figure 12.4. The queue length has been set to 20. Twenty successive entries are made into the queue; subsequently all are flushed out sequentially. The time interval of arrival of successive entrants to the queue is a random number – in the range 0 to 7. Simulation results are in Table 12.4. The queue service details are also displayed. It includes the mean interval time of arrivals of entrants to the queue.

```
module dem_qg;
integer alpha, beta ,gama,i,j,n,qlen,mnit,xql; reg[2:0] aa[22:1]; reg[2:0] b;
initial begin
        $display ("addition\ni\tbeta\tb");
        $q_initialize (1,1,20,alpha); //start a fifo queue no:1
        n=$random(22);
        for(j=1;j<21;j=j+1) aa[j]=3+($random%4);
        for(i=1;i<21;i=i+1)          begin
                beta=50 + $random%50;
                b=aa[i];
```

continued

continued

```
        #b        $q_add (1,i,beta,alpha);
                  $display ("%0d\t%0d\t%d",i,beta,b);
                                end
        $display;
        $q_exam(1,1,qlen,alpha);$q_exam(1,2,mnit,alpha);
        $q_exam(1,3,xql,alpha);
        $display ("current queue length = %0d,  mean interval time =
%d",qlen,mnit);
        $display ("maximum queue length = %0d", xql);
        $display("removal");
        $display ("i\tgama");
        for(i=1;i<21;i=i+1)          begin
                  $q_remove (1,i,gama,alpha);
                  $display ("%0d\t%0d ",i,gama);
                                end

        #500 $stop;
        end
endmodule
```

Figure 12.5 The module of Figure 12.4 modified to show queue statistics.

Table 12.4 Results of simulating the module pair in Figure 12.5

#addition						# removal			
i	beta	b	i	beta	b	i	gama	i	gama
# 1	48	3	# 11	27	5	# 1	48	# 11	27
# 2	89	0	# 12	90	4	# 2	89	# 12	90
# 3	86	0	# 13	32	4	# 3	86	# 13	32
# 4	4	2	# 14	86	3	# 4	4	# 14	86
# 5	22	4	# 15	64	4	# 5	22	# 15	64
# 6	7	4	# 16	25	1	# 6	7	# 16	25
# 7	84	0	# 17	26	0	# 7	84	# 17	26
# 8	47	1	# 18	35	1	# 8	47	# 18	35
# 9	33	4	# 19	15	4	# 9	33	# 19	15
# 10	43	4	# 20	37	2	# 10	43	# 20	37

current queue length = 20, mean interval time = 2
maximum queue length = 20

12.3   PROGRAMMABLE LOGIC DEVICES (PLDs)

All logic functions can be realized in the sum of products or the product of sums form. The practice is to express it in terms of minterms or maxterms and express the function in terms of prime implicants [Micheli]. If its size is small, the function can be realized in terms of a few SSIs or MSIs. But if the number of ICs required increases beyond a limit, economic alternatives are provided by PLDs.

PLDs have circuit structures to realize combinational circuits; they also have flip-flops and allow realization of sequential circuits. Often sequential circuits or finite state machines with tens of states can be realized with them.

The term PLD refers to families of devices which can be programmed to carry out different functions. Figure 12.6 shows the circuit arrangement of such a device in simplified form.

The device accepts three logic inputs – x, y, and z – and through buffers makes them as well as their complements available on six signal lines inside. These form the inputs to a set of three AND gates – a1, a2, and a3 – with outputs $p1$, $p2$, and $p3$. Borrowing a term from combinational circuits, $p1$, $p2$ and $p3$ are called the "Product Terms." The signal lines $p1$, $p2$, and $p3$ form possible inputs to the following set of OR gates – r1, r2, and r3. $s1$, $s2$, and $s3$ are the OR gate outputs. These can be made available as possible outputs of the device – designated $g1$, $g2$, and $g3$. The device acts purely as a general-purpose combinatorial circuit with programming facility. Alternately, $s1$, $s2$, and $s3$ can be the inputs to the flip-flops – ff1, ff2, and ff3. The flip-flop outputs can be the chip outputs. The product term $p1$ can be formed by selectively connecting inputs to the AND gate a1. The dots at the crossings signify such a connection. With the dots shown, we get

$$p1 = x\,y'\,z'$$

and

$$p2 = z'$$

where x', y', and z' signify the complements of x, y, and z, respectively. Selectively establishing the connections to $p1$, $p2$, and $p3$ forms the first level of programming of the PLD.

The inputs to the OR gates r1, r2, and r3 can be selected from amongst $p1$, $p2$, and $p3$. The dots at the crossings signify such connections. With the dots shown for r3, we get

$$s3 = x\,y'\,z' + z.$$

Deciding the inputs to the OR gates constitutes the second level of programming of the PLD.

The OR gate outputs can be made available directly at the output side as $q1$, $q2$, and $q3$ respectively. Alternately, they can be loaded into the respective flip-flops at the clock edges. The selection between these alternatives constitutes the third level of programming of the PLD. In short, a combinational or a sequential circuit is realized using a PLD, through three levels of programming.

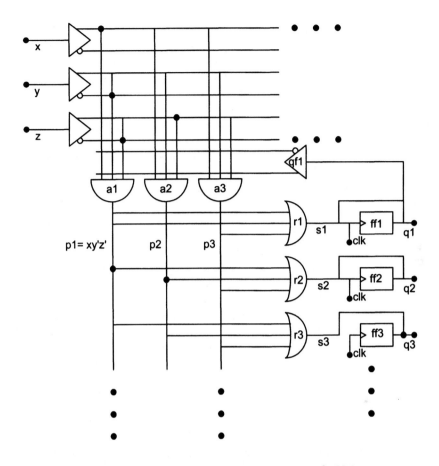

Figure 12.6 A diagram illustrating the structure of a PLA.

What is described here constitutes the general structure of a PLD. Details of individual devices and device families differ from manufacturer to manufacturer.

Observations:

- In mask programmable PLDs programming is done at the mask level at the manufacturing stage itself.
- In erasable PLDs (EPLDs), programming is done electrically at the customer's site. It is carried out with a particular sequence of electrical voltage pulses applied to selected pins of the device. The program can be erased by subjecting the device to UV rays for a specified period (~20 minutes) and the device can be reprogrammed.

- An electrically erasable PLD (E^2PL) is programmed like an EPLD by applying a specific electrical voltage sequence to it. The erasure is also carried out electrically by applying another set of electrical voltage pulses. The electrically erased device can be programmed again.
- EPLDs and E^2PLDs are also known as field programmable devices.
- Programmable array logic (PAL) is a class of PLD. The OR gates are fixed here and not programmable. Only the AND gates can be programmed.
- A programmable logic array (PLA) is another class of PLD. The inputs to their OR gates are also programmable; this is in addition to the programming facility at the AND gate level.
- Normally, the size of a PLD is referred to as $a \times b \times c$, where a, b, and c are the number of input lines, the number of product terms, and the number of output lines respectively.
- Manufacturers offer different families of PLDs; normally, a family has the size fixed with the number of its inputs and outputs changing. Depending on the application, a family and a device in the family can be selected.
- The flip-flop outputs are normally available as additional input lines to the product terms of the device. One can use them judiciously and realize sequential circuits and different finite state machines.

12.3.1   Programming of PLD in VERILOG

Verilog provides a family of system tasks to simulate different types of PLDs. Figure 12.7 shows the structure of the tasks. The keyword for the task is in three parts; the alternatives for each are shown in the figure. Possible combinations of the alternatives lead to a total of 16 such tasks. Their use is illustrated here through an example.

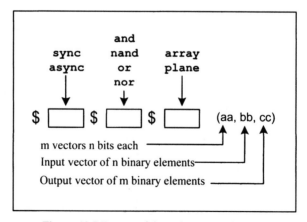

Figure 12.7 Format of the tasks to model PLAs.

Example 12.4 A 2 × 2 Multiplier

Figure 12.8 shows a module to realize a 2 × 2 multiplier as a PLA. A test bench for the multiplier is also included in the figure. Multipliers of much larger sizes can be described and realized; the one considered here is purely illustrative in nature.

a and b form the two 2-bit inputs to the multiplier and c is its 4-bit output. The multiplier has been realized in a sum-of-products form. The input combinations, the outputs, and the relevant product terms are shown in Table 12.5.

```
module mltp_a(a1,a0,b1,b0,c1,c2,c3,c4);//PLA based 2-bit multiplier
input a1,a0,b1,b0; output c1,c2,c3,c4;
reg[1:4] mand[1:9]; reg[1:9] mor[1:4]; reg p1,p2,p3,p4,p5,p6,p7,p8,p9,c1,c2,c3,c4;
initial begin
        mand[1]=4'b1010; mand[2]=4'b0110; mand[3]=4'b1110;
mand[4]=4'b1001;
        mand[5]=4'b0101; mand[6]=4'b1101; mand[7]=4'b1011;
mand[8]=4'b0111;
        mand[9]=4'b1111; mor[1]=9'b1_0100_0101; mor[2]=9'b0_1110_1110;
        mor[3]=9'b0_0001_1010; mor[4]=9'b0_0000_0001;
        $async$and$array(mand,{a0,a1,b0,b1},{p1,p2,p3,p4,p5,p6,p7,p8,p9});
        $async$or$array(mor,{p1,p2,p3,p4,p5,p6,p7,p8,p9},{c1,c2,c3,c4});

        end
endmodule

module mltp_a_tst;
reg a1,a0,b1,b0; reg[1:4] n; integer i;wire c4,c3,c2,c1;
mltp_a mm(a1,a0,b1,b0,c1,c2,c3,c4);
always  begin
        n=4'b0000;
        for(i=0;i<17;i=i+1) begin
                {a1,a0,b1,b0}=n;
        #1      n=n+1'b1;
                        end
        end
initial $monitor("%b\t%b\t%b",{a1,a0},{b1,b0},{c4,c3,c2,c1});
initial begin
        $display("a \tb\t a*b");
#15     $stop;
        end
endmodule
```

Figure 12.8 A 2 × 2 multiplier module and its test bench.

Table 12.5 Details of the multiplier module in Figure 12.8

Bits of input a		Bits of input b		Product	Bits of output c			
a1	a0	b1	b0	p	c3	c2	c1	c0
0	0	0	0	–	0	0	0	0
0	0	0	1	–	0	0	0	0
0	0	1	0	–	0	0	0	0
0	0	1	1	–	0	0	0	0
0	1	0	0	–	0	0	0	0
0	1	0	1	p1	0	0	0	*1*
0	1	1	0	p2	0	0	*1*	0
0	1	1	1	p3	0	0	*1*	*1*
1	0	0	0	–	0	0	0	0
1	0	0	1	p4	0	0	*1*	0
1	0	1	0	p5	*1*	0	0	0
1	0	1	1	p6	*1*	*1*	0	0
1	1	0	0	–	0	0	0	0
1	1	0	1	p7	0	0	*1*	*1*
1	1	1	0	p8	*1*	*1*	0	0
1	1	1	1	p9	*1*	0	0	*1*

The output bits affected by the product terms are shown in bold italics in the table. The two PLA task statements in Figure 12.8 together realize the multiplier in asynchronous form. Simulation results are reproduced in Table 12.6.

12.4   DESIGN OF FINITE STATE MACHINES

A finite state machine (FSM) is the most basic form of describing a digital system. Properly carried out, it forms the optimal and compact representation. Knowledge of design of FSMs will remain a basic one; hence its importance [Comer, Devadas *et al.*] An FSM is characterized by the following:

- A set of finite states
- A set of logic inputs

Table 12.6 Simulation results of the module set in Figure 12.8

a	b	a*b	a	b	a*b
# 00	00	0000	# 01	11	0011
# 00	01	0000	# 10	00	0000
# 00	10	0000	# 10	01	0010
# 00	11	0000	# 10	10	0100
# 01	00	0000	# 10	11	0110
# 01	01	0001	# 11	00	0000
# 01	10	0010	# 11	01	0011

- A set of logic outputs
- A set of logic equations connecting the next state to the present state and present input vectors
- A set of logic equations connecting the next output state to the present-state input values

In general, a state description of a state machine consists of descriptions of the state transitions, the output functions, and the next-state register functions. Because the next-state functions call for memory or register based operations, an **always** block is an appropriate way to describe it in Verilog. If-else-if or Case statements, the 3-operand operator, or proper usage of combinational and sequential UDPs perform the state transition and output function descriptions. In case all the possible states are not defined and the Case statement is used to realize the FSM, it should always have a default statement to ensure that the state machine does not go into an undefined state. In case the If-else-if construct is used, at the beginning of the always block an asynchronous reset or clear can be used to bring the machine to a known initialized state.

The instant of transition from the present to the next can be completely controlled by a clock; additionally, changes in the inputs may also dictate such transitions. FSMs can be broadly classified into two categories – Moore machines and Mealy machines. Design of both types is discussed and illustrated in the sequel.

12.4.1   Moore Machine

The Moore model of the FSM is shown in Figure 12.9 in block diagram form. The input vector A and the present state vector S_p together form the input to a

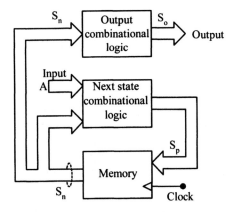

Figure 12.9 Moore machine in block diagram form.

combinational circuit block. Its output vector S_o forms the address input to the memory block. At the active clock edge the memory location is accessed and the next state vector output from it. At any moment a second combinational logic block defines the output in terms of the present state vector. Two aspects characterize a Moore machine:

- Next state of the output is decided fully by the present state.
- All changes in the output are brought about only at the active edge of the clock. Hence the Moore machine is inherently synchronous.

Edge-triggered flip-flops, synchronous counters, *etc.*, are typical examples of Moore machines. Design of Moore machines can be described in various ways. Since the active edge of the clock is pivotal to its operation, the circuit design block can be directly activated at the active edge of the clock. Figure 12.10 shows two approaches to the Moore machine design. Other combinations of procedural and continuous assignments to S_o and S_p are also possible. All the approaches are characterized by the following:

- Assignments to S_n are procedural and at the active edge of the clock.
- Assignments within the procedural block are of the nonblocking type, since the updating of values depends on the previous state.

The step-by-step procedure for the design of a synchronous (Moore) machine is as follows:

- Generate a state diagram from the problem statement.
- Minimize the number of states.
- Select a binary encoding for the states.
- Generate an encoded state table.
- Select the memory device –T flip-flop or D flip-flop.
- Generate a next-state K map for each memory device.
- Generate a K map for each output.
- Implement memory and combinational logic using PLAs or other devices.

always@(posedge clk) begin Sn <= F1(Sp); Sp <= F2(Sn, A); So <= F3(Sn); End	Assign Sp = F2(Sn, A); always@(posedge clk) begin Sn <= F1(Sp); So <= F3(Sn); End

Figure 12.10 Two of the possible approaches to a Moore machine description.

Example 12.5

A sequence generator is to sequence through eight distinct states. The states are represented by a set of four binary variables – W, X, Y, and Z. The states and the sequence are as follows (the 4 bits represent values of W, X, Y, and Z, respectively):
$1000 \rightarrow 1100 \rightarrow 0100 \rightarrow 0110 \rightarrow 0010 \rightarrow 0011 \rightarrow 0001 \rightarrow 1001 \rightarrow 1000 \ldots..$
Each transition is to take place at the positive edge of the clock. Since the scheme has no external primary input to affect the output, it is realized as a Moore machine. A 3-bit state machine suffices to generate the eight independent states specified. The step-by-step implementation of the FSM is on the following lines:

- The binary encoding and the corresponding state assignment are shown in Table 12.7 with the 3 bits being designated as Q_a, Q_b, and Q_c; the states are designated as S_0, S_1, S_2, S_3, S_4, S_5, S_6, and S_7, respectively. The binary sequence of the FSM, the next state, and the set of outputs for each of the states are given in the table.
- The outputs W, X, Y, and Z are expressed as functions of Q_a, Q_b, and Q_c in the Sum of Products form and the respective Karnaugh maps are given in Table 12.8 to Table 12.11.
- The outputs W, X, Y, and Z are given in minimized form as Equations (12.1), (12.2), (12.3) and (12.4).
- The next state variables Q_a, Q_b, and Q_c are implemented using T-flip-flops designated T_a, T_b and T_c respectively.
- The inputs to the flip-flops Q_a, Q_b, and Q_c are designated as T_a, T_b, and T_c; Karnaugh maps for the respective functions are in Tables 12.12, 12.13 and 12.14.
- Equations (12.5), (12.6) and (12.7) represent the minimized functional form of T_a, T_b, and T_c.

Table 12.7 State assignments and transitions

Present state		Next state	
State designation	$Q_aQ_bQ_c$	$Q_aQ_bQ_c$	Outputs $WXYZ$
S_0	000	001	1000
S_1	001	010	1100
S_2	010	011	0100
S_3	011	100	0110
S_4	100	101	0010
S_5	101	110	0011
S_6	110	111	0001
S_7	111	000	1001

Table 12.8 Karnaugh map for FSM output W (Example 12.5)

		Q_aQ_b			
		00	00	00	00
Q_c	0	1	1	0	0
	1	0	0	1	0

Table 12.9 Karnaugh map for FSM output X (Example 12.5)

		Q_aQ_b			
		00	01	11	10
Q_c	0	0	1	1	1
	1	0	0	0	0

Table 12.10 Karnaugh map for FSM output Y (Example 12.5)

		Q_aQ_b			
		00	01	11	10
Q_c	0	0	0	1	0
	1	1	1	0	0

Table 12.11 Karnaugh map for FSM output Z (Example 12.5)

		Q_aQ_b			
		00	01	11	10
Q_c	0	0	0	0	0
	1	0	1	1	1

Table 12.12 Karnaugh map for T_a (Example 12.5)

		Q_aQ_b			
		00	01	11	10
Q_c	0	0	0	1	0
	1	0	0	1	0

Table 12.13 Karnaugh map for T_b (Example 12.5)

		Q_aQ_b			
		00	01	11	10
Q_c	0	0	1	1	0
	1	0	1	1	0

Table 12.14 Karnaugh map for T_c (Example 12.5)

		Q_aQ_b			
		00	01	11	10
Q_c	0	1	1	1	1
	1	1	1	1	1

$$W = \overline{Q}_x\overline{Q}_y + Q_xQ_yQ_z \tag{12.1}$$

$$X = \overline{Q}_xQ_z + \overline{Q}_xQ_y \tag{12.2}$$

$$Y = Q_x\overline{Q}_y + \overline{Q}_xQ_yQ_z \tag{12.3}$$

$$Z = Q_xQ_z + Q_xQ_y \tag{12.4}$$

$$T_a = Q_bQ_c \tag{12.5}$$

$$T_b = Q_c \tag{12.6}$$

$$T_c = 1 \tag{12.7}$$

The above can be realized directly in a programmable logic device; but with the increase in the number of states, the state machine becomes too big to be manually designed. In such cases the design description can be done in Verilog.

12.4.1.1   Design Realization: Version 1

Figure 12.11 shows a module to realize the machine under discussion; a test bench is also shown in the figure. There are two always block in the state machine description. The outputs change as per the description of the first always blocks. The change in the state occurs at the clock transitions as described in the second always block. Thus the outputs are steered through whenever a state transition occurs. Simulation results are shown in Figure 12.12 as waveforms of the signals. The synthesized circuit is shown in Figure 12.13.

```
//sequence generator
//moore machine_a

`define s0 3'b000//wxyz=1000
`define s1 3'b001//wxyz=1100
`define s2 3'b010//0100
`define s3 3'b011//0110
`define s4 3'b100//0010
`define s5 3'b101//0011
`define s6 3'b110//0001
`define s7 3'b111//1001
module a_seqmoorev(clr,clk,w,x,y,z);
input clr,clk;
output w,x,y,z;
reg w,x,y,z;
reg [2:0]present_state;

always@(present_state)
begin
case(present_state)
                `s0: {w,x,y,z}=4'b1000;
                `s1: {w,x,y,z}=4'b1100;
                `s2: {w,x,y,z}=4'b0100;
                `s3: {w,x,y,z}=4'b0110;
                `s4: {w,x,y,z}=4'b0010;
                `s5: {w,x,y,z}=4'b0011;
                `s6: {w,x,y,z}=4'b0001;
                `s7: {w,x,y,z}=4'b1001;
        endcase
end
```

continued

continued

```
always@(posedge clk)
begin
if (clr) present_state =`s0;
        else      begin
                           case(present_state)
                           `s0: present_state=`s1;
                           `s1: present_state=`s2;
                           `s2: present_state=`s3;
                           `s3: present_state=`s4;
                           `s4: present_state=`s5;
                           `s5: present_state=`s6;
                           `s6: present_state=`s7;
                           `s7: present_state=`s0;
                           default: present_state=`s0;
                           endcase
               end
end
endmodule

//test-bench
//In a moore machine the next_state logic is independent of primary inputs and
hence
module test_a_seqmoorev();
reg clr,clk;
wire w,x,y,z;
a_seqmoorev vv(clr,clk,w,x,y,z);
initial begin clk=1'b0;clr=1'b1; #3 clr =1'b0; #50 $stop; end
always  #2 clk = ~clk;
endmodule
```

Figure 12.11 Module Version 1 of the design for the FSM of Example 12.5; a test bench for the design is also shown in the figure.

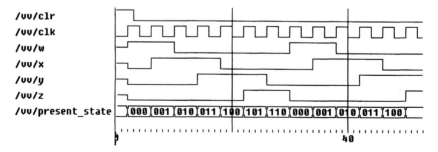

Figure 12.12 Simulation results of the set of modules in Figure 12.11.

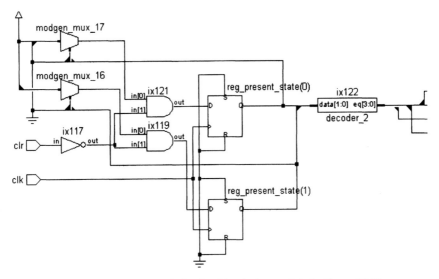

Figure 12.13 Synthesized circuit of the design module in Figure 12.11.

12.4.1.2   Design Realization: Version 2

An alternate design module is shown in Figure 12.14. The transition output variables and the state transitions of the next-state variables have been combined in a common always block: The outputs are steered through at the clock edge; hence as many extra latches are inferred as the number of outputs. The simulation output waveforms are identical to those for Version 1 shown in Figure 12.12 and are not reproduced. Six latches are implied – 2 for the state variables and four for the 4 output variables. The synthesized circuit in Figure 12.15 too confirms the same. In contrast, the coding in Version 1 above does not explicitly imply latches for the output variables.

```
//sequence generator – moore machine
`define s0 3'b000//wxyz=1000
`define s1 3'b001//wxyz=1100
`define s2 3'b010//0100
`define s3 3'b011//0110
`define s4 3'b100//0010
`define s5 3'b101//0011
`define s6 3'b110//0001
`define s7 3'b111//1001
```

continued

continued

```
module seqmoorev2(clr,clk,w,x,y,z);
input clr,clk;
output w,x,y,z;
reg w,x,y,z;
reg [2:0]present_state;

//mainblock
always@(posedge clk or posedge clr)
begin
        if (clr)
        present_state =`s0;
        else
        begin
                case(present_state)
                `s0:begin present_state=`s1; {w,x,y,z}=4'b1000; end
                `s1:begin present_state=`s2; {w,x,y,z}=4'b1100; end
                `s2:begin present_state=`s3; {w,x,y,z}=4'b0100; end
                `s3:begin present_state=`s4; {w,x,y,z}=4'b0110; end
                `s4:begin present_state=`s5; {w,x,y,z}=4'b0010; end
                `s5:begin present_state=`s6; {w,x,y,z}=4'b0011; end
                `s6:begin present_state=`s7; {w,x,y,z}=4'b0001; end
                `s7:begin present_state=`s0; {w,x,y,z}=4'b1001; end
                default:  present_state=`s0;
                endcase
        end
 end
endmodule


//test-bench
//In a moore machine the next_state logic is independent of primary inputs
module test_seqmoorev2();
reg clr,clk;
wire w,x,y,z;
seqmoorev2 vv(clr,clk,w,x,y,z);
initial begin clk=1'b0;clr=1'b1; #3 clr = 1'b0; #50 $stop; end
always #2 clk = ~clk;
endmodule
```

Figure 12.14 Module Version 2 of the design for the FSM of Example 12.5; a test bench for the design is also shown in the figure.

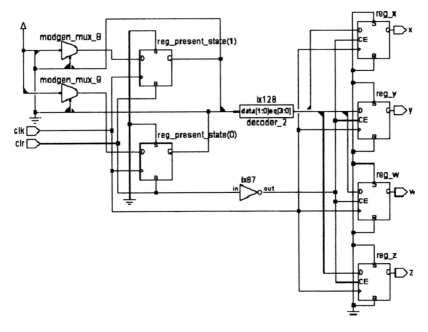

Figure 12.15 Synthesized circuit of the design module in Figure 12.14.

12.4.2   Mealy Machine

The Mealy machine is shown in block diagram form in Figure 12.17. It differs from the Moore machine at the output stage. The inputs can affect the outputs directly. Thus for the output S_o we have

$$S_o = F4(Sn, A)$$

The changes in the input A reflect as corresponding changes in the outputs without the clock being directly involved: To that extent the behavior is asynchronous. Counters with asynchronous Preset and Clear and Shift Registers with Preset are examples of Mealy machines. A Mealy machine has to respond to changes in input in addition to the response to the active edges of the clock. The same can be accommodated in various ways. Figure 12.16 shows two possible realizations.

always@ (negedge clk or A) assignments;	Continuous assignments; **always@ (negedge** clk or A) Assignments;

Figure 12.16 Two possible approaches to Mealy machine description.

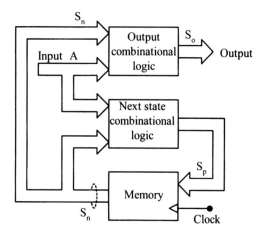

Figure 12.17 A Mealy machine in block diagram form.

Example 12.6

A sequence generator is to have four binary outputs designated W, X, Y, and Z. They are to follow either of two sequences depending on the value of a Boolean variable A:

- If $A = 0$, the sequence to be followed is

$1000 \rightarrow 1100 \rightarrow 0100 \rightarrow 0110 \rightarrow 0010 \rightarrow 0011 \rightarrow 0001 \rightarrow 1001 \rightarrow 1000 \ldots$

where W is the most significant bit and Z the least significant bit.

- If $A = 1$, the sequence to be followed is

$1001 \rightarrow 0001 \rightarrow 0011 \rightarrow 0010 \rightarrow 0110 \rightarrow 0100 \rightarrow 1100 \rightarrow 1000 \rightarrow 1001 \ldots$

The encoding used for the Moore machine has been retained. The design is done on the following lines:

- The encoding, the set of present states, corresponding next states and respective outputs are shown in Table 12.15 for the sequence when $A = 0$: Table 12.16 shows the same when $A = 1$.
- Each of the outputs is a function of 4 variables, namely, Q_a, Q_b, Q_c, and A. The Karnaugh map representation of the functions is in Tables 12.17 to 12.20. The minimized functions are given by Equations (12.8) to (12.11).
- Each next state variable is also a function of Q_a, Q_b, Q_c, and A; the functions are represented in Karnaugh map form in Tables 12.21 to 12.24. The respective minimized functions are given by Equations (12.12) to (12.14).

Table 12.15 State transition details for Example 12.6 for $A = 0$

Present state		Next state for $A = 0$	
State designation	$Q_aQ_bQ_c$	$Q_aQ_bQ_c$	Outputs $WXYZ$
S_0	000	001	1000
$S1$	001	010	1100
S_2	010	011	0100
$S3$	011	100	0110
$S4$	100	101	0010
$S5$	101	110	0011
$S6$	110	111	0001
$S7$	111	000	1001

Table 12.16 State transition details for Example 12.6 for $A = 1$

Present state		Next state for $A = 1$	
State designation	$Q_aQ_bQ_c$	$Q_aQ_bQ_c$	Outputs $WXYZ$
S_0	000	111	0001
$S1$	111	110	0011
S_2	110	101	0010
$S3$	101	100	0110
$S4$	100	011	0100
$S5$	011	010	1100
$S6$	010	001	1000
$S7$	001	000	1001

Table 12.17 Karnaugh map for FSM output W (Example 12.6)

		Q_bQ_c			
		00	01	11	10
	00	1	1	0	0
AQ_a	01	0	0	1	0
	11	0	0	1	1
	10	1	0	0	0

Table 12.18 Karnaugh map for FSM output X (Example 12.6)

		Q_bQ_c			
		00	01	11	10
	00	0	1	1	1
AQ_a	01	0	0	0	0
	11	1	1	0	1
	10	0	0	0	0

Table 12.19 Karnaugh map for FSM output Y (Example 12.6)

		Q_bQ_c			
		00	01	11	10
	00	0	0	1	0
AQ_a	01	1	1	0	0
	11	1	0	0	0
	10	0	0	1	1

Table 12.20 Karnaugh map for FSM output Z (Example 12.6)

		Q_bQ_c			
		00	01	11	10
	00	0	0	0	0
AQ_a	01	0	1	1	1
	11	0	0	0	0
	10	1	1	0	1

$$W = \overline{A}\,\overline{Q}_a\overline{Q}_b + Q_aQ_bQ_c + AQ_aQ_b\overline{Q}_c + \overline{Q}_a\overline{Q}_b\overline{Q}_c \tag{12.8}$$

$$X = \overline{A}\,\overline{Q}_a\overline{Q}_c + A\overline{Q}_a\overline{Q}_b + AQ_a\overline{Q}_c + AQ_a\overline{Q}_b \tag{12.9}$$

$$Y = Q_a\overline{Q}_b\overline{Q}_c + \overline{A}\,Q_a\overline{Q}_b + \overline{Q}_aQ_bQ_c + AQ_a\overline{Q}_b \tag{12.10}$$

$$Z = \overline{A}\,Q_aQ_c + \overline{A}\,Q_aQ_b + AQ_a\overline{Q}_b + A\overline{Q}_a\overline{Q}_c \tag{12.11}$$

**Table 12.21 Karnaugh map for T_a
(Example 12.6)**

		Q_bQ_c			
		00	01	11	10
	00	0	0	1	0
AQ_a	01	0	0	1	0
	11	1	1	1	1
	10	1	1	1	1

**Table 12.22 Karnaugh map for T_b
(Example 12.6)**

		Q_bQ_c			
		00	01	11	10
	00	0	1	1	0
AQ_a	01	0	1	1	0
	11	1	1	1	1
	10	1	1	1	1

**Table 12.23 Karnaugh map for T_c
(Example 12.6)**

		Q_bQ_c			
		00	01	11	10
	00	1	1	1	1
AQ_a	01	1	1	1	1
	11	1	1	1	1
	10	1	1	1	1

$$T_a = YZ + A \tag{12.12}$$

$$T_b = \overline{A}Z + A \tag{12.13}$$

$$T_c = 1 \tag{12.14}$$

The above FSM can be realized using a PLD. However, here it is realized through a Verilog module. The module and its test bench are in Figure 12.18. The simulation waveforms are in Figure 12.19. The synthesized circuit is shown in Figure 12.20.

```
//sequence generator – mealy machine
`define s0 3'b000//wxyz=1000
`define s1 3'b001//wxyz=1100
`define s2 3'b010//0100
`define s3 3'b011//0110
`define s4 3'b100//0010
`define s5 3'b101//0011
`define s6 3'b110//0001
`define s7 3'b111//1001
module p_seqmealy(a,clk,w,x,y,z,state);
input a,clk; output w,x,y,z; output [2:0]state;
reg w,x,y,z; reg [2:0] state,next_state;
```

continued

continued

```
initial begin
        if (!x)
                begin
                state=`s0;
                next_state=`s0;
                end
        else    begin
                state =`s7;
                next_state =`s7;
                end
        end
always@(posedge clk)
state = next_state;
always@(state)
        begin
                case(state)
                `s0: {w,x,y,z}=4'b1000;
                `s1: {w,x,y,z}=4'b1100;
                `s2: {w,x,y,z}=4'b0100;
                `s3: {w,x,y,z}=4'b0110;
                `s4: {w,x,y,z}=4'b0010;
                `s5: {w,x,y,z}=4'b0011;
                `s6: {w,x,y,z}=4'b0001;
                `s7: {w,x,y,z}=4'b1001;
                endcase
        end

//mainblock
always@(clk)
        begin
        case(state)
                `s0:    next_state=a?`s7:`s1;
                `s1:    next_state=a?`s0:`s2;
                `s2:    next_state=a?`s1:`s3;
                `s3:    next_state=a?`s2:`s4;
                `s4:    next_state=a?`s3:`s5;
                `s5:    next_state=a?`s4:`s6;
                `s6:    next_state=a?`s5:`s7;
                `s7:    next_state=a?`s6:`s0;
        endcase
        end
endmodule
```

continued

continued

```
//test-bench
module p_tst_seqmealy();
reg a, clk; wire w,x,y,z; wire [2:0]state;
p_seqmealy sm (a,clk,w,x,y,z,state);
initial begin clk=1'b0;a=1'b0; #150 $stop; end
always #2 clk = ~clk;
always #60 a=~a;
endmodule
```

Figure 12.18 A design module for the FSM of Example 12.6. A test bench is also included in the listing.

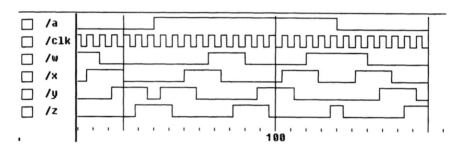

Figure 12.19 Waveforms of the variables during the simulation of the module in Figure 12.18.

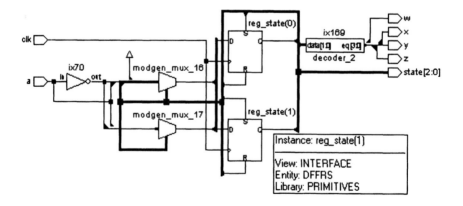

Figure 12.20 Synthesized circuit of the FSM module in Figure 12.18.

12.5   EXERCISES

A microcontroller is shown in block diagram form in Figure 12.21.   Its organization and working are briefly on the following lines:

- A clock unit generates a nonoverlapping two-phase clock – designated *alpha* and *beta*.
- The program memory is of 1 Kb.
- The program counter (PC) is 10 bits wide. Its output forms the address for the program memory.
- The instruction to be executed is fetched from the program memory and loaded into the instruction register (IR). The program counter content forms address of the instruction.
- The PC incrementer can increment the program counter by 1, 2, or 3 as desired; the incrementing is synchronous with one of the clocks.
- The instruction decoder decodes the instruction and provides the control outputs to the different units.
- A versatile register file (VRF), an ALU, a RAM, a serial I/O unit, a clock scaler and a move block byte (MBB) are the other units of the Microcontroller.
- Instructions are fetched from the program memory and executed successively and cyclically.

The microcontroller is to be developed in steps through the exercises given below.

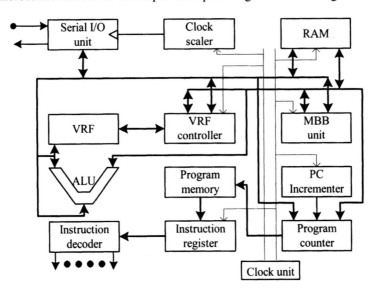

Figure 12.21 A microcontroller in block diagram form.

1. The instruction decoder block of a microcontroller is shown in Figure 12.22(a). Its clock – called '*alpha*' – is to work with output as shown in Figure 2.22(b). Details are as given below:

   a. Widths of register outputs are given in the figure.
   b. The program memory is to be 1 Kb wide.
   c. At every positive edge of the clock the program counter (PC) has to increment by one.
   d. The PC output forms the address to the program memory: Data at the location are transferred to the instruction register (IR) at the following negative edge of *alpha* clock.
   e. The IR output is decoded by the instruction decoder.

   The instruction decoder output is 24 bits wide – designated *c0* to *c23*. Set up the scheme and test it through a test bench. The *alpha* clock is to be generated; the 8-bit IR content is to be output. load the program memory with 1 Kb of random data. Bring them out as IR output successively in the test-bench. The instruction decoder may be ignored for the present.

2. Modify the scheme in Exercise 1 above by interposing a "PC Incrementer" between the clock and the PC as shown in Figure 12.23 to form the instruction decoder (ID) module. The control bits *c22* and *c23* decide its role:

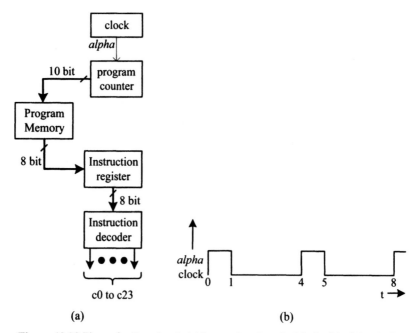

(a)   (b)

Figure 12.22 Figure for Exercise 1: (a) Instruction decoder block. (b) *alpha* clock.

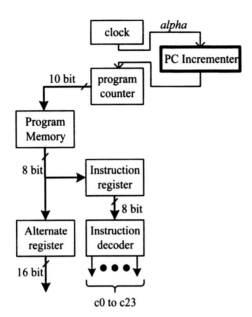

Figure 12.23 Figure for Exercise 2: A modified version of the instruction decoder block in Figure 12.22.

    a. If $c22\ c23 = 00$, increment the PC by one byte.

    b. If $c22\ c23 = 01$, increment the PC by two bytes; load the following byte to the alternate register (immediate memory addressing).

    c. If $c22\ c23 = 10$, increment the PC by three bytes; load the following two bytes to the alternate register (immediate memory addressing – two bytes of data).

    d. If $c22\ c23 = 11$, increment the PC by three bytes; load the following two bytes back to the PC (unconditional jump).

3. Set up a versatile register file (VRF) on the following lines:

    a. 16 numbers of 8-bit registers addressed by 4 bits

    b. A control vector c of 13 bits:

       i. $c0 - c3$ address A

      ii. $c5 - c8$ address B

     iii. $c9 - c11$ steer vector (SV)

     iv. $c12$ enable bit

    c. Two buses aa and bb – 8 bits wide

    d. Clock waveform (clock $beta$) as in Figure 12.24(a).

    e. Carry out instructions as in Table 12.24. If $c12 = 1$, VRF is active. If $c12 = 0$, VRF is inactive.

The VRF will have a VRF controller – a decoder, a mux and a demux combined – [Figure 12.24(b)].

Realize the module, form a test bench, and test the module for all its functions.

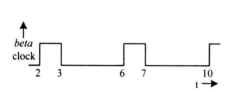

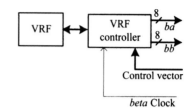

Figure 12.24 (a) *beta* clock waveform.

Figure 12.24 (b) The VRF in Exercise 3 in block diagram form.

Table 12.24 Instructions linked to VRF in Exercise 3

SV value	Instruction
000	Clear VRF.
001	Load data on bus *ba* to register of address *A* (8-bit store).
010	Clear register of address *A*.
011	Read from register of address *A* into bus *bb* (8 bit load).
100	Transfer content of register of address *A* into register of address *B*; content of register of address *A* to remain unaffected.
101	Swap contents of register of address *A* and register of address *B*.
110	Load from *ba* to register of address *A* and *bb* to register of address *B* (16-bit store).
111	Load from register of address *A* to *ba* and register of address *B* to *bb* (16-bit Load).

4. Combine the ID and the VRF modules in the two previous exercises as shown in Figure 12.25. Details of Instructions are in Table 12.25. The 8-bit content of IR is the instruction to be carried out. For the set of values of IR content given in Table 12.25, the control lines active are $c0 - c11$, $c22$ and $c23$. All other control lines are zero. Note that the IR values given in Table 12.25 are all Opcodes. The instruction decoder has to accept the 8-bit content of IR and generate all the necessary control signal values through appropriate combinational logic.

5. Combine the IR-VRF of the last exercise with an ALU. *A* and *B* will be the source and destination addresses of the ALU operation for all the ALU instructions. The ALU operation is specified by the bits $b0-b2$ of IR; the additional commitment of IR bits is follows:

   a. $b3-b4$ specify *A* address

   b. $b5-b6$ specify *B* address

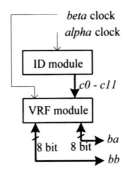

Figure 12.25 Combining ID and VRF units as in Exercise 12.4.

Table 12.25 Instructions for the unit in Figure 12.25

IR content	Instruction to be carried out	No. of bytes by which the PC is to be incremented
00H	No operation	1
01H	Specify address of register A; load following byte into Ra	2
02H	Specify addresses of registers A and B;; load following word into the pair.	3
03H	Specify addresses of registers A and B; load the content of the pair into the PC.	–
04H	Clear VRF	1
05H to 0bH	All register based instructions in Exercise 3 above, as detailed in Table 12.24; every instruction is a 2 byte instruction with the 2nd byte specifying the A and B addresses	2

   c. $b7 = 1$ for ALU operation and 0 for other operations

   d. Carry, borrow, half-carry, and zero bits are to be loaded into the 0Fh register of VRF.

   Prepare a test bench for the design and test all assigned functions.

6. The above compact processor has some (many) limitations. When $b7 = 1$, 127 Opcode possibilities exist; only eight of them have been used. Others cannot be used. Why?

7. Add a serial transmitter module to the unit in Exercise 5 as shown in Figure 12.26. Assign addresses 0Eh and 0Dh of VRF to RcR and TxR. Specific details are as follows:

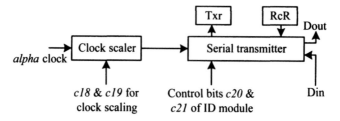

Figure 12.26 Block diagram of the serial I/O unit to be added to the microcontroller.

    a. In the Opcode $b7\ b6\ b5 = 000$ & $b4 = 1$ for serial transmission.

    b. When IR = 10H transmit TxR content serially to Dout.

    c. When IR = 11H, receive Din data and load it into RcR.

    d. IR = 12H transmit as with IR = 10H but at half the clock rate.

    e. IR = 13H receive as with IR = 11H but at half the clock rate.

    f. IR = 14H transmit as with IR = 10H but at 1/4th the clock rate.

    g. IR = 15H receive as with IR = 11H but at 1/4th the clock rate.

    h. IR = 14H transmit as with IR = 10H but at 1/8th the clock rate.

    i. IR = 15H receive as with IR = 11H but at 1/8th the clock rate.

Four possible clock values are specified here; use control bits $c18$ & $c19$ to select clock scale factor.

8. Form a processor by combining the Memory block of Exercise 5 in Chapter 7 with ID, VRF, and serial I/O block above. Have the following additional instructions:

    a. For IR content 20H to 2fH, load the following two bytes from the program memory into the MAR at *beta* clock. In the following *alpha* clock, do memory read and load into the register in VRF with address as the 2nd nibble of the Opcode.

    b. For IR content 31H to 3fH load the following two bytes from the program memory into the MAR at *beta* clock. In the following *alpha* clock, take the content of register from VRF with address as the 2nd nibble of the Opcode and do memory write at the address in MAR.

9. The processor built up above has room for only a limited set of Opcodes. It is due to the limited IR width and the constraint imposed on it that $b7 = 1$ means ALU operation. The constraint was imposed to make the instruction decoder simple. Remove this constraint and absorb further decoding onto it. With such a change, many more instructions can be accommodated as shown in Table 12.26.

Table 12.26 Part of a compact Instruction set for the processor

Opcode	Operation
00H to 0bH	All the instructions in Exercise 5 above linking ID unit and the VRF unit.
10H – 1fH	Immediate-type ALU instructions; result to be put in the VRF register of address 0H.
20 H– 2fH	Load and store-type instructions between RAM, VRF, and the immediate bytes.
30H – 3fH	Fetch from memory, ALU operation and back to VRF.
40H – 4fH	Fetch from VRF, ALU operation, and back to memory.
50 onward	Serial transmit, serial receive, block move; provision for additions to the instruction set.

10. Add reset input to the processor in the above example. All the registers and all the locations in the memory block should get reset and remain so as long as the reset line is high. It includes the *alpha* and *beta* clocks also. The reset input should remain low for 100 ns and revert to the high state automatically. As soon as it goes high the processor should start working. It should start with $t = 0$ ns with the *alpha* clock waveform. The program memory is to remain undisturbed during the reset mode (Note that the reset line is to be added to all the modules as an additional input). Test the reset function through a test-bench.

11. Add "hold" input to the processor. Normally the hold input should be low. As long as it remains high, the processor operation will remain suspended; PC, IR, all registers, memory locations, *etc.*, will retain their stored values (Hint: Disable the clocks to all the blocks). Test the hold function through a test bench.

12. "*Move Bulk Byte*" (MBB) *unit*: The unit is shown in block diagram form in Figure 12.27. The MBB block is to move a block of data bytes to the memory block from *ba* bus; or it is to move a block of data bytes from the memory block to bus *ba*. The movement is effected byte-wise on successive clock pulses. MBB is a skeletal DMA unit. Its activity is decided by the contents of three registers within. They decide the starting address of the block (register *strt*), the size of the block (register *sz*), and the mode of transfer (bits M1 and M2). Form the unit with the following functions:

a. The block is selected when the Select input goes high; any activity connected with the block is done with the select line held high.

b. When Load1 input goes high, the byte on the *ba* bus will be loaded into register *strt* (more significant byte of the starting address). It will happen at the following positive edge of *beta*.

c. When Load2 goes high, the byte on the *ba* bus will be loaded into register *sz* (size) inside. It will happen at the following positive edge of *beta*.

d. During the beta clock pulse following Load2, bits M1 and M2 will be loaded through the 0th and 1st bits on *ba* bus. With such three successive load operations, MBB is ready for data transfer.

e. M1 decides mode and M2 decides the operation. M1M2 = 00 means the MBB is to do read operation with the processor in HOLD mode. MM2 = 01 means the MMB is to do write operation in hold mode. MM2 = 10 means that doing read operation in "cycle steal" mode. MM2 = 11 means write operation in "cycle steal" mode.

Specify the sequence of activities for each of the operations. The bits M1, M2, Load1, Load2, and clocks alpha and beta are all to be used as inputs to a finite state machine and outputs WR, RD, SD (selective disable), and EBM (enable bulk move) generated to conform to the above requirements. Design the FSM and test it through a test-bench. Integrate it with the registers *sz* and *strt* to complete the MBB. Prepare a test-bench for the MBB and test its functioning.

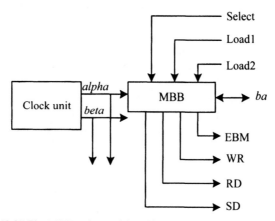

Figure 12.27 The MBB unit considered in Exercise 12 in block diagram form.

13. Form two single byte ports and link them to the MBB; one is to the input port for data being written to the MBB and the other the output port for data read from it. Form the composite module; test it with a test bench.

14. Designate two registers 0ch and 0bh in VRF as input and output ports. Assign two flags – data input flag (DIF) and data output flag (DOF) in the Status register (*b6* and *b5*) dedicated to I//O operation. Whenever data are written into the DIF, *b6* of status register is set. Whenever data in DIF is read, *b6* of status register is reset. Similarly, whenever data are written into the DOF, *b6* of status register is set. Whenever data in DOF are read, b6 of status register is reset. Through a suitable test bench, test port allocations and assignments.

15. Integrate the MBB and the memory block of Exercise 5 in Chapter 7. Note that the data selected for movement of data start at the beginning of pages 128 bytes apart. Test the operation of the composite unit through a test-bench.

16. Integrate the MBB with the processor; add the reset input to the MBB also. Note that Load1, Load2, Enable, M1 and M2 can be combined and output as one byte on the bb bus from the ID block of the processor. Depending on the activity desired, the select bit can be made 1 or 0. c17 of the ID unit is to be connected to the select input of the MBB block. Load EBM into *b7* of the status byte (0th byte of the VRF block).

   a. Ready the processor for bulk writing into memory by writing into *strt* and *sz* registers: Hold mode is to be used.

   b. Ready the processor for bulk reading from memory by writing into *strt* and *sz* registers: Hold mode is to be used.

   c. Ready the processor for bulk writing into memory by writing into *strt* and *sz* registers: Cycle steal mode is to be used.

   d. Ready the processor for bulk reading from memory by writing into *strt* and *sz* registers: Cycle steal mode is to be used.

   Assign Opcodes to the above instructions. Test the functions through a test bench.

17. In the processor two registers of the VRF were designated as input and output registers, respectively. Instead use separate ports as the basis: Attach an input port and an output port to the processor. Use *c15* and *c16* of the ID module to read from the input port and write to the output port, respectively.

APPENDIX A

Keywords and Their Significance

always	Start of a continuous type of behavior activity flow
and	Instantiation of an AND gate primitive
assign	Assign a value or an expression to a net or a variable
automatic	The qualified function / task is of a reentrant type
begin	Start of a block of statements
buf	Instantiation of a buffer primitive
bufif0	Instantiation of a tri-state buffer primitive; On when the control input is at 0 state
bufif1	Instantiation of a tri-state buffer primitive; On when the control input is at 1 state
case	Start of a multiway decision statement
casex	Start of a multiway decision statement: **x** and **z** values are don't cares
casez	Start of a multi-way decision statement: **z** values are don't cares
cell	Design element such as module, primitive, *etc.*
cmos	Instantiation of a CMOS switch primitive
config	Configuration for instantiation
deassign	Termination of a procedural continuous assignment
default	Unspecified instances in a configuration
defparam	Modified value of parameter(s) follows.
design	Library and cell of the top level module
disable	Termination of a concurrent activity
edge	Type of edge for timing checks
else	Alternative in a conditional assignment
end	Termination of a block definition
endcase	Termination of a case statement
endconfig	Termination of configuration for instantiation
endfunction	Termination of a function definition
endgenerate	Termination of multiple instantiations
endmodule	Termination of a module definition

`endprimitive`	Termination of an UDP definition
`endspecify`	Termination of a specify block definition
`endtable`	Termination of a table definition
`endtask`	Termination of a task definition
`event`	A flag
`for`	Control execution of statement(s) by a three-step process
`force`	An overriding assignment on a variable / variables
`forever`	Continuous execution of statement (s)
`fork`	Statements for concurrent execution
`function`	Start of definition of a function
`generate`	Multiple instantiation
`genvar`	The index variable in a generate loop
`highz0`	0 state of a net with high source impedance
`highz1`	1 state of a net with high source impedance
`if`	Conditional operator
`ifnone`	Default state-dependent path delay
`include`	Inclusion of specified file
`initial`	Start of an "only once" type of behavior activity flow
`inout`	Declaration of port(s) of input or output types
`input`	Declaration of input port(s)
`instance`	A specific instance (to be followed by an expansion clause)
`integer`	A variable of type integer
`join`	Termination of list of statements for concurrent execution
`large`	Largest value of charge strength on a net
`liblist`	A set of libraries to be searched for the instance
`library`	A logical collection of cells
`localparam`	Parameter value not alterable externally
`macromodule`	Start of a module definition
`medium`	Medium value of charge strength on a net
`module`	Start of a module definition
`nand`	Instantiation of a NAND gate primitive
`negedge`	Falling edge of a net or variable
`nmos`	Instantiation of an NMOS switch primitive
`nor`	Instantiation of a NOR gate primitive
`noshowcancelled`	In case of anomalous delay specifications, avoid output transition to x state
`not`	Instantiation of a NOT gate primitive
`notif0`	Instantiation of a tri-state NOT gate primitive: ON when control input is at 0 state
`notif1`	Instantiation of a tri-state NOT gate primitive: ON when control input is at 1 state
`or`	Alternative event in sensitivity list

`output`	Declaration of output port(s)
`parameter`	Declaration of a constant / constants
`pmos`	Instantiation of a PMOS switch primitive
`posedge`	Rising edge of a net or variable
`primitive`	Start of an UDP definition
`pull0`	Strength value of net (s) at 0 state
`pull1`	Strength value of net (s) at 1 state
`pulldown`	A resistive connection of a net to logic 0
`pullup`	A resistive connection of a net to logic 1
`pulsestyle_onevent`	Possible transition of output to **x** state on event
`pulsestyle_ondetect`	Possible transition of output to **x** state on detection
`rcmos`	Instantiation of a resistive CMOS switch primitive
`real`	A variable or a constant of the real number type
`realtime`	Numerical value of simulation time
`reg`	A data storage element
`release`	Termination of an overriding assignment on a variable / variables
`repeat`	Execute a statement / statements a fixed number of times
`rnmos`	Instantiation of a resistive NMOS switch primitive
`rpmos`	Instantiation of a resistive PMOS switch primitive
`rtran`	A bi-directional resistive pass switch primitive
`rtranif0`	A bi-directional resistive pass switch primitive; it is ON when control input is 0.
`rtranif1`	A bidirectional resistive pass switch primitive; it is ON when control input is 1.
`scalared`	No restriction on operation on vectors specified
`showcancelled`	Output transition to **x** state bypassing possible anomalous delay specifications
`signed`	The qualified variable has a sign associated
`small`	Smallest value of charge strength on a net
`specify`	Specific value assignments to parameters follow
`specparam`	Specifies values for the parameters that follow
`strong0`	Strength value of net when in 0 state
`strong1`	Strength value of net when in 1 state
`supply0`	Connection to logic 0 supply
`supply1`	Connection to logic 1 supply
`table`	Beginning of state table of a UDP
`task`	Start of a task definition
`time`	Time variable
`tran`	A bi-directional pass switch primitive
`tranif0`	A bi-directional pass switch primitive; it is ON when control input is 0.

tranif1	A bi-directional pass switch primitive; it is ON when control input is 1.
tri	A net driven from multiple sources
tri0	A net driven from multiple sources with resistive pulldown
tri1	A net driven from multiple sources with resistive pullup
triand	A net (nets) driven by multiple sources with AND-type output in case of conflict
trior	A net (nets) driven by multiple sources with OR-type output in case of conflict
trireg	A capacitive type net which can store charge
use	A binding for the cell specified
vectored	Restricted operation on vectors specified
wait	Wait for an expression to be true to start execution
wand	A net (nets) driven by multiple sources with AND-type output in case of conflict
weak0	Strength value of net (s) at 0 state
weak1	Strength value of net (s) at 1 state
while	Execute a statement / statements until an expression becomes false
wire	A type of net
wor	A net (nets) driven by multiple sources with OR type output in case of conflict
xnor	Instantiation of an XNOR gate primitive
xor	Instantiation of an XOR gate primitive

APPENDIX B

Truth Tables of Gates and Switches

The truth tables for gates are given with two inputs each; it remains the same for multiple inputs as well. The inputs are designated as 'Input 1' and 'Input 2'; the output values are in the respective cells of the table.

Table B.1 Truth table of AND gate

Input 2 \ Input 1	0	1	x	z
0	0	0	0	0
1	0	1	x	x
x	0	x	x	x
z	0	x	x	x

Table B.2 Truth table of OR gate

Input 2 \ Input 1	0	1	x	z
0	0	1	x	x
1	1	1	1	1
x	x	1	x	x
z	x	1	x	x

Table B.3 Truth table of NAND gate

Input 2 \ Input 1	0	1	x	z
0	1	1	1	1
1	1	0	x	x
x	1	x	x	x
z	1	x	x	x

Table B.4 Truth table of NOR gate

Input 2 \ Input 1	0	1	x	z
0	1	0	x	x
1	0	0	0	0
x	x	0	x	x
z	x	0	x	x

Table B.5 Truth table of XOR gate

Input 2 \ Input 1	0	1	x	z
0	0	1	x	x
1	1	0	x	x
x	x	x	x	x
z	x	x	x	x

Table B.6 Truth table of XNOR gate

Input 2 \ Input 1	0	1	x	z
0	1	0	x	x
1	0	1	x	x
x	x	x	x	x
z	x	x	x	x

Table B.7 Truth table for Buffer and NOT gates

Input	**buf** output	NOT output
0	0	1
1	1	0
x	x	x
z	x	x

Table B 8 Truth table of bufif0 gate

		Data input			
		0	1	x	z
Control input	0	0	1	x	x
	1	z	z	z	z
	x	L	H	x	x
	z	L	H	x	x

Table B 9 Truth table of bufif1 gate

		Data input			
		0	1	x	z
Control input	0	z	z	z	z
	1	0	1	x	x
	x	L	H	x	x
	z	L	H	x	x

Table B 10 Truth table of notif0 gate

		Data input			
		0	1	x	z
Control input	0	1	0	x	x
	1	z	z	z	z
	x	H	L	x	x
	z	H	L	x	x

Table B 11 Truth table of notif1 gate

		Data input			
		0	1	x	z
Control input	0	z	z	z	z
	1	1	0	x	x
	x	H	L	x	x
	z	H	L	x	x

Table B 12 Truth table of pmos and rpmos gates

		Data input			
		0	1	x	z
Control input	0	0	1	x	x
	1	z	z	z	z
	x	L	H	x	x
	z	L	H	x	x

Table B 13 Truth table of nmos and rnmos gates

		Data input			
		0	1	x	z
Control input	0	z	z	z	z
	1	0	1	x	x
	x	L	H	x	x
	z	L	H	x	x

REFERENCES

Arnold MG (1998) *Verilog Digital Computer Design*. Prentice-Hall, Englewood Cliffs, NJ.

Baker RJ, Li HW, Boyce DE (1998) *CMOS Circuit Design, Layout and Simulation*. IEEE, New York.

Bhaskar J (1997) *Verilog HDL Primer*. Star Galaxy Press, Allentown, PA.

Bignell J, Donovan R (2000) *Digital Electronics*, 4th ed. Thomson Learning, New York.

Bogart Jr., TF (1992) *Introduction to Digital Circuits*. McGraw-Hill, New York.

Ciletti MD (1999) *Modeling, Synthesis and Rapid Prototyping with the Verilog HDL*. Prentice-Hall, Englewood Cliffs, NJ.

Comer DJ (1995) *Digital Logic and State Machine Design*. 3rd ed. Saunders College Publishing, New York.

Devadas S, Ghosh A, Keutzer K (1994) *Logic Synthesis*. McGraw-Hill, New York.

Gopalan KG (1996) *Introduction to Digital Microelectronic Circuits*. McGraw-Hill, New York.

Gottfried BS (1990) *Programming with C*. McGraw-Hill, New York.

Heuring VP, Jordan HJ (1997) *Computer Systems Design and Architecture*. Addison-Wesley, Menlo Park, CA.

Hill FJ, Peterson G (1987) *Digital Systems: Hardware Organization and Design*. 3rd ed. John Wiley & Sons, New York.

IEEE (2001) *IEEE Standard Verilog Hardware Description Language*. IEEE, New York

Lee JM (1997) *Verilog Quickstart*. Kluwer Academic Publishers, Norwell, Ma.

Micheli GD (1994) *Synthesis and Optimization of Digital Circuits*. McGraw-Hill, New York.

Navabi Z (1999) *Verilog Digital System Design*. McGraw-Hill, New York.

Oldfield JV, Dorf RC (1995) *Field Programmable Gate Arrays*. Wiley Interscience, New York.

Palnitkar S (1996) *Verilog HDL: A Guide to Digital Design and Synthesis*. Prentice-Hall, Engelwood Cliffs, NJ.

Proakis JG (2001) *Digital Communications*. 4th ed. McGraw-Hill, New York.

Roth Jr., Charles H (1998) *Digital Systems Design Using VHDL.* PWS Publishing, Boston.

Rabaey JM (1996) *Digital Integrated Circuits – A Design Perspective.* Prentice-Hall, Engelwood Cliffs, NJ.

Sedra AS, Smith KC (1998) *Microelectronic Circuits,* 4th ed. Oxford University Press, New York.

Smith DJ (1996) *A Practical Guide for Designing, Synthesizing and Simulating ASICs and FPGAs using VHDL or Verilog.* Doone Publications, Madison, AL.

Smith MJ (1997) *Application-Specific Integrated Circuits.* Addison-Wesley Longman, Reading, MA.

Sutherland S (2001) *Verilog 2001.* Kluwer Academic Publishers, Norwell, MA.

Thomas DE, Moorby PR (1996) *The Verilog Hardware Description Language.* 3rd ed. Kluwer Academic Publishers, Norwell, MA.

Tocci RJ, Widmer NS (2001) *Digital Systems – Principles and Applications.* 8th ed. Pearson Education, Singapore.

Wai-Kai Chen (ed) (2000) *VLSI Handbook,* CRC Press, Florida.

Wakerly JF (2000) *Digital Design – Principles and Practices.* Prentice-Hall, Engelwood Cliffs, NJ.

Wolf W (1998) *Modern VLSI Design – Systems on Silicon.* Prentice-Hall, Engelwood Cliffs, NJ.

INDEX

A

Adder, 239-244, 277-283, 290, 291
    BCD, 147, 148
ALU, 116-123, 143-145, 210-212
always, 161, 168,169
and, 48-50
Array of instances, 66
ASIC, 4
assign, 128, 225-230
Assignment, 12
    blocking, 201-204
    concurrent, 201
    continuous, 127-130
        and delays, 133, 134
        and nets, 131
        and strengths, 132
    nonblocking, 201-204
        and delays, 204, 205
    procedural, 160, 161
        with delays, 184-187
    procedural continuous, 227, 228
    sequential, 161
    vector, 135
automatic, 285, 286

B

begin, 28, 161-163
Bidirectional pass switch, 328-329
Block
    disabling, 244-249
    named, 163
    nesting, 163
buf, 51-52
Buffer, 51-52

bufif0, 64, 65
bufif1, 64, 65
Bus switcher, 151, 152, 329-331

C

C language, 13, 16, 159, 219
case, 205-210
Case sensitivity, 31
casex, 210-212
casez, 210-212
Clock, 184, 254, 255
cmos, 318-321
CMOS
    NOR gate, 312-314
    switch, 318-321
Comment, 33, 34
Comparator, 67-69
Compiler directive, 385-392
    define, 385
Concatenation, 135
Contention resolution, 102-109,
    334-337
Counter, 170-179, 224, 225, 232,
    234
    ring, 152-156

D

Data types, 40, 41
deassign, 225-230
default, 205-210
defparam, 42, 340
Delay, 15, 28, 91-102, 133, 134
    assigning, 184-187, 191
    conditional, 359-361

distributed, 348
gate, 94-99
intra-assignment, 187, 188
net, 92-94
path, 348-371
pin to pin, 348-371
propagation, 97
with tri-state gates, 99-102
zero, 191
Demux, 222-224, 227, 228
Design description
levels, 11
behavioral, 13, 14
circuit, 11
data flow, 12
gate, 12
disable, 244-249
display, 44, 374-378
Distributed delay, 348
Dynamic shift register, 325-327

E
else, 219-225
end, 28, 161-163
endcase, 205-210
endfunction, 274
endmodule, 18, 50
endprimitive, 293
endspecify, 348-361
endtable, 293
endtask, 286-287
Escaped identifiers, 32, 33
Event, 38, 169, 170, 266-268
Expression
bit width, 150, 151

F
Finish, 29, 45, 381
Finite State Machine, 418-432
Mealy machine, 427-432
Moore machine, 419-427
Flip-flop
clocked, 181, 182
D, 86-88, 183, 228-235

edge-triggered, 88-91, 153, 154,
300, 301
RS, 83-86
for, 238-244
loop flowchart, 238
force, 261-265
forever, 254-258
fork, 258-261
FPGA, 7
Full adder, 71-72, 294-296
function, 273, 274
Function, 274
recursive, 284-286
scope, 284

G
Gates
resistive, 308
tri-state, 64-66

H
Half adder, 70-71
Hardware
trade off with speed, 283, 284
HDL, 3, 9
Hierarchical access, 14, 56-62,
361-362, 393-404
Hierarchical name, 375-377

I
Identifier, 32
if, 219-225
loop flowchart, 220
if-else, 219-225
loop flowchart, 221
initial, 28, 161, 164, 165
inout, 17, 50
input, 17, 50
Instantiation, 19-21
integer, 34-36
Inverter
CMOS, 311, 312
NMOS, 317, 318

J
join, 258-261

K
Keyword, 31, 32

L
large, 106
latch, 152, 153, 183, 298, 299
    with NAND gates, 81
Level
    behavioral level, 159, 160
    data flow level, 127
    gate level, 47-80
    RTL, 14
    switch level, 305
Lexical tokens, 17, 31
    comment, 33, 34
    identifier, 32, 33
    keyword, 31, 32
    number, 34-38
    operator, 43
    string, 36-38
    white space, 33
Logic values, 38

M
Macromodule, *see* **module**
medium, 39, 106
Memory, 43
Microcontroller design, 433-441
Modeling, *see* Level
module, 16-18, 50
Module
    path, 349-371
    stimulus, 18, 54, 55
    structure, 16-21, 50
    test bench, 18
monitor, 29, 44, 380, 381
Multiple always blocks, 194-197
Multiple initial blocks, 167
Mux, 73-79, 122, 146

N
nand, 51, 52
negedge, 169, 170
Net, 40, 131
    charge, 106, 107
    types, 109-115
nmos, 306
NMOS inverter, 317, 318
nor, 51, 52
not, 51, 52
notif0, 64, 65
notif1, 64, 65
Number, 34-38
    integer, 34-36
    real, 36
    sign, 35, 36

O
Operand, 160
Operation, 160
Operator, 43, 136-150, 160
    algebraic
    arithmetic, 137, 138
    binary, 137, 138
    equality, 139
    logical, 138-140
    modulus
    precedence, 148-150
    relational, 138, 139
    shift, 141
    ternary, 141-143
    unary, 137
or, 51, 52
output, 17, 50

P
Parallel blocks, 258-261
parameter, 42, 341-347, 372
Parameter, 42, 43, 339, 340
    assignment, 341-347
    declaration, 341-347

module, 371-373
over-riding, 342-347
type, 347
Parity bit generation, 274- 277
Path delay, 348-371
conditional, 359-361
edge sensitive, 364-366
PLA, 416
PLD, 414-418
PLI, 16
pmos, 307
Port, 16, 21
posedge, 169, 170
primitive, 293
Primitive, 12
gate, 47-52, 81
user defined, 292-302
Programmable Logic Device,
    414-418
programming, 416-418
Programmable Logic Interface, 16
pull0, 39
pull1, 39
pulldown, 309-311
pullup, 309-311
Pulse filtering, 367-371

Q
Queue, 407-413

R
RAM cell, 321-325, 331-333
rcmos, 328
real, 41
realtime, 388, 389
reg, 41
release, 261-265
repeat, 236, 237
Ring counter, 152-156
rnmos, 308
rpmos, 308
RS latch, 83-84
Rtran, 328
rtranif0, 328

rtranif1, 328

S
Scalar, 41,42
Shift register, 179-181
dynamic, 325-327
Simulation, 7, 24, 25, 28, 214, 215
concurrency, 13
small, 39, 106
specify, 348-361
specparam, 340, 351-352
stop, 45, 165, 381
Stratified event queue, 215, 216
Strength, 38-40, 102-109, 132
task for display, 377, 378
String, 36-38
Strobe, 378, 379
strong0, 39
strong1, 39
supply0, 39, 115
supply1, 39, 115
Switch primitive
bi-directional, 328-333
CMOS, 311-312, 318-321
NMOS, 306
PMOS, 307
resistive, 308
with delay, 333, 334
Synthesis, 7, 14, 25-27
System function, 16, 381-383
file related, 383-385
for random number, 381-383
System task, 16, 44, 45, 374-381
file related, 383-385
for display, 44, 45
for output, 44, 45
timescale related, 386-392

T
table, 293
task, 286-287
enabling, 286
structure, 287
Test

functional, 14, 15
timing, 14
Test bench, 14, 18, 27, 54, 55
time, 388, 389
simulation time
Time scale, 386-392
default, 388
Time step, 28
Tokens, *see* Lexical tokens
tran, 328
tranif0, 328
tranif1, 328
tri, 40, 113
tri0, 114
tri1, 114
triand, 113
trior, 113
trireg, 38, 106

U
UDP, *see* User Defined Primitive
User Defined Primitive
combinational, 292-294
instantiation, 295
with delay, 302
sequential, 297, 298

V
Variable, 41
local, 164
Vector, 41, 42, 135
VHDL, 9

W
wait, 192-195
wand, 109, 110
weak0, 39
weak1, 39
while, 249-254
loop flowchart, 249
White space, 33
wire, 40
wor, 111-113
write, 224, 374

X
xnor, 51, 52
xor, 51, 52

Z
Zero delay, 191

CPSIA information can be obtained at www.ICGtesting.com
Printed in the USA
BVOW042015121112

305373BV00002B/2/P